Antibacterial Coatings and Biofilm

Antibacterial Coatings and Biofilm

Editor

Dan Cristian Vodnar

Basel • Beijing • Wuhan • Barcelona • Belgrade • Novi Sad • Cluj • Manchester

Editor
Dan Cristian Vodnar
Department of Food Science
University of Agricultural
Sciences and Veterinary
Medicine
Cluj-Napoca
Romania

Editorial Office
MDPI AG
Grosspeteranlage 5
4052 Basel, Switzerland

This is a reprint of articles from the Special Issue published online in the open access journal *Coatings* (ISSN 2079-6412) (available at: https://www.mdpi.com/journal/coatings/special_issues/antibacterial_coating).

For citation purposes, cite each article independently as indicated on the article page online and as indicated below:

Lastname, A.A.; Lastname, B.B. Article Title. *Journal Name* **Year**, *Volume Number*, Page Range.

ISBN 978-3-7258-2085-6 (Hbk)
ISBN 978-3-7258-2086-3 (PDF)
doi.org/10.3390/books978-3-7258-2086-3

© 2024 by the authors. Articles in this book are Open Access and distributed under the Creative Commons Attribution (CC BY) license. The book as a whole is distributed by MDPI under the terms and conditions of the Creative Commons Attribution-NonCommercial-NoDerivs (CC BY-NC-ND) license.

Contents

Dan Cristian Vodnar and Bernadette-Emőke Teleky
Recent Trends in Antibacterial Coatings and Biofilm
Reprinted from: *Coatings* **2023**, *13*, 255, doi:10.3390/coatings13020255 1

Catalin Vitelaru, Anca C. Parau, Adrian E. Kiss, Iulian Pana, Mihaela Dinu, Lidia R. Constantin, et al.
Silver-Containing Thin Films on Transparent Polymer Foils for Antimicrobial Applications
Reprinted from: *Coatings* **2022**, *12*, 170, doi:10.3390/coatings12020170 4

Marco Consumi, Kamila Jankowska, Gemma Leone, Claudio Rossi, Alessio Pardini, Eric Robles, et al.
Non-Destructive Monitoring of *P. fluorescens* and *S. epidermidis* Biofilm under Different Media by Fourier Transform Infrared Spectroscopy and Other Corroborative Techniques
Reprinted from: *Coatings* **2020**, *10*, 930, doi:10.3390/coatings10100930 17

Steluta Carmen Ciobanu, Simona Liliana Iconaru, Daniela Predoi, Alina Mihaela Prodan and Mihai Valentin Predoi
Physico-Chemical Properties and In Vitro Antifungal Evaluation of Samarium Doped Hydroxyapatite Coatings
Reprinted from: *Coatings* **2020**, *10*, 827, doi:10.3390/coatings10090827 32

Dorota Kowalczuk
FTIR Characterization of the Development of Antimicrobial Catheter Coatings Loaded with Fluoroquinolones
Reprinted from: *Coatings* **2020**, *10*, 818, doi:10.3390/coatings10090818 56

Nicole Ciacotich, Lasse Kvich, Nicholas Sanford, Joseph Wolcott, Thomas Bjarnsholt and Lone Gram
Copper-Silver Alloy Coated Door Handles as a Potential Antibacterial Strategy in Clinical Settings
Reprinted from: *Coatings* **2020**, *10*, 790, doi:10.3390/coatings10080790 70

Santiago Arango-Santander, Carolina Gonzalez, Anizac Aguilar, Alejandro Cano, Sergio Castro, Juliana Sanchez-Garzon and John Franco
Assessment of *Streptococcus Mutans* Adhesion to the Surface of Biomimetically-Modified Orthodontic Archwires
Reprinted from: *Coatings* **2020**, *10*, 201, doi:10.3390/coatings10030201 79

Katalin Szabo, Bernadette-Emoke Teleky, Laura Mitrea, Lavinia-Florina Călinoiu, Gheorghe-Adrian Martău, Elemer Simon, et al.
Active Packaging—Poly(Vinyl Alcohol) Films Enriched with Tomato By-Products Extract
Reprinted from: *Coatings* **2020**, *10*, 141, doi:10.3390/coatings10020141 90

Yuanzhe Li, Boyang Luo, Claude Guet, Srikanth Narasimalu and Zhili Dong
Preparation and Formula Analysis of Anti-Biofouling Titania–Polyurea Spray Coating with Nano/Micro-Structure
Reprinted from: *Coatings* **2019**, *9*, 560, doi:10.3390/coatings9090560 108

Bianca Eugenia Ștefănescu, Carmen Socaciu and Dan Cristian Vodnar
Recent Progress in Functional Edible Food Packaging Based on Gelatin and Chitosan
Reprinted from: *Coatings* **2022**, *12*, 1815, doi:10.3390/coatings12121815 123

Silvia Amalia Nemes, Katalin Szabo and Dan Cristian Vodnar
Applicability of Agro-Industrial By-Products in Intelligent Food Packaging
Reprinted from: *Coatings* **2020**, *10*, 550, doi:10.3390/coatings10060550 . . . . **149**

Editorial

Recent Trends in Antibacterial Coatings and Biofilm

Dan Cristian Vodnar [1,2,*] and Bernadette-Emőke Teleky [2]

1. Faculty of Food Science and Technology, University of Agricultural Sciences and Veterinary Medicine Cluj-Napoca, Calea Mănăştur 3-5, 400372 Cluj-Napoca, Romania
2. Institute of Life Sciences, University of Agricultural Sciences and Veterinary Medicine Cluj-Napoca, Calea Mănăştur 3-5, 400372 Cluj-Napoca, Romania
* Correspondence: dan.vodnar@usamvcluj.ro; Tel.: +40-747341881

In modern society, the growing use of plastic packaging has innumerable and unquestionable consequences. Even though this type of packaging plays an essential role in the global economy and presents numerous advantages, it also has several disadvantages, especially regarding environmental pollution. As conventional/synthetic materials account for over 50% of the materials used for packaging in the food industry, it is crucial to produce eco-friendly, active/edible, and intelligent packaging with antimicrobial properties [1,2]. Another essential aspect of the adoption of biodegradable, eco-friendly packaging is the reduction in greenhouse gas emissions and the transition from a "linear economy" (where the use of fossil fuels and unrecyclable materials is prevalent) to a "circular economy" (no/or minimal waste, by-product reutilization) [3]. This transition requires a reassessment of plastic's life cycle from its base materials to production and recirculation [4,5].

The most important aspect of food packaging is maintaining foodstuffs' appearance and quality, thereby significantly increasing perishable commodity foods' shelf life [6]. Foodborne pathogenic microorganisms are responsible for several diseases worldwide, with the most widely known being *Salmonella* spp., *Pseudomonas aeruginosa*, *Staphylococcus aureus*, *Listeria monocytogenes*, and *Escherichia coli*. Bio-based biodegradable materials can successfully be used as a substitute for non-renewable fossil-based polymers by integrating antimicrobial compounds. These can be organic materials, such as enzymes, proteins, and polymers, or bioactive compounds extracted from food by-products. The integration of bioactive compounds (i.e., macronutrients, phytochemicals) in these eco-friendly materials can confer antimicrobial and antioxidant properties and improve the biofilm's physical–chemical properties [7–10]. As a consequence, productive packaging is vital to protect food from the surrounding environment throughout transportation, storage, and distribution [11].

Increased attention is given to the production of polysaccharide-based edible materials such as alginate, pectin, chitosan, carrageenan, cellulose, xanthan, and other polymers. These materials have gained particular interest in the last few years as a consequence of being natural, regenerable, and abundant [12,13]. Nevertheless, synthetic materials, such as polyvinyl alcohol (PVA) and polylactic acid (PLA), are intensively studied in the production of marketable food packaging materials [14,15]. The integration of bioactive molecules, including carotenoids, vitamins, polyphenols, essential oils, and other compounds, especially those loaded in nanocarriers, is vital to conferring these natural or synthetic materials' antibacterial and antifungal qualities [14,16]. Subsequently, these intelligent/active packaging materials diminish lipid oxidation, prevent the growth of microorganisms, can preserve quality, and increase the shelf life of food products.

In addition to natural bioactive compounds, inorganic antimicrobial compounds comprising metal or metal oxide nanoparticles, such as silver or hydroxyapatite or other metal oxides, have been intensively studied, with outstanding results [14,17,18]. Silver nanoparticles can be integrated in TiO_2 and SiO_2 translucent matrices with the help of high-power impulse magnetron sputtering and radio frequency, and display efficient antimicrobial

properties against *E. coli* [17]. Antibacterial TiO_2 was also applied to produce a titania-polyurea coating that presented efficient activity against *P. aeruginosa*. Besides using these biopolymers in food packaging, they can also be applied as antimicrobial coatings in sectors such as the medical field, pharmaceutics, electronics, and the cosmetics industry [17–20]. Recently, several studies incorporated antibacterial substances in biopolymers, for instance, fluoroquinolone antibiotics; these can also be efficiently used in the medical field [20,21]. Copper–silver alloy coatings can also be used effectively on door handles or other surfaces as they inhibit the growth of *S. aureus, P.aeruginosa, E. coli*, and *Enterobacter aerogenes* [22].

Another relevant aspect in producing these antibacterial coatings and films is the integration of various sensors that can distinguish deficiencies through packaging, supervise the food's quality, and specify the freshness of a product [23]. Furthermore, the sensory characteristics should be considered, as several antimicrobial agents confer an unacceptable odor on food products. Additionally, conventional preservation methods negatively affect the nutritional quality of foods. Thus, the shift from synthetic antimicrobials to natural ones is inevitable [24]. To adopt and efficiently commercialize intelligent/active/smart packaging, further research is necessary to find appropriate materials that maintain their distinct features throughout handling, transportation, and storage. Thereby, attention should be given to the technological features, the interaction of the food with the packaging material, sensorial qualities, and cost reduction.

Conflicts of Interest: The authors declare no conflict of interest.

References

1. Nemes, S.A.; Szabo, K. Applicability of Agro-Industrial By-Products in Intelligent Food Packaging. *Coatings* **2020**, *10*, 550. [CrossRef]
2. Szabo, K.; Teleky, B.E.; Mitrea, L.; Călinoiu, L.F.; Mărtău, G.A.; Simon, E.; Varvara, R.A.; Vodnar, D.C. Active packaging-poly (vinyl alcohol) films enriched with tomato by-products extract. *Coatings* **2020**, *10*, 141. [CrossRef]
3. Mitrea, L.; Leopold, L.F.; Bouari, C.; Vodnar, D.C. Separation and Purification of Biogenic 1,3-Propanediol from Fermented Glycerol through Flocculation and Strong Acidic Ion-Exchange Resin. *Biomolecules* **2020**, *10*, 1601. [CrossRef] [PubMed]
4. Teleky, B.-E.; Vodnar, D.C. Recent Advances in Biotechnological Itaconic Acid Production, and Application for a Sustainable Approach. *Polymers* **2021**, *13*, 574. [CrossRef]
5. Socaciu, M.-I.; Semeniuc, C.; Vodnar, D. Edible Films and Coatings for Fresh Fish Packaging: Focus on Quality Changes and Shelf-life Extension. *Coatings* **2018**, *8*, 366. [CrossRef]
6. Stefănescu, B.E.; Socaciu, C.; Vodnar, D.C. Recent Progress in Functional Edible Food Packaging Based on Gelatin and Chitosan. *Coatings* **2022**, *12*, 1815. [CrossRef]
7. Teleky, B.-E.; Mitrea, L.; Plamada, D.; Nemes, S.A.; Călinoiu, L.-F.; Pascuta, M.S.; Varvara, R.-A.; Szabo, K.; Vajda, P.; Szekely, C.; et al. Development of Pectin and Poly(vinyl alcohol)-Based Active Packaging Enriched with Itaconic Acid and Apple Pomace-Derived Antioxidants. *Antioxidants* **2022**, *11*, 1729. [CrossRef]
8. Mitrea, L.; Călinoiu, L.-F.; Mărtău, G.-A.; Szabo, K.; Teleky, B.-E.E.; Mureșan, V.; Rusu, A.-V.V.; Socol, C.-T.T.; Vodnar, D.-C.C.; Mărtău, G.A.; et al. Poly(vinyl alcohol)-based biofilms plasticized with polyols and colored with pigments extracted from tomato by-products. *Polymers* **2020**, *12*, 532. [CrossRef]
9. Mihalca, V.; Kerezsi, A.D.; Weber, A.; Gruber-Traub, C.; Schmucker, J.; Anca, F.; Vodnar, D.C.; Dulf, F.V.; Socaci, S.A.; Muresan, C.I.; et al. Protein-Based Films and Coatings for Food Industry Applications. *Polymers* **2021**, *23*, 769. [CrossRef]
10. Zhang, W.; Rhim, J.W. Functional edible films/coatings integrated with lactoperoxidase and lysozyme and their application in food preservation. *Food Control* **2022**, *133*, 108670. [CrossRef]
11. Al-Tayyar, N.A.; Youssef, A.M.; Al-hindi, R. Antimicrobial food packaging based on sustainable Bio-based materials for reducing foodborne Pathogens: A review. *Food Chem.* **2020**, *310*, 125915. [CrossRef] [PubMed]
12. Mărtău, G.A.; Mihai, M.; Vodnar, D.C. The Use of Chitosan, Alginate, and Pectin in the Biomedical and Food Sector—Biocompatibility, Bioadhesiveness, and Biodegradability. *Polymers* **2019**, *11*, 1837. [CrossRef] [PubMed]
13. Consumi, M.; Jankowska, K.; Leone, G.; Rossi, C.; Pardini, A.; Robles, E.; Wright, K.; Brooker, A.; Magnani, A. Non-Destructive Monitoring of P. fluorescens and S. epidermidis Biofilm under Different Media by Fourier Transform Infrared Spectroscopy and Other Corroborative Techniques Marco. *Coatings* **2020**, *10*, 930. [CrossRef]
14. Pascuta, M.S.; Vodnar, D.C. Nanocarriers for sustainable active packaging: An overview during and post COVID-19. *Coatings* **2022**, *12*, 102. [CrossRef]
15. Martau, G.-A.; Unger, P.; Schneider, R.; Venus, J.; Vodnar, D.C.; Pablo, L.-G. Integration of Solid State and Submerged Fermentations for the Valorization of Organic Municipal Solid Waste. *J. Fungi* **2021**, *7*, 766. [CrossRef] [PubMed]

16. Szabo, K.; Mitrea, L.; Călinoiu, L.F.; Teleky, B.-E.; Martău, G.A.; Plamada, D.; Pascuta, M.S.; Nemes, S.-A.; Varvara, R.-A.; Vodnar, D.C. Natural Polyphenol Recovery from Apple-, Cereal-, and Tomato-Processing By-Products and Related Health-Promoting Properties. *Molecules* **2022**, *27*, 7977. [CrossRef]
17. Vitelaru, C.; Parau, A.C.; Kiss, A.E.; Pana, I.; Dinu, M.; Constantin, L.R.; Vladescu, A.; Tonofrei, L.E.; Adochite, C.S.; Costinas, S.; et al. Silver-Containing Thin Films on Transparent Polymer Foils for Antimicrobial Applications. *Coatings* **2022**, *12*, 170. [CrossRef]
18. Carmen Ciobanu, S.; Liliana Iconaru, S.; Predoi, D.; Mihaela Prodan, A.; Valentin Predoi, M. Physico-chemical properties and in vitro antifungal evaluation of samarium doped hydroxyapatite coatings. *Coatings* **2020**, *10*, 827. [CrossRef]
19. Li, Y.; Luo, B.; Guet, C.; Narasimalu, S.; Dong, Z. Preparation and formula analysis of anti-biofouling Titania-polyurea spray coating with nano/micro-structure. *Coatings* **2019**, *9*, 560. [CrossRef]
20. Kowalczuk, D. FTIR characterization of the development of antimicrobial catheter coatings loaded with fluoroquinolones. *Coatings* **2020**, *10*, 818. [CrossRef]
21. Arango-Santander, S.; Gonzalez, C.; Aguilar, A.; Cano, A.; Castro, S.; Sanchez-Garzon, J.; Franco, J. Assessment of streptococcus mutans adhesion to the surface of biomimetically-modified orthodontic archwires. *Coatings* **2020**, *10*, 201. [CrossRef]
22. Ciacotich, N.; Kvich, L.; Sanford, N.; Wolcott, J.; Bjarnsholt, T.; Gram, L. Copper-Silver alloy coated door handles as a potential antibacterial strategy in clinical settings. *Coatings* **2020**, *10*, 790. [CrossRef]
23. Soltani Firouz, M.; Mohi-Alden, K.; Omid, M. A critical review on intelligent and active packaging in the food industry: Research and development. *Food Res. Int.* **2021**, *141*, 110113. [CrossRef] [PubMed]
24. Rout, S.; Tambe, S.; Deshmukh, R.K.; Mali, S.; Cruz, J.; Srivastav, P.P.; Amin, P.D.; Gaikwad, K.K.; Andrade, E.H.d.A.; de Oliveira, M.S. Recent trends in the application of essential oils: The next generation of food preservation and food packaging. *Trends Food Sci. Technol.* **2022**, *129*, 421–439. [CrossRef]

Disclaimer/Publisher's Note: The statements, opinions and data contained in all publications are solely those of the individual author(s) and contributor(s) and not of MDPI and/or the editor(s). MDPI and/or the editor(s) disclaim responsibility for any injury to people or property resulting from any ideas, methods, instructions or products referred to in the content.

Article

Silver-Containing Thin Films on Transparent Polymer Foils for Antimicrobial Applications

Catalin Vitelaru [1], Anca C. Parau [1], Adrian E. Kiss [1], Iulian Pana [1], Mihaela Dinu [1], Lidia R. Constantin [1], Alina Vladescu [1,2,*], Lavinia E. Tonofrei [3], Cristina S. Adochite [4,*], Sarah Costinas [4], Liliana Rogozea [4], Mihaela Badea [4] and Mihaela E. Idomir [4]

[1] National Institute for Research and Development in Optoelectronics INOE 2000, Strada Atomiștilor 409, 077125 Măgurele, Romania; catalin.vitelaru@inoe.ro (C.V.); anca.parau@inoe.ro (A.C.P.); kadremil@yahoo.com (A.E.K.); iulian.pana@inoe.ro (I.P.); mihaela.dinu@inoe.ro (M.D.); lidia.constantin@inoe.ro (L.R.C.)
[2] Research Center for Physical Materials Science and Composite Materials, Research School of Chemistry & Applied Biomedical Sciences, Tomsk Polytechnic University, Lenin's Avenue 43, 634050 Tomsk, Russia
[3] ATS Novus SRL, 030352 Bucharest, Romania; lavinia.voinea@gmail.com
[4] Faculty of Medicine, Transilvania University of Brasov, B-dul Eroilor nr 29, 500036 Brașov, Romania; sarah.costinas@gmail.com (S.C.); r_liliana@yahoo.com (L.R.); mihaela.badea@unitbv.ro (M.B.); mihaela.idomir@unitbv.ro (M.E.I.)
* Correspondence: alinava@inoe.ro (A.V.); cristina.adochite@unitbv.ro (C.S.A.)

Abstract: The increasing occurrence of infections caused by pathogens found on objects of everyday use requires a variety of solutions for active disinfection. Using active materials that do not require daily maintenance has a potential advantage for their acceptance. In this contribution, transparent films, with silver as the main antimicrobial agent and a total thickness of a few tens of nm, were deposited on flexible self-adhesive polymer foils used as screen protectors. TiO_2 and SiO_2 were used as transparent matrix to embed the Ag nanoparticles, ensuring also their mechanical protection and controlled growth. HiPIMS (High-Power Impulse Magnetron Sputtering) was used for the sputtering of the Ag target and fine control of the Ag amount in the layer, whereas TiO_2 and SiO_2 were sputtered in RF (Radio Frequency) mode. The thin film surface was investigated by AFM (Atomic Force Microscopy), providing information on the topography of the coatings and their preferential growth on the textured polymer foil. XRD (X-Ray Diffraction) revealed the presence of specific Ag peaks in an amorphous oxide matrix. UV-Vis-NIR (Ultraviolet-Visible-Near Infrared) spectroscopy revealed the presence of nanostructured Ag, characterized by preferential absorption in the 400 to 500 nm spectral range. The antimicrobial properties were assessed using an antimicrobial test with the *Escherichia coli* strain. The highest efficiency was observed for the Ag/SiO_2 combination, in the concentration range of 10^4–10^5 CFU/mL.

Keywords: antimicrobial; silver; magnetron sputtering; polymer foils

Citation: Vitelaru, C.; Parau, A.C.; Kiss, A.E.; Pana, I.; Dinu, M.; Constantin, L.R.; Vladescu, A.; Tonofrei, L.E.; Adochite, C.S.; Costinas, S.; et al. Silver-Containing Thin Films on Transparent Polymer Foils for Antimicrobial Applications. *Coatings* **2022**, *12*, 170. https://doi.org/10.3390/coatings12020170

Academic Editor: Jeff Rao

Received: 13 December 2021
Accepted: 25 January 2022
Published: 28 January 2022

Publisher's Note: MDPI stays neutral with regard to jurisdictional claims in published maps and institutional affiliations.

Copyright: © 2022 by the authors. Licensee MDPI, Basel, Switzerland. This article is an open access article distributed under the terms and conditions of the Creative Commons Attribution (CC BY) license (https://creativecommons.org/licenses/by/4.0/).

1. Introduction

Pathogenic infectious agents capable of forming biofilms survive on surfaces for long periods of time. It is known that some multidrug-resistant bacterial strains such as MRSA (Methicillin-resistant *Staphylococcus aureus*) and VRE (Vancomycin-resistant *Enterococci*) can survive for weeks on various surfaces in the hospital [1–3]. The use of disinfectants for various surfaces is the most common solution but is associated with some disadvantages related to the fact that they must be used in certain concentrations with varying degrees of toxicity. If applied incorrectly, their bactericidal activity is short-lived, recontamination being possible in a few minutes [4]. Therefore, it is desired to introduce films with long-term antibacterial properties on frequently touched surfaces, such as the screens of various devices (telephones, medical equipment, computers, etc.).

Ever since cellular phone introduction in the 1980s, there has been a continuous increase in its popularity around the world, with an estimated number of 4.5 billion people (60% of the global population) owning such a device between 2016 and 2017. The increase was accelerated after the introduction of smartphones in 2012, with an estimated 2.3 billion owners [5]. While using a mobile phone, the device comes into contact with contaminated areas of the human body, like the hands, or more sensitive areas like the mouth, nose, and eyes [6]. The mobile phone, which has become an essential part of human life, is therefore a very sensitive device when it comes to possible contaminations, and the use of touch screens only prolongs the contact with the skin and increases the probability of contamination [7].

Silver nanoparticles are widely used as potent antimicrobial agents in various applications [8,9]. By embedding them into a transparent dielectric matrix, it is possible to obtain coatings that are both transparent and have antimicrobial properties [10,11]. Chemical methods such as in situ reduction [12] and electrochemical deposition [13] are used for fixing nanoparticles onto dielectric surfaces. Among the physical methods used for deposition of thin films, magnetron sputtering stands as one of the most versatile, being intensively used in both research and industry [14]. Compared to chemical methods it is considered cleaner and more environmentally friendly, leaving no chemical residues or hazardous compounds. The use of high-power pulses in magnetron sputtering, as defined in HiPIMS [15,16], allows adding supplementary control over the properties of the thin film by using highly ionized fluxes that can be directed toward the substrate at variable fluxes and energies [17]. Compared to magnetron sputtering it has the advantage of providing precise control of the quantity by using the additional temporal parameters added by the pulsed regime, namely, the frequency and pulse duration.

Such thin films can be deposited onto self-adhesive polymer foils, adding additional functionalities such as antibacterial properties to the existing mechanical protection provided by such foils. In order to keep the transparency of the coatings, one needs to limit the amount of silver on the surface and to avoid coalescence by confining the metal only in isolated nanosized particles. On the other hand, for achieving antimicrobial efficacy of these layers it is usually necessary to ensure a prolonged release of the silver biocide at a concentration level (0.1 ppb) [18] needed to provide a sufficient amount.

2. Materials and Methods

The method we used for obtaining the thin films was magnetron sputtering. A confocal configuration with 3 targets of 1 inch diameter was used. This configuration allowed us to deposit each material individually, to make multilayer structures, or to co-deposit 2 or 3 materials and mix them at the same time.

For the deposition, we used silicon oxide, titanium oxide, and silver targets. The substrates were made of thin self-adhesive polyurethane foils. The oxide targets were operated in RF sputtering conditions, at 50 W applied power on each of the targets. The deposition was performed in argon at 6 mTorr pressure, and the deposition rates were chosen so that the oxide remained the main material and the silver was only a small addition. The typical deposition time was 30 min, yielding film thicknesses in the 30 to 35 nm range.

For the fine tuning of the deposition rate for Ag, HiPIMS was used to sputter the Ag. The pulse characteristics selected for the deposition of Ag are illustrated in Figure 1. The pulse voltage was set at 650 V, with a peak current of 1.5 A. Pulse duration was 50 μs, and the repetition frequency was varied in the 1 to 10 Hz range.

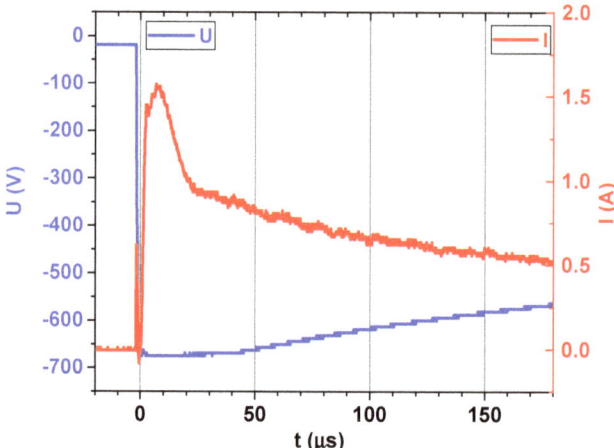

Figure 1. Typical HiPIMS voltage and current pulse shapes used for the deposition of Ag under HiPIMS conditions.

The topography of thin films was analyzed based on AFM images acquired with the AFM/STM Microscopy System (INNOVA VEECO, Berlin, Germany) working in taping mode. UV-spectrophotometry was performed using a Jasco V-670 UV-Vis/NIR Spectrophotometer (Tokyo, Japan). A SEM-Hitachi Tabletop Scanning Electron Microscope (TM3030, Tokyo, Japan) system equipped with Energy Dispersive X-ray spectrometer (EDS, QUANTAX 70, Bruker, MS, USA) was used for the investigation of the composition of the thin films. The crystalline structure was examined using a SmartLab (RIGAKU, Tokyo, Japan) diffractometer equipped with a Cu rotating anode (9 kW) and vertical goniometer with 5 axes. High resolution optics of the incident beam was used to obtain Cu Ka1 (λ = 1.540597 Å) radiation. The measurements were performed in the 2θ interval 10–100°, using an incidence angle of 1°, at a scanning speed of 3°/min. The grain sizes of the investigated samples were calculated using the Scherrer Equation (1):

$$\text{Grain Size} = \frac{0.9\,\lambda}{\beta\,\cos\theta} \qquad (1)$$

where λ represents the diffraction wavelength; β represents the full width at half maximum intensity of the peak (in radians); θ represents the Bragg angle (in degrees).

The antibacterial performance of films was evaluated against *Escherichia coli* strains (ATCC 25922) according to ISO 22196:2011 [19] but with some modifications made depending on the type of films tested as described in the following. The strains were cultured on blood agar for 24 h at 37 °C before use. The films were cut into round shapes (Ø 15 mm) and sterilized on both sides, for 15 min on each side at 60 cm distance using UVC light irradiation. The prepared inoculum of 10^5 and 10^4 CFU/mL (113 µL) was poured on each film. The contact time of the film with inoculum was performed at room temperature for 30 min. Afterward, the suspension and the film were washed in 3 mL of saline solution. The washed solution was dispersed on a Petri dish with blood agar. Incubation of plates was then carried out at 37 °C for 24 and 48 h. The number of *Escherichia coli* colonies was quantified using the automatic analyzer (InterScience Scan 300-Soft Scan Saint Nom la Brétèche, France). The number of colonies were reported as colony forming units (CFUs). Duplicate measurements were performed.

The results were then tabulated and the means and standard deviations between the two experiments were calculated, respectively, and the antibacterial efficiency compared to the films on which the bacteria were deposited using Equation (2) [20]:

$$\text{Antimicrobial Activity (\%)} = \frac{N_c - N_s}{N_c} \cdot 100 \qquad (2)$$

where N_c represents the number of colonies on the control films; N_s represents the number of colonies on the tested films.

3. Results
3.1. Thin Film Deposition and Characterization
3.1.1. Sample Preparation

The polyurethane foils were cleaned in isopropyl alcohol prior to deposition and a plasma cleaning procedure was carried out prior to each deposition. The plasma cleaning consisted of applying an RF bias on the substrate, with 50 V DC self-bias, at 6 mTorr of Argon pressure, for a duration of 15 min.

The topography of the foils before cleaning showed a nanopatterning with roughly 500 nm radius holes distributed on the surface, a few nanometers deep (Figure 2). After the cleaning procedures, one can see that the nanopattern was kept, with a small increase in surface roughness in between the holes. Additional testing with higher voltages and longer processing duration was performed. It was found that if the voltage was too high and/or the duration too long, the surface of the polymer changed dramatically. Therefore, the cleaning conditions for all the deposited samples was kept at 50 V of DC self-bias and 15 min process duration.

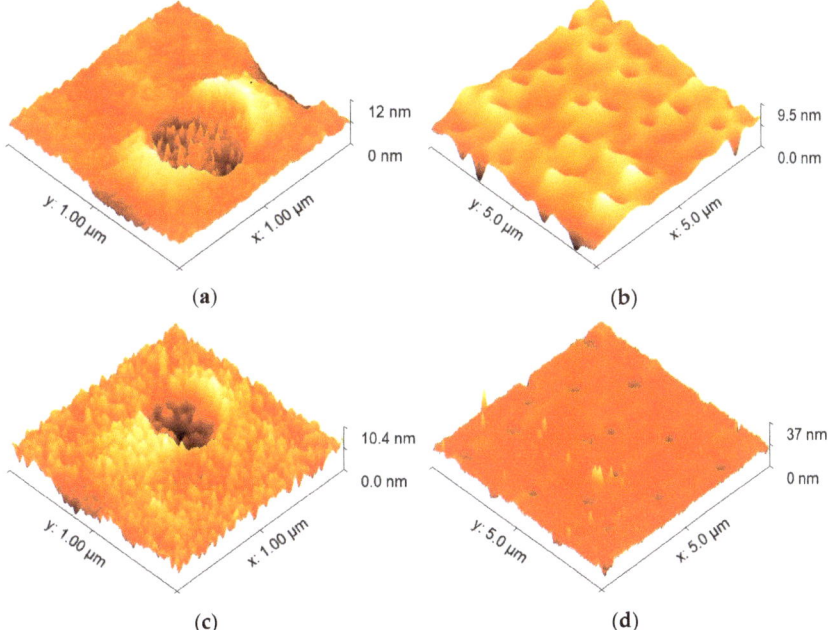

Figure 2. AFM images of the polymer surface before (**a**,**b**) and after the plasma cleaning procedure (**c**,**d**). The images correspond to the same sample, at different scanning areas.

Regarding the depth of the holes, it was observed that after cleaning, the depth slightly increased from 4.16 ± 0.74 to 6.44 ± 2.66 nm. These values were calculated by averaging the depth of all holes found on surfaces recorded on 25 µm² (Figure 2b–d). The same result can also be seen in the case of holes presented in Figure 2a–c, indicating that the cleaning process affected the surface of foils by deepening the already existing holes.

The changes induced in the optical properties of the polymer were investigated by UV-Vis Spectrophotometry. The transmittance, reflectivity, and absorbance of the samples before and after cleaning are represented in Figure 3. The transparency of the foil was diminished from 90 to 85%, whereas the reflectivity remained almost constant. This means that the appearance of the foil did not change significantly. The reduction of transmittance was almost entirely due to the increase in absorption.

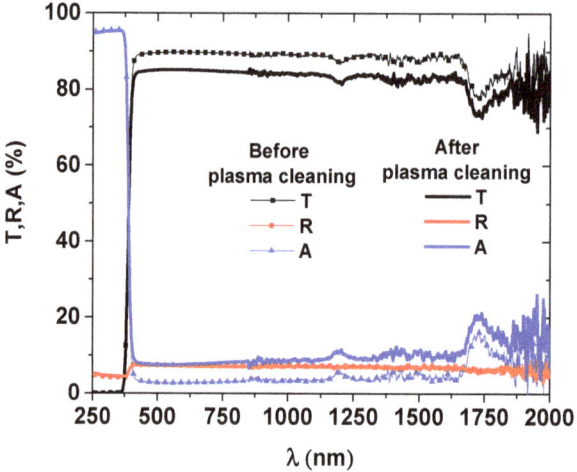

Figure 3. Spectrophotometric curves of the polymer foil sample before and after plasma cleaning at 50 V for 15 min (T–transmittance, R–reflectivity, A–absorbance).

3.1.2. Deposition Process and Physical Characterization

The main challenge for the deposition of transparent coatings with sufficient antimicrobial activity is to find the balance between the maximum quantity of metal atoms that do not reduce drastically the transmittance and the minimum quantity required to obtain the antimicrobial effect. Therefore, the deposition rates of the materials to be deposited and the deposition time should be adapted so that the total quantity of metal atoms remains within these constrains. The deposition rate of SiO_2 and TiO_2 was ~1 nm/min, and the process duration was fixed at 30 min.

In order to find the optimum quantity of silver to be embedded in the oxide matrix, the frequency of the HiPIMS pulses was changed between 1 and 10 Hz, and the Ag films were deposited on the polymer substrate for a total duration of 30 min. The AFM images of the Ag thin films deposited on the polymer surface are represented in Figure 4. One important feature that emerges from these images is that the deposition followed the topography of the polymer foil, with preferential growths in the places where the circular patterns were present. The roughness of the surface increased with increasing frequency, with a visible increase in the grain size on the surface. At a frequency of 10 Hz, the Ag films covered almost entirely the initial pattern on the surface, creating a continuous film. The thickness of the Ag layer obtained at 10 Hz was 30 nm, making it equal to the envisaged thickness of the oxide deposited in a similar duration. The thicknesses of the samples obtained at 1 and 3 Hz, respectively, were not directly evaluated, being too small to obtain reliable results. Nevertheless, a direct relation with the frequency should be considered, accounting for an equivalent deposition rate of 3 nm per 1 Hz.

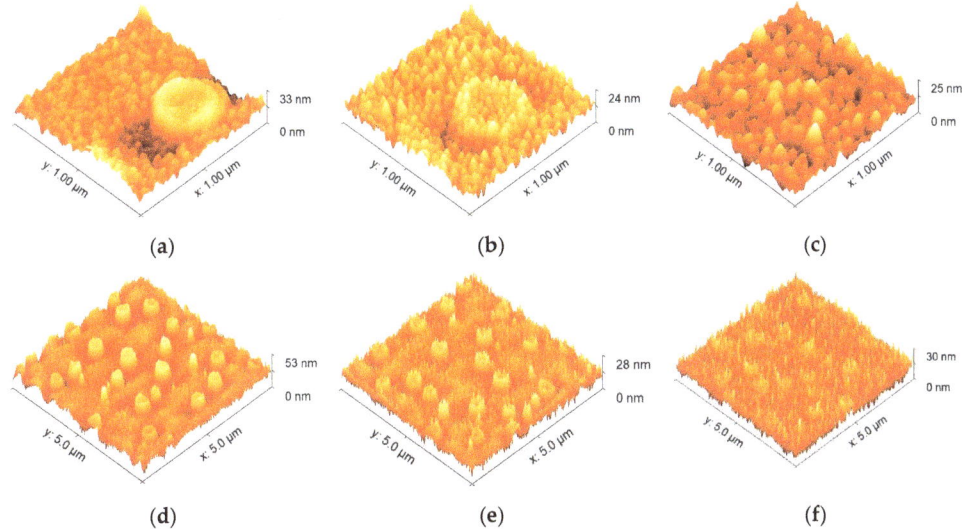

Figure 4. AFM images of the polymer foils deposited with Ag films, for a period of 30 min and at different frequencies: 1 Hz (**a,d**), 3 Hz (**b,e**), and 10 Hz (**c,f**). The images are grouped by deposition conditions corresponding to the same sample and were obtained at different scanning areas.

The XRD spectra obtained for the same samples are presented in Figure 5, demonstrating the presence of crystalline Ag (identified based on JCDPS No.1-071-4613) with cubic structure and (111) preferential orientation. There is a clear dependency of the crystallinity on the deposition frequency, since by increasing the frequency, the crystalline films were obtained. This finding was well evident for the film deposited at 10 Hz. For the films prepared at 1 Hz, the peak located at 38.3° broadened, indicating a low crystallinity, tending to be amorphous. One may also observe an increase in grain size from 1.07 nm (for Ag at 1 Hz) up to 9.42 nm (for Ag at 10 Hz), as seen in Table 1. This increase in peak intensity and crystallinity can be attributed to the film thickness effects [21].

The next experimental step was to deposit simultaneously Ag and two oxides, SiO_2 and TiO_2, respectively, which was done because of mechanical stresses caused by cell compression and the oxidation occurring on the surface of silver. Because of the constrains related to the transparency of the film, the quantity of Ag must be limited to a minimum. Therefore, the HiPIMS sputtering was set at 1 Hz and the total process duration at 30 min. It was seen in Figure 6 that when Ag was found in an amorphous matrix of SiO_2 and TiO_2, respectively, there was a shift of the peak maximum toward higher values, indicating the existence of compressive stress. Taking also into account the oxide matrix surrounding the Ag clusters and the atomic radius of O_2 (0.73 Å) and Ag (1.445 Å) [22], one may conclude that a partial oxidation of the silver occurred, the elementary cell being compressed by the forming compound. Moreover, there were shifts of 2.47° and 1.66° between the standard peak and the one registered for Ag + SiO_2 and Ag + TiO_2, respectively. Although the quantity of Ag embedded in the thin film was very small, comparable with the one corresponding to the 1 Hz deposition, the grain sizes of Ag embedded in the SiO_2 and TiO_2 matrices were significantly higher than the ones obtained for bare silver at 1, 3, and 10 Hz deposition frequencies. According to the data in literature, an amorphous matrix allows Ag to diffuse more easily and to form larger size Ag nanoparticles [23]. Considering that only the maximum Ag peak could be identified in the Ag + SiO_2 and Ag + TiO_2 films, it results that the Ag was finely dispersed in the amorphous matrix [24]. According to Adochite et al., Ag was finely dispersed in the amorphous matrix, since only the maximum Ag peak could be identified in the Ag + SiO_2 and Ag + TiO_2 films [24]. Moreover, the fact that Ag prevented the crystallization of anatase phase, as indicated by XRD patterns of as-deposited Ag + TiO$_2$ nanocomposite films in the mentioned study, also represented an indication of Ag dispersion. This effect was attributed by Okumu et al. [25] to the impinging energetic oxygen ions which were formed during sputtering of Ag:TiO$_2$ films.

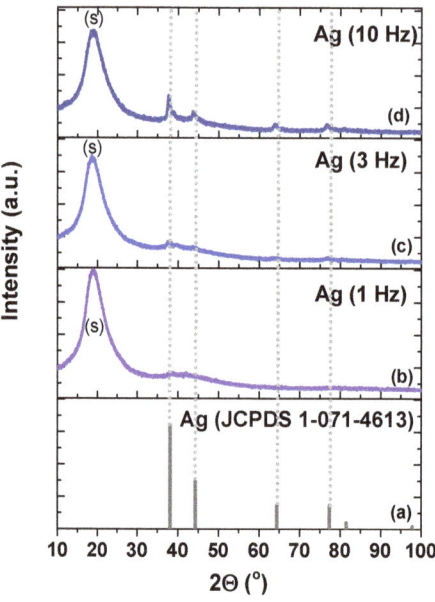

Figure 5. XRD spectra for the Ag thin films obtained at different pulse frequencies 1 Hz (**b**), 3 Hz (**c**), 10 Hz(**d**). The position of Ag peaks in the JCPDS card is depicted in (**a**).

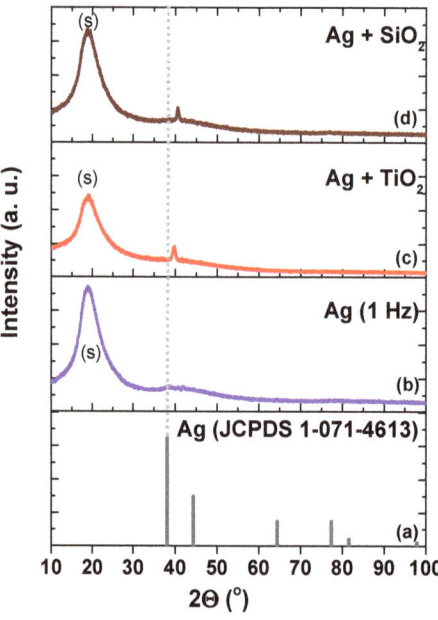

Figure 6. XRD spectra of thin films obtained during 30 min sputtering of Ag at 1 Hz, without any oxide (**b**), with TiO$_2$ (**c**), and with SiO$_2$ (**d**), respectively. The position of Ag peaks in the JCPDS card is depicted in (**a**).

Table 1. Maximum peak position for Ag (111) (according to JCDPS no. 1-071-4613 card), the equivalent position of measured maxima and the grain size of the investigated samples.

Sample	Peak Position (111) (°)	Grain Size (nm)
Ag (JCPDS 01-071-4613)	38.1	NA
Ag (1 Hz, 30 min)	38.3	1.07
Ag (3 Hz, 30 min)	37.64	1.53
Ag (10 Hz, 30 min)	37.63	9.42
Ag + SiO_2	40.57	16.6
Ag + TiO_2	39.76	12.12

For the silver-containing films, only 10% of equivalent Ag thickness was used, compared with the oxide thin film, i.e., ~3 vs. 30 nm. A version of only the oxide thin film was also made by combining two oxides in equal amounts to form a ~30 nm thick film, obtained during a 15 min deposition process. The AFM images of these 3 types of thin films are presented in Figure 7. These functional films combine two materials but keep the ensemble thin enough to maintain the initial surface features. It can be clearly seen that there is a preferential growth that follows the initial pattern of the polymer film.

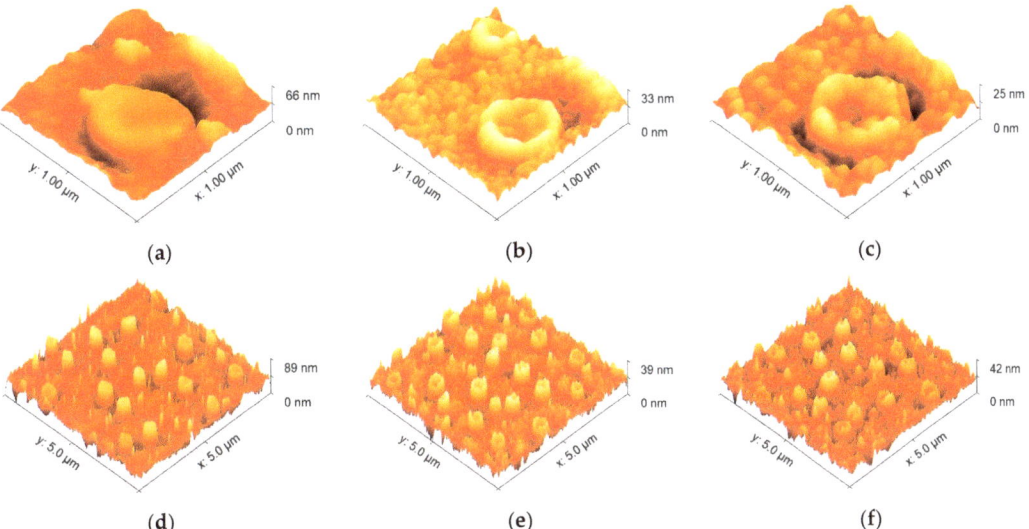

Figure 7. AFM images of the polymer foils deposited with TiO_2 + Ag films (a,d), SiO_2 + TiO_2; (b,e), SiO_2 + Ag; (c,f); all films have a total thickness of ~33 nm. The images grouped by type of thin film correspond to the same sample and were obtained at different scanning areas.

The optical properties of the resulting structure are very important for this application since the foil is applied directly on the phone screen and should keep its transparency and general appearance. From the transmission and reflection spectrum presented in Figure 8, one can see that the SiO_2 film had the best transparency of all. Titanium oxide, on the other hand, had the lowest transparency, most probably because of an insufficient amount of oxygen, leading to sub-stoichiometric composition.

From the absorption spectra, it is evident that an absorption peak around 450–500 nm was present for the silver-containing films. This was specific to the absorption on silver nanoparticles, associated with the surface plasmon resonance phenomena [26]. The position of the peak depends on the matrix, showing a potential way of tuning the size and density of particles by embedding them in different matrices. The presence of this peak confirmed the Ag nanoparticle were finely dispersed in the oxide matrix.

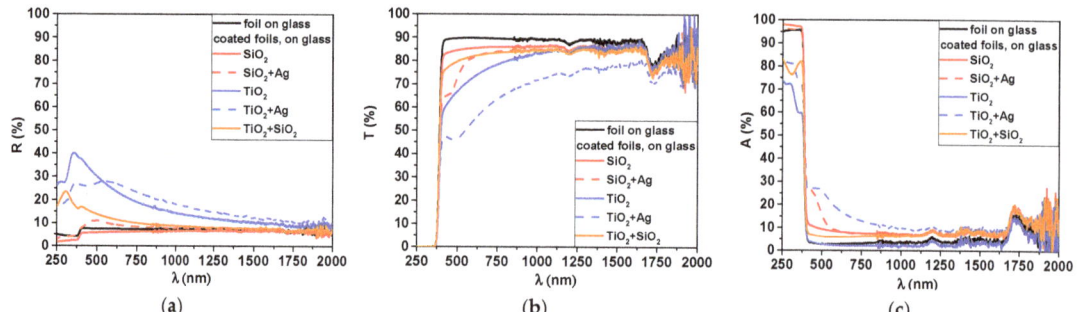

Figure 8. Spectrophotometric curves, reflectivity (**a**), transmittance (**b**) and absorbance (**c**) of the uncoated foil and the foils coated with SiO_2, SiO_2 + Ag, TiO_2, TiO_2 + Ag, TiO_2 + SiO_2.

The film composition was assessed by EDX (Figure 9). Although very thin, a measurable amount for Ag was detected, accounting for only 0.1 to 0.2% of the total. Nevertheless, the silver represented 23 and 28% as compared with Si and Ti, respectively. In Figure 9, the elemental distribution is also presented. One can see that each constitutive element of each investigated layer was well distributed on the investigated area, indicating that the layers were homogeneously coated on the whole surface of the foil.

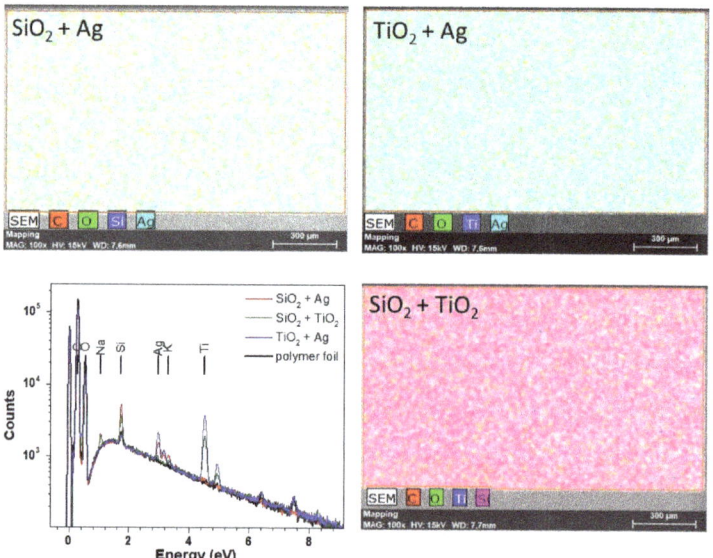

Figure 9. EDX spectra and element distribution of the polymer foils and the foils deposited with TiO_2 + Ag films, SiO_2 + TiO_2, SiO_2 + Ag; all films have a total thickness of ~33 nm.

3.2. Antimicrobial Activity

The antimicrobial activity of all three types of layers was assessed using the methodology described in the Materials and Methods section. The graphs in Figure 10 show the difference between the results read at 24 and 48 h, respectively, for all of the investigated layer types in contact with *Escherichia coli* suspensions of concentrations 10^5 CFU/mL and 10^4 CFU/mL. A maximum efficiency of almost 97% was obtained for samples deposited with SiO_2 + Ag at a concentration of 10^4 CFU/mL. Silver's antibacterial properties can be explained by the following mechanism: owing to the sulfur–Ag affinity, the bacterial cell membrane is enriched with sulfur-containing proteins, which could be favored locations for Ag particle attachment. Hence, silver nanoparticles can damage or affect the structure of bacteria by attaching to the bacterial cell membrane [27,28].

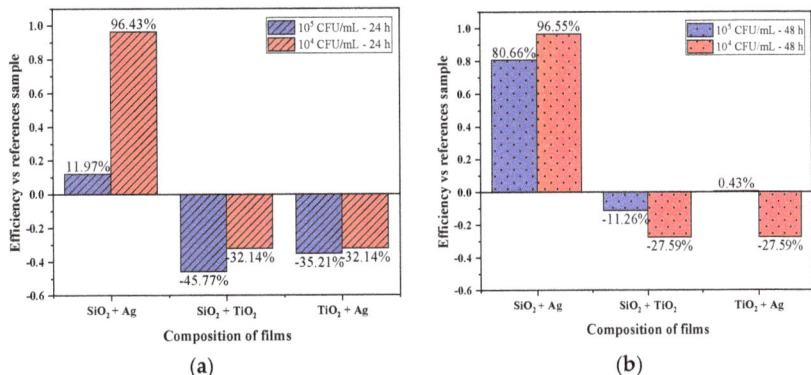

Figure 10. Antimicrobial efficiency of SiO$_2$ + Ag, SiO$_2$ + TiO$_2$, TiO$_2$ + Ag films against *Escherichia coli*, at two concentrations of 10^5 and 10^4 CFU/mL and two incubation times of 24 (**a**) and 48 h (**b**).

On the other hand, for the other two films submitted with SiO$_2$ + TiO$_2$ and TiO$_2$ + Ag under the tested conditions, there was an increased growth compared to the reference sample, the polymer foil without any deposited layer. At 24 h of incubation, for the film submitted with SiO$_2$ + TiO$_2$, an increase of 45.77% in the growth of bacterial colonies was observed compared to the reference sample for the concentration of 10^5 CFU/mL, while the concentration of 10^4 CFU/mL for the same experimental conditions recorded a percentage of 32.14% over the number of colonies in the reference sample. Studies show low antibacterial properties for TiO$_2$ and SiO$_2$. Two very important factors influencing the antibacterial grade are the size of the nanoparticles used [29] and the inoculation method, which, being photocatalytic materials, has been shown in the presence of UV-TiO$_2$ radiation to enhance the antibacterial efficiency for Gram-negative bacterial species [30].

Because TiO$_2$ is a photocatalytic material that has activity in UV-radiation, to make it active and in the visible spectrum, it must be doped with transition metals such as Ag, V, Cr, Mn, Fe, Co, or N. In the case of films deposited with TiO$_2$ + Ag, bacterial activity increased in both *Escherichia coli* concentrations [31].

Figure 10b indicates abundant bacterial growth resulting in a percentage of 80.66% for the antimicrobial efficiency of the sample deposited with SiO$_2$ + Ag at the concentration of 10^5 CFU/mL. For the concentration of 10^4 CFU/mL the increasing trend was maintained, nearly maximum antibacterial efficiency, but the samples deposited with SiO$_2$ + TiO$_2$ had the same proliferative effect as reading the plates at 24 h. As a result, we observed that at the concentration of 10^4 CFU/mL there were determined percentages that indicated an increased efficiency compared to the experiments performed at the concentration of 10^5 CFU/mL. In the case of blood agar plates incubated and evaluated after 48 h, given the determined antibacterial efficacy, early senescence phenomena could be observed, phenomena that needed to be deepened to find a mechanism between the variability of the incubation time and the materials deposited on the polymer films.

4. Discussion

To summarize, we evaluated three types of thin films to be applied on self-adhesive polymer foils for smartphone applications. Among these, the one consisting of the combination of silicon oxide and silver was the most efficient.

We managed to maintain the surface characteristics of the original foil, obtaining a good transparency, and made evident the presence of nanostructured Ag nanoparticles embedded in the silicon oxide matrix.

For films deposited with TiO$_2$ + SiO$_2$ and those with TiO$_2$ + Ag, the microbiological activity of *Escherichia coli* indicated an increased growth of both bacterial concentrations. SiO$_2$ + Ag films had an elevated antibacterial activity of 10^4 CFU/mL, indicating according to Zhao et al. [32] that Ag can form complexes with DNA and RNA resulting in DNA condensation and replication inhibition, as established in other investigations. Silver interaction with thiol groups in proteins can result in inactivation of respiratory enzyme function [33]. Antimicrobial actions can be represented by free radicals produced on silver particles, particularly reactive oxygen species (ROS) [34].

SiO$_2$ + Ag films exhibited a high effect on the *Escherichia coli* bacteria. The efficiency could usually be related to the grain size, with size effects that were identified both at micro and nano scales. For the microscale range, it has already been demonstrated that bacterial attachment depends significantly on the size of particles [35]. Several researchers have reported that bacteria tend to attach to larger particles because they exhibit a larger surface that provides more attachment sites, which permit more attached cells. For example, Soupir et al. reported that at least 60% of attached *Escherichia coli* and enterococci were related to particles smaller than 62 µm [36]. Jeng et al. demonstrated that more than 90% of *Escherichia coli* were attached to particles with sizes ranging from 5 to 30 µm [37]. A similar result was published by George et al., who showed that the percentage of attached *Escherichia coli* bacteria is related to particles larger than 5 µm and increases with particle concentration [38]. Moreover, Oliver et al. reported that the *Escherichia coli* bacteria prefers attachment to particles ranging from 16 to 32 µm, without explaining this mechanism [39].

Related to the nanoscale range, on the other hand, it was observed that the lower values, up to 5 nm, were more likely for the inhibition of the *Escherichia coli* bacteria. Ahmed et al. found that the inactivation of *Escherichia coli* under ultraviolet light irradiation of the TiO$_2$ surface can be attributed to the reduction of the crystallite size from the 30 to 5 nm [40]. Furthermore, the study of Ahmed et al. was in good agreement with others who agreed that the degradation efficiency of TiO$_2$ nanoparticles depends on their morphology, preparation methods, and especially the size of the particles [41]. The challenge in preparation of nanoparticles of nanoscale size is to obtain smaller nanoparticles with homogeneous size distribution [40].

In the present paper it was shown that by using magnetron sputtering co-deposition, this homogeneity can be achieved, with grain sizes in the range of ~10 nm scale. The larger grain size of Ag nanoparticles embedded in SiO$_2$ as compared to the ones embedded in TiO$_2$ appears to be beneficial for the increase in bactericidal effect. In perspective, the combination of these thin films with embedded Ag nanoparticles, with the effect of UV-radiation, can be envisaged as a way to increase the antibacterial effect.

Author Contributions: Conceptualization, C.V. and M.B.; methodology, A.C.P., M.D., M.E.I. and M.B.; formal analysis, A.C.P., M.D., L.R.C., I.P., S.C. and C.S.A.; investigation, A.C.P., M.D., I.P., A.E.K., L.R.C., L.E.T., C.S.A., S.C. and L.R.; writing—original draft preparation, C.V., C.S.A. and M.B.; writing—review and editing A.V., M.E.I., C.S.A. and M.B.; supervision, C.V. and M.B.; project administration, C.V. and M.B.; funding acquisition, C.V. and M.B. All authors have read and agreed to the published version of the manuscript.

Funding: This work was funded by the Romanian Ministry of Education and Research, CCCDI-UEFISCDI, Project No. PN-III-P2-2.1-PED-2019-4966 PED 489 and No. PN-III-P1-1.1-TE-2019-1924 TE 105, within PNCDI III.

Acknowledgments: A.V. thanks Tomsk Polytechnic University within the framework of the Tomsk Polytechnic University–Competitiveness Enhancement Program grant. INOE co-authors also want to thank to Core Program, Project No. 18N/2019.

Conflicts of Interest: The authors declare no conflict of interest. The funders had no role in the design of the study; in the collection, analyses, or interpretation of data; in the writing of the manuscript, or in the decision to publish the results.

References

1. Neely, A.N.; Maley, M.P. Survival of enterococci and staphylococci on hospital fabrics and plastic. *J. Clin. Microbiol.* **2000**, *38*, 724–726. [CrossRef] [PubMed]
2. Dancer, S.J. Importance of the environment in meticillin-resistant *Staphylococcus aureus* acquisition: The case for hospital cleaning. *Lancet Infect. Dis.* **2008**, *8*, 101–113. [CrossRef]
3. Martinez, J.A.; Ruthazer, R.; Hansjosten, K.; Barefoot, L.; Snydman, D.R. Role of environmental contamination as a risk factor for acquisition of vancomycin-resistant enterococci in patients treated in a medical intensive care unit. *Arch. Intern. Med.* **2003**, *163*, 1905–1912. [CrossRef]
4. Perez-Gavilan, A.; de Castro, J.V.; Arana, A.; Merino, S.; Retolaza, A.; Alves, S.A.; Francone, A.; Kehagias, N.; Sotomayor-Torres, C.M.; Cocina, D.; et al. Antibacterial activity testing methods for hydrophobic patterned surfaces. *Sci. Rep.* **2021**, *11*, 6675. [CrossRef]
5. Graveto, J.M.; Costa, P.J.; Santos, C.I. Cell Phone Usage By Health Personnel: Preventive Strategies to Decrease Risk of Cross Infection In Clinical Context. *Texto Contexto Enferm.* **2018**, *27*, e5140016. [CrossRef]
6. Elmanama, A.; Hassona, I.; Marouf, A.; Alshaer, G.; Ghanima, E.A. Microbial Load of Touch Screen Mobile Phones Used by University Students and Healthcare Staff. *J. Arab Am. Univ.* **2015**, *1*, 1–18. [CrossRef]
7. Kister, M.P.; Borowska, K.; Jodłowska-Jędrych, B.; Kister, K.A.; Drop, B. The potential role of cell phones in dissemination of bacteria in a healthcare setting. *Our Dermatol. Online.* **2016**, *7*, 219–224. [CrossRef]
8. Anees Ahmad, S.; Sachi Das, S.; Khatoon, A.; Tahir Ansari, M.; Afzal, M.; Saquib Hasnain, M.; Kumar Nayak, A. Bactericidal activity of silver nanoparticles: A mechanistic review. *Mater. Sci. Energy Technol.* **2020**, *3*, 756–769. [CrossRef]
9. Le Ouay, B.; Stellacci, F. Antibacterial activity of silver nanoparticles: A surface science insight. *Nano Today* **2015**, *10*, 339–354. [CrossRef]
10. Ivanova, T.; Harizanova, A.; Koutzarova, T.; Vertruyen, B. Optical and structural characterization of TiO_2 films doped with silver nanoparticles obtained by sol-gel method. *Opt. Mater.* **2013**, *36*, 207–213. [CrossRef]
11. Varghese, S.; Elfakhri, S.; Sheel, D.W.; Sheel, P.; Bolton, F.J.; Foster, H.A. Novel antibacterial silver-silica surface coatings prepared by chemical vapour deposition for infection control. *J. Appl. Microbiol.* **2013**, *115*, 1107–1116. [CrossRef] [PubMed]
12. Salem, M.A.; Elsharkawy, R.G.; Ayad, M.I.; Elgendy, M.Y. Silver nanoparticles deposition on silica, magnetite, and alumina surfaces for effective removal of Allura red from aqueous solutions. *J. Solgel Sci. Technol.* **2019**, *91*, 523–538. [CrossRef]
13. Yin, D.; Liu, Y.; Chen, P.; Meng, G.; Huang, G.; Cai, L.; Zhang, L. Controllable synthesis of silver nanoparticles by the pulsed electrochemical deposition in a forced circulation reactor. *Int. J. Electrochem. Sci.* **2020**, *15*, 3469–3478. [CrossRef]
14. Baptista, A.; Silva, F.J.G.; Porteiro, J.; Míguez, J.L.; Pinto, G.; Fernandes, L. On the Physical Vapour Deposition (PVD): Evolution of Magnetron Sputtering Processes for Industrial Applications. *Proc. Procedia Manuf.* **2018**, *17*, 746–757. [CrossRef]
15. Anders, A. Tutorial: Reactive high power impulse magnetron sputtering (R-HiPIMS). *J. Appl. Phys.* **2017**, *121*, 171101. [CrossRef]
16. Britun, N.; Minea, T.; Konstantinidis, S.; Snyders, R. Plasma diagnostics for understanding the plasma-surface interaction in HiPIMS discharges: A review. *J. Phys. D Appl. Phys.* **2014**, *47*, 224001. [CrossRef]
17. Greczynski, G.; Lu, J.; Jensen, J.; Bolz, S.; Kölker, W.; Schiffers, C.; Lemmer, O.; Greene, J.E.; Hultman, L. A review of metal-ion-flux-controlled growth of metastable TiAlN by HIPIMS/DCMS co-sputtering. *Surf. Coat. Technol.* **2014**, *257*, 15–25. [CrossRef]
18. Kumar, R.; Münstedt, H. Silver ion release from antimicrobial polyamide/silver composites. *Biomaterials* **2005**, *26*, 2081–2088. [CrossRef]
19. ISO. *Measurement of Antibacterial Activity on Plastics and Other Non-Porous Surfaces*; ISO 22196; ISO: Geneva, Switzerland, 2011.
20. Al-Sharqi, A.; Apun, K.; Vincent, M.; Kanakaraju, D.; Bilung, L.M. Enhancement of the antibacterial efficiency of silver nanoparticles against gram-positive and gram-negative bacteria using blue laser light. *Int. J. Photoenergy* **2019**, *2019*, 2528490. [CrossRef]
21. Jalili, S.; Hajakbari, F.; Hojabri, A. Effect of silver thickness on structural, optical and morphological properties of nanocrystalline Ag/NiO thin films. *J. Theor. Appl. Phys.* **2018**, *12*, 15–22. [CrossRef]
22. Guo, S.; Liu, C.T. Phase stability in high entropy alloys: Formation of solid-solution phase or amorphous phase. *Prog. Nat. Sci. Mater.* **2011**, *21*, 433–446. [CrossRef]
23. Navabpour, P.; Ostovarpour, S.; Hampshire, J.; Kelly, P.; Verran, J.; Cooke, K. The effect of process parameters on the structure, photocatalytic and self-cleaning properties of TiO_2 and Ag-TiO_2 coatings deposited using reactive magnetron sputtering. *Thin Solid Films* **2014**, *571*, 75–83. [CrossRef]
24. Adochite, R.C.; Munteanu, D.; Torrell, M.; Cunha, L.; Alves, E.; Barradas, N.P.; Cavaleiro, A.; Riviere, J.P.; Le Bourhis, E.; Eyidi, D.; et al. The influence of annealing treatments on the properties of Ag:TiO_2 nanocomposite films prepared by magnetron sputtering. *Appl. Surf. Sci.* **2012**, *258*, 4028–4034. [CrossRef]
25. Okumu, J.; Dahmen, C.; Sprafke, A.N.; Luysberg, M.; Von Plessen, G.; Wuttig, M. Photochromic silver nanoparticles fabricated by sputter deposition. *J. Appl. Phys.* **2005**, *97*, 094305. [CrossRef]
26. Xu, G.; Tazawa, M.; Jin, P.; Nakao, S.; Yoshimura, K. Wavelength tuning of surface plasmon resonance using dielectric layers on silver island films. *Appl. Phys. Lett.* **2003**, *82*, 3811–3813. [CrossRef]
27. Sarkheil, M.; Sourinejad, I.; Mirbakhsh, M.; Kordestani, D.; Johari, S.A. Application of silver nanoparticles immobilized on TEPA-Den-SiO_2 as water filter media for bacterial disinfection in culture of Penaeid shrimp larvae. *Aquac. Eng.* **2016**, *74*, 17–29. [CrossRef]

28. Zhang, H.; Chen, G. Potent antibacterial activities of Ag/TiO$_2$ nanocomposite powders synthesized by a one-pot sol-gel method. *Environ. Sci. Technol.* **2009**, *43*, 2905–2910. [CrossRef]
29. Adams, L.K.; Lyon, D.Y.; McIntosh, A.; Alvarez, P.J.J. Comparative toxicity of nano-scale TiO$_2$, SiO$_2$ and ZnO water suspensions. *Water Sci. Technol* **2006**, *54*, 327–334. [CrossRef]
30. Levchuk, I.; Kralova, M.; Rueda-Márquez, J.J.; Moreno-Andrés, J.; Gutiérrez-Alfaro, S.; Dzik, P.; Parola, S.; Sillanpää, M.; Vahala, R.; Manzano, M.A. Antimicrobial activity of printed composite TiO$_2$/SiO$_2$ and TiO$_2$/SiO$_2$/Au thin films under UVA-LED and natural solar radiation. *Appl. Catal.* **2018**, *239*, 609–618. [CrossRef]
31. Ahmad Barudin, N.H.; Sreekantan, S.; Thong, O.M.; Sahgal, G. Antibacterial activity of Ag-TiO$_2$ nanoparticles with various silver contents. *Proc. Mater. Sci. Forum* **2013**, *756*, 238–245. [CrossRef]
32. Zhao, J.X.; Zhang, B.P.; Li, Y.; Yan, L.P.; Wang, S.J. Optical and photocatalytic properties of TiO$_2$/Ag-SiO$_2$ nanocomposite thin films. *J. Alloys Compd.* **2012**, *535*, 21–26. [CrossRef]
33. Suwanchawalit, C.; Wongnawa, S.; Sriprang, P.; Meanha, P. Enhancement of the photocatalytic performance of Ag-modified TiO$_2$ photocatalyst under visible light. *Ceram. Int.* **2012**, *38*, 5201–5207. [CrossRef]
34. Kim, J.S.; Kuk, E.; Yu, K.N.; Kim, J.H.; Park, S.J.; Lee, H.J.; Kim, S.H.; Park, Y.K.; Park, Y.H.; Hwang, C.Y.; et al. Antimicrobial effects of silver nanoparticles. *Nanomed. Nanotechnol. Biol. Med.* **2007**, *3*, 95–101. [CrossRef] [PubMed]
35. Wu, T.; Zhai, C.; Zhang, J.; Zhu, D.; Zhao, K.; Chen, Y. Study on the attachment of *Escherichia coli* to sediment particles at a single-cell level: The effect of particle size. *Water* **2019**, *11*, 819. [CrossRef]
36. Soupir, M.L.; Mostaghimi, S.; Dillaha, T. Attachment of *Escherichia coli* and Enterococci to Particles in Runoff. *J. Environ. Qual.* **2010**, *39*, 1019–1027. [CrossRef]
37. Jeng, H.C.; England, A.J.; Bradford, H.B. Indicator organisms associated with stormwater suspended particles and estuarine sediment. *J. Environ. Sci. Health-Part A Toxic/Hazard. Subst. Environ. Eng.* **2005**, *40*, 779–791. [CrossRef]
38. George, I.; Anzil, A.; Servais, P. Quantification of fecal coliform inputs to aquatic systems through soil leaching. *Water Res.* **2004**, *38*, 611–618. [CrossRef]
39. Oliver, D.M.; Clegg, C.D.; Heathwaite, A.L.; Haygarth, P.M. Preferential attachment of *Escherichia coli* to different particle size fractions of an agricultural grassland soil. *Water Air Soil Pollut.* **2007**, *185*, 369–375. [CrossRef]
40. Ahmed, F.; Awada, C.; Ansari, S.A.; Aljaafari, A.; Alshoaibi, A. Photocatalytic inactivation of *Escherichia coli* under UV light irradiation using large surface area anatase TiO$_2$ quantum dots. *R. Soc. Open Sci.* **2019**, *6*, 191444. [CrossRef]
41. Benabbou, A.K.; Derriche, Z.; Felix, C.; Lejeune, P.; Guillard, C. Photocatalytic inactivation of *Escherischia coli*. *Appl. Catal. B Environ.* **2007**, *76*, 257–263. [CrossRef]

Article

Non-Destructive Monitoring of *P. fluorescens* and *S. epidermidis* Biofilm under Different Media by Fourier Transform Infrared Spectroscopy and Other Corroborative Techniques

Marco Consumi [1,*,†], Kamila Jankowska [1,†], Gemma Leone [1], Claudio Rossi [1], Alessio Pardini [1], Eric Robles [2], Kevin Wright [2], Anju Brooker [2] and Agnese Magnani [1,*]

1. Department of Biotechnology, Chemistry and Pharmacy, University of Siena, Via A. Moro 2, 53100 Siena, Italy; Kamila.Jankowska@student.unisi.it (K.J.); gemma.leone@unisi.it (G.L.); claudio.rossi@unisi.it (C.R.); alessio.pardini@unisi.it (A.P.)
2. Procter & Gamble Ltd.—Newcastle Innovation Centre, Whitley Road, Longbenton, Newcastle upon Tyne NE12 9BZ, UK; robles.es@pg.com (E.R.); wright.ki@pg.com (K.W.); brooker.am@pg.com (A.B.)
* Correspondence: marco.consumi@unisi.it (M.C.); agnese.magnani@unisi.it (A.M.)
† Both authors contributed equally to this work.

Received: 4 September 2020; Accepted: 25 September 2020; Published: 28 September 2020

Abstract: In the present study, the early stage of bacteria biofilm formation has been studied as a function of different nutrients. Infrared spectra of *Pseudomonas fluorescens* (PF) and *Staphylococcus epidermidis* (SE), on germanium ATR crystal, were collected under deionized water H_2O, phosphate buffered solution (PBS) and PBS with glucose (PBS-G). In H_2O, protein bands of PF increased while, no difference in PBS and PBS-G were observed until 135 min. SE strain showed a low sensitivity to PBS composition starting to expose proteins on surfaces after 120 min. SE shows a low polysaccharides increase in H_2O while, in bare and enriched PBS their intensity increases after 120 and 75 min. in PBS and PBS-G respectively. PF exhibits a peculiar behavior in H_2O where the saccharide bands increased strongly after 100 min, while under all the other conditions, the intensity of polysaccharide bands increased up to the plateau probably because the layer of the biofilm exceeded the penetration capability of FTIR technique. All data suggest that, under lack of nutrients, both the bacteria tend to firmly anchor themselves to the support using proteins.

Keywords: Fourier transform infrared spectroscopy; attenuated total reflectance; bacteria; in-situ analysis; ATR-FTIR; vibrational spectroscopy; 2nd order derivative method

1. Introduction

Most bacteria, when propagated in static liquid culture, grow within the broth phase or quickly sediment to the bottom. External environmental conditions deeply affect the success in surface colonization and the consequent pathogenesis of biofilm related infections [1–5]. Bacteria are equipped to live at the solid–liquid interface using their flagella, pili, exopolysaccharides, and other adhesive components, often using also external environmental substances. Indeed, bacteria are able to respond to environmental stimuli, by appropriately changing their metabolism and producing extracellular polymeric substances (EPS) that permit a prompt formation of surface-related bacterial communities [6,7]. Metals [8], nanoparticles [9], extracellular DNA [10], and different media [6,7] induce changes in polysaccharides, proteins, and extracellular DNA composing the EPS [11]. Several studies have highlighted the influence of external environment also on attached bacteria in terms of metabolic cascades and cell–cell communications that differ significantly from what observed in a planktonic state [12,13]. Even though the extensive scientific literature on the influence of external environment in several biofilm processes, the signals that influence the system at a very beginning state, promoting surface attachment and growing, are not yet clear. When attached to the surfaces, bacteria can be seen as highly heterogeneous communities that exhibit complex biochemical processes. Their development involves initial attachment of the microorganism to the surface, the formation of microcolonies, and, finally, differentiation of microcolonies into exopolysaccharide-encased mature biofilms [7,8]. Limited to the early stage of bacteria attachment and EPS development, several components play a key role as polysaccharides, proteins and lipids [14]. Polysaccharides play a fundamental role in the biofilm's matrix. Some of the most common polysaccharides are cellulose, Psl, Pel, alginate, and the staphylococcal polysaccharide intercellular adhesin [15–19]. Proteins also play a critical role and, in some cases, are present at higher concentrations than polysaccharides [20,21]. Common proteins present in the matrix are amyloid fiber [22]. Knowing how proteins and polysaccharides change under different stimuli is fundamental to understand the biofilm evolution. In fact, biofilm phenotype is not only different from the planktonic phenotype, but it changes during the biofilm development as a function of external physico-chemical stimuli [7–10,23]. The whole picture of biofilm formation is extremely challenging because of the inherent complexity and the multifactorial dependence of bacterial biofilm. In fact, not only different bacterial species may form different biofilm structures under identical conditions, but also the same bacterial species may form different biofilm structures under different environmental conditions. *Pseudomonas aeruginosa* forms mushroom-shaped microcolonies when it grows in the presence of glucose medium, whereas it forms flat biofilms when in citrate medium [24].

ATR-FTIR spectroscopy has proven to be a useful analytical tool for monitoring biofilms in situ, being non-destructive, in real-time and under fully hydrated conditions technique [11,19,25–29]. Boualam et al. [30] showed that ATR-FTIR technique permitted to differentiate biofilms as a function of water samples containing variable quantities of biodegradable organic matter.

This work focus on the study of the early stages of bacteria attachment to germanium crystal surface, using attenuated total reflectance Fourier transform infrared (ATR-FTIR) technique. A detailed analysis of the whole spectral profile of bacteria under study has been performed to gain more insight at the molecular level into biochemical and physiological changes during the early stages of biofilm development. In fact, through the analysis of the whole spectra of the bacteria, it is possible to monitor spectra changes, even weak, induced by metabolic changes induced by environmental changes [22]. *Pseudomonas fluorescens* (PF) and *Staphylococcus epidermidis* (SE) bacteria were chosen as a model because of their inclination to be surface bound in diverse environmental conditions. In particular, PF, a Gram-negative bacterium, was chosen as the reference Gram-negative bacterium for the following reasons: (i) It is present in drinking water distribution networks [31], (ii) can be grown in low-nutrient situations [32], (iii) is used in a standard procedure for measurement of assimilable organic carbon in water [32], and (iv) has been widely used in model bacterial surface colonization studies [2,15,26,33]. SE, a Gram-positive bacterium, was picked because: (i) It is an opportunistic agent possessing an intrinsic pathogenic character [34], (ii) is part of the normal mucosa and skin microflora, and (iii) is the causative agent in numerous invasive and toxigenic diseases [34].

2. Materials and Methods

2.1. Materials

All reagents, salts, nutrients, agar (analytical grade), and solvents (HPLC grade) were purchased from Sigma-Aldrich (Milan, Italy) and used without any additional purification. Non-pathogenic strain of Pseudomonas fluorescens was purchased from the Leibniz Institute DSMZ-German Collection of Microorganisms and Cells Cultures (Braunschweig, Germany). Staphylococcus epidermidis NCTC 11,047 Lenticule® discs were purchased from Sigma Aldrich company.

2.2. Media Used and Their Preparation

H_2O: Deionized water, freshly made by Acquinity P/7 apparatus (MembraPure GmbH, Berlin, Germany)

PBS: Phosphate Buffer Saline PBS for microbiology (catalogue n° P3813 Sigma Aldrich) was dissolved in deionized water according to the instruction. The final pH was 7.5.

PBS-G: Was prepared by adding glucose to a PBS solution to a final concentration of 2.5 g/L. The final pH was 7.5.

All media were autoclaved at 121 °C for 21 min prior to use.

2.3. Bacteria Cultivation

Lyophilized bacteria pellets were rehydrated adding 4 mL of sterile Tryptic soy broth (TSB) media. After 30 min, the bacteria suspensions were gently mixed. 0.5 mL of each bacterial suspension were collected and transferred to Petri dishes containing nutrient agar to permit the growth of bacterial colonies. PF was incubated at room condition (21 ± 2 °C) under laminar chamber whereas SE was incubated at 36 °C till colonies formation on the agar plate are evident. Then, a loopful of bacterial biomass was transferred from the nutrient agar plate to fresh 100% TSB medium and incubated for 24 h. Optical density for Pseudomonas strain at 570 nm (OD 570) and Staphylococcus at 600 nm (OD 600) have been measured to monitor bacterial growth.

2.4. Planktonic P. fluorescens and S. epidermidis Preparation for ATR Analysis

TSB media bacteria suspension were centrifuged for 10 min at 4000 rpm. The supernatant was discarded, the pellet was resuspended in desired medium and the optical density was reduced to 0.23 by adding appropriate volume of media. The final suspension has been used for experiment without any additional treatment.

Comparison of bacteria at the beginning and after 180 min in contact with the media of interest have been performed drying the bacteria on the ATR crystal using a homemade top, equipped with a diffusor able to flow dry nitrogen homogeneously onto ATR crystal surface. Drying take about 5 min at a temperature of 23 ± 2 °C. Three repetitions of ATR spectra measurement were performed for each bacteria and media system.

2.5. Spectroscopy Study

All the samples were analyzed using a Nicolet IS50 FTIR spectrophotometer (Thermo Nicolet Corp., Madison, WI, USA), equipped with nine-reflection germanium ATR crystal (Pike 16154, Pike Technologies) and a deuterated-triglycine sulphate (DTGS) detector [35]. Before each experiment Ge crystal was treated with 70% ethanol overnight and dried under a flow of sterilized nitrogen. Typically, 128 scans at a resolution of 4 cm^{-1} in the range of 4000–800 cm^{-1} were recorded. The frequency scale was internally calibrated with a helium–neon reference laser to an accuracy of 0.01 cm^{-1}. OMNIC software (OMNIC software system Version 9.8 Thermo Nicolet) was used for spectra manipulation. MinMax normalization was applied to the spectra where they are first offset-corrected by setting the minimum intensity of the whole spectrum, or of a defined spectral region, to zero. Spectra are then scaled with the maximum intensity value equaling to one and a piecewise baseline correction was performed. Baseline is obtained by several user-defined points which are connected by straight lines. Correction is achieved by subtracting the baseline from the sample spectrum. Chosen points are the same for all the spectra [36]. The second derivative of the FTIR spectra were calculated using the Savitsky–Golay method (29 points and third-degree polynomial) by OMNIC software (Thermo Nicolet Corp., Madison, WI, USA).

2.6. Adhesion Assay

Adhesion of bacteria was quantified by using a crystal violet method (CV). The experiments have been performed on the same bacterial suspension used for spectroscopy studies and using the same experimental conditions, together with control samples (TSB medium). CV has been made as described by Genevaux et al. [37] with minor modifications. Briefly, each bacterial strain suspended in the media of interest was monitored for 240 min (early stage) to follow biofilm growth using a rapid screening method in 96-well microliter plates. At predefined time, unbound cells were removed by vigorous washing for 5 times with PBS buffer. Then, 200 µL of 0.1% aqueous CV was added to each well and the plate was left to stand for 30 min, then, 180 µL of each well was aspirated again and each well was washed 5 times with large amount of water. Finally, CV bound biofilm was eluted by adding 200 µL of 96% ethanol to each well and left to stand for 30 min before reading it with a micro-plate reader at 540 nm using Biotrak II (Amersham Bioscience).

O-safranin assay (SN) was carried out at the end of spectroscopy study. The amount of formed biofilm was quantified by staining samples with 0.1% of safranin (incubation time: 10 min). Then, each sample was vigorously washed 5 times with PBS and photographed by a digital camera (Olympus Camedia C2000Z, Olympus Corporation Tokyo, Japan)

3. Results and Discussion

3.1. Adhesion Test

Relationship between time and adhesion of the two strains, to polystyrene micro-titer plates as a function of environmental conditions is depicted in Figure 1. Both the bacteria strains adhered to the plate surface and the amount of EPS slowly increased. Small differences in adhesion properties among the different media can be seen after 30 min of contact with the surface. The control samples grew faster than the others because of Tryptic soy broth (TSB), a complex, general purpose medium, that is routinely used as a culture broth. Indeed, it offers a high nutritional environment to bacteria stimulating their proliferation.

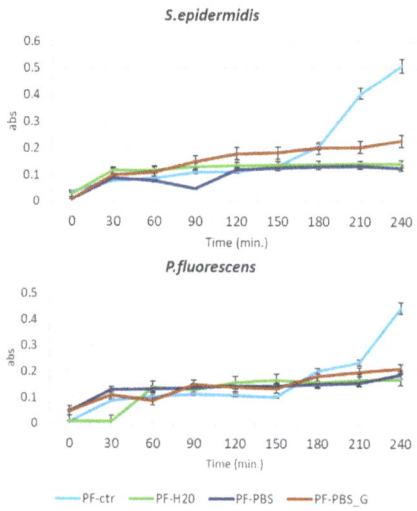

Figure 1. Relationship between time and adhesion of the two strains, to polystyrene micro-titer plates as function of environmental conditions.

3.2. ATR-FTIR Spectra of SE and PF Bacteria

Considering the average size of bacteria, that oscillates from 0.7 to 2 µm, the very first layer of attached bacteria can be analyzed by ATR technique [26–28]. The infrared spectra exhibited typical bands corresponding to carboxyl, amide, ester, phosphate, and carbohydrate moieties [38–48] (Table 1).

Table 1. Assignment of main infrared vibration bands in the 3000–900 cm^{-1} region of ATR-FTIR spectrum of *Pseudomonas fluorescens* pellet harvested by centrifugation. Key: ν—stretching, δ—bending, τ—twist, a-asymmetric, s—symmetric, LPS—lipopolysaccharides.

Wavenumber (cm^{-1})	Tentative Assignment of the Main Band to the Relevant Functional Groups			Ref
	Assignment	Principal Compounds	Main Corresponding Cellular Compounds	
1736	νC=O	Esters from lipids	Membranes	[30–33,44]
1713	νC=O	Esters, carboxylic acids	Nucleoid, ribosomes	
1700–1580	νC=O, νC=N, νC=C, δNH	DNA, RNA bases	Nucleoid, ribosomes	
1693–1627	Amide I (νC=O coupled with δN-H), δH2O	Proteins, water (1640cm^{-1})	Membranes, cytoplasm, flagella, pili, ribosomes	[41]
1568-1531	Amide II (δN-H coupled with νC-N)	Proteins	Membranes, cytoplasm, flagella, pili, ribosomes	[41]
1468,1455	δCH2, δCH3	Lipids	Membranes	
1400	νsCOO-	Amino acids, fatty acids chains	Capsule, peptidoglycan	[26–28]
1317;1281	τCH2; Amide III(νC-N coupled with δN-H)	Fatty acids chains, proteins	Membranes, cytoplasm, flagella, pili, ribosomes	
1238	νa PO2-	Phosphodiester, phospholipids, LPS, nucleic acids, ribosomes	Membrane, nucleoids, ribosomes	
1220	νC-O-C	-	Capsule, storage inclusion	[16,25,33,44]
1200–900	νC-O, νC-C, δC-O-H,	Polysaccharides	-	
-	δC-O-C	-	-	
1172,1153	νsC-OH, νC-O	Proteins, carbohydrates, esters	-	
1118	νsCC	Phosphodiester, phospholipids	Nucleoids, ribosomes	[20,28,40–48]
1086	νsPO2-	LPS, nucleic acids	Membranes, nucleoid, ribosomes	[28,48]
1058	νsC-O-C,	Polysaccharides	Capsule, peptidoglycan	[16,33,40,44]
-	νsP-O-C (R-O-P-O-R')	-	-	
1041	νO-H coupled with δC-O	Polysaccharides	Capsule, peptidoglycan	[42–45]
1026	CH2OH	Carbohydrates	Storage inclusion	
993		Ribose skelet (ARN)	Ribosomes	[28,40,47]
970	νC-C, νP-O-P	RNA backbone	Ribosomes	

The infrared spectra of SE and PF bacteria were reported in Figure 2.

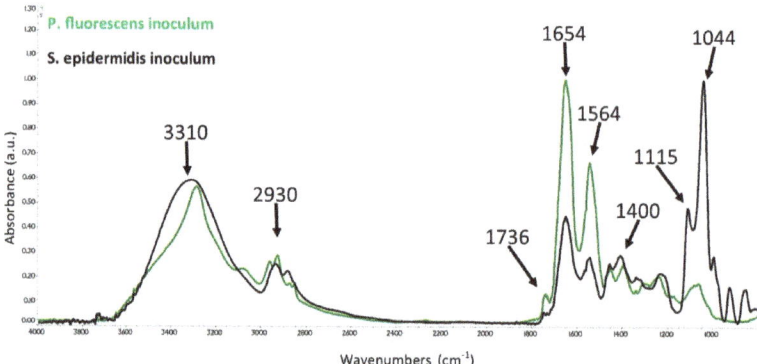

Figure 2. FTIR spectra of *P. fluorescens* (green) and *Staphylococcus epidermidis* (black).

Typical spectra of bacteria can be divided in 3 main zones:

Zone (I) proteins (1700–1500 cm^{-1}) identified by the N-H stretching and the vibrations of the peptide linkage, the Amide I band arising from backbone amide C=O stretching vibrations and the amide II band arising from the out-of-phase combination of the NH in-plane bending and the CN stretching vibrations [38].

Zone (II) phospholipids and nucleic acids (1220–1260 cm^{-1}) characterized mainly by PO2—asymmetric stretching vibrations of phosphate groups [39,40].

Zone (III) polysaccharides (1200–900 cm^{-1}) mainly localized in the fingerprint region with broad band [41–44].

Figure 2 shows the starting spectra of the two bacterial cells biofilm formed on Ge crystal, dried under a flow of dry nitrogen. Both strains biofilm reveals all the typical bands corresponding to bacterial macrocomponents (proteins, polysaccharides, lipids). A strong difference in proteins/polysaccharide ratio is found. SE infrared spectrum is characterized by strong bands in Zone III due to polysaccharides, while, conversely, PF shows strong bands in Zone I, related to protein absorption, and a weak polysaccharides absorption. The phospholipids and nucleic acid bands are similar for both SE and PF, as expected, because we started from about equal number of bacteria cells and cell division and proliferation generally started only 6–8 h after inoculation.

3.3. Changing of Amide Bands as a Function of Time and Media

Figure 3 shows the changes of Amide bands intensities over the time, for both bacterial species under the media of interest. SE shows a rapid increase of Amide II band intensity after 45 min while PF shows the same behavior only after 75 min.

The Amide band of SE has a lower intensity than that of PF. This can be explained considering that in H$_2$O, PF increased its protein production to favor its attachment to the surface. No difference in PF growth in PBS and PBS-G is observed for the first 60 min, but, after 135 min, additional proteins are exposed. PF in PBS media behaves as SE in PBS-G, while SE strain shows a lower sensitivity to PBS composition starting to expose proteins on surfaces after 120 min.

The collected data suggest that, under lack of nutrients and in presence of osmotic shock, both the bacteria tend to firmly anchor themselves to the support. The osmotic shock seems to be the predominant parameter affecting bacterial attachment to the surface. Indeed, in the presence of bare PBS (physiological ionic strength without nutrients) the bacteria need to produce and spread a lower amount of proteins on the surface. The addition of a nutrient, as glucose, provokes only a slight effect on protein expression from bacteria.

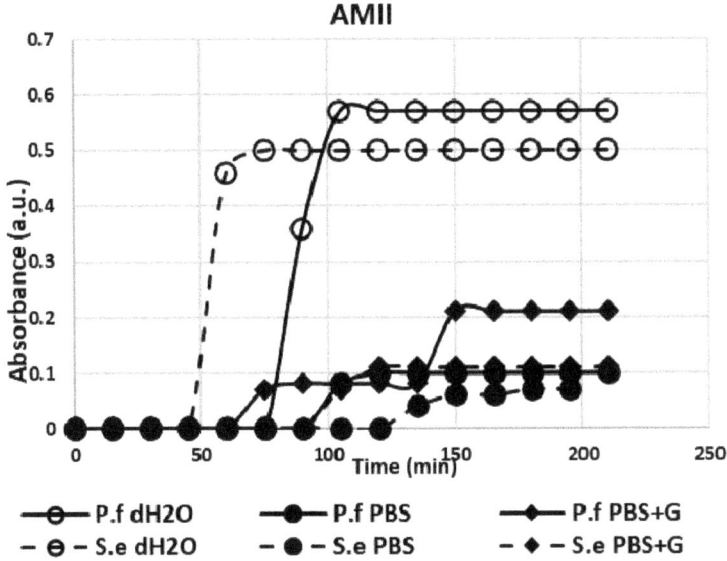

Figure 3. The evolution of Amide II (AMII) intensity band over the time.

3.4. Changing of Polysaccharides Bands as a Function of Time and Media

Figure 4 shows the changing of intensity of the polysaccharide bands over a period of about 200 min. Polysaccharide bands intensity starts to be detectable only after 50 min. At the inoculum, SE (Gram-positive) shows polysaccharide bands more intense than those of PF (Gram-negative) because of differences in their membrane structures (see Figure 2). In fact, the Gram-positive cell wall is primarily made up of peptidoglycan (ca. 40–80% of the dry weight of the wall), a polymer of N-acetylglucosamine and N-acetylmuramic acid, whereas the cell walls of Gram-negative bacteria are more complex due to the presence of an outer membrane, that does not contain teichoic or teichuronic acids but rich of proteins [45].

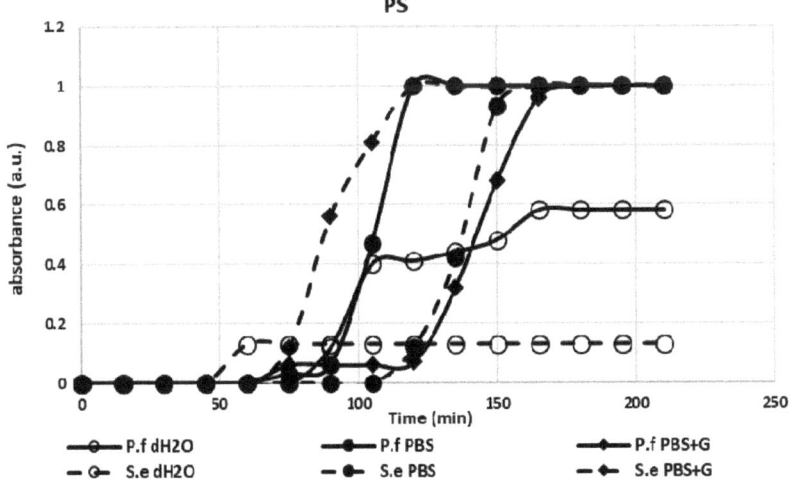

Figure 4. Evolution of polysaccharide (PS) intensity band over the time.

SE shows a low increase of polysaccharide bands in H_2O medium (osmotic shock). A different behavior is found in bare and enriched PBS. Indeed, polysaccharide bands intensity increases after 75 min in PBS with glucose, whereas without glucose it increases only after 120 min.

PF exhibits a peculiar behavior in H_2O where the saccharide bands increase strongly after 100 min without reaching the plateau. In all the other conditions, the intensity of polysaccharide bands increases to the plateau probably because the layer of the biofilm exceeded the penetration capability of FTIR technique. In particular, PF shows an increase of the polysaccharide bands intensity after 90 min in PBS and only after 120 min in the presence of glucose. This behavior suggests that in all the conditions, both the bacteria species start to increase their polysaccharides content, probably because of biofilm needs to counteract the lack of nutrients.

3.5. FTIR Spectra Changes as a Function of Media Composition

3.5.1. S. epidermidis

Figure 5 shows the IR spectra of the SE bacteria after 180 min in H_2O, PBS, and PBS-G. Comparing them to inoculum FTIR spectrum, a significant increase of protein band intensity in the absence of nutrients (H_2O) is observed. The spectra change from a prevalence of polysaccharide absorption band to a prevalence of amide absorption bands, confirming the presence of high level of proteins in contact with the surface, under osmotic stress conditions. This finding underline how the lack of nutrients in the medium encourages bacterial attachment to a surface.

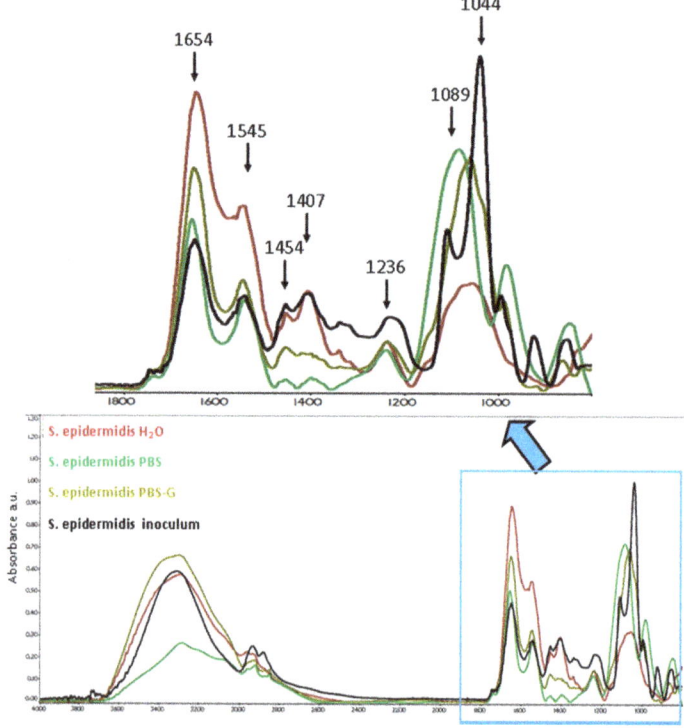

Figure 5. FTIR spectra of *S. epidermidis* after 180 min in H_2O, phosphate buffered solution (PBS), and PBS with glucose (PBS-G) compared with inoculum.

This behavior could be attributed to the needs of resources to produce proteins. The addition of glucose in the media (PBS-G) allows to bacteria to produce a higher amount of proteins in EPS whereas a smaller reduction of carboxylate group is observed than in PBS alone. This could again indicate the lack of the osmosis stimulus but in this case, glucose could also be considered as a source of material for EPS construction.

In summary, under osmotic stress and without nutrients, bacteria are stimulated to produce more proteins on the surface to firmly attach to it, probably due to the reduction in the uronic based biomolecules. The presence of ions mitigates this behavior, so that the formation of proteins and reduction of uronic based biomolecules are less evident, whereas the presence of glucose helps the synthesis of uronic base biomolecules together with the protein formation.

3.5.2. *P. fluorescens*

Figure 6 shows the spectra of PF in different media. The inoculum IR spectra are mainly dominated by Amide bands absorption due to proteins whereas the polysaccharides contribution to the infrared spectrum is very low. After 180 min in contact with different media, the spectra show an increase of polysaccharide bands greater in PBS and PBS-G than in H_2O. Unlike the others, the spectrum of PF in PBS shows a decrease of the band at 1400 cm^{-1} suggesting the reduction of carboxylate group probably due to the presence of low level of uronic acid and its derivates.

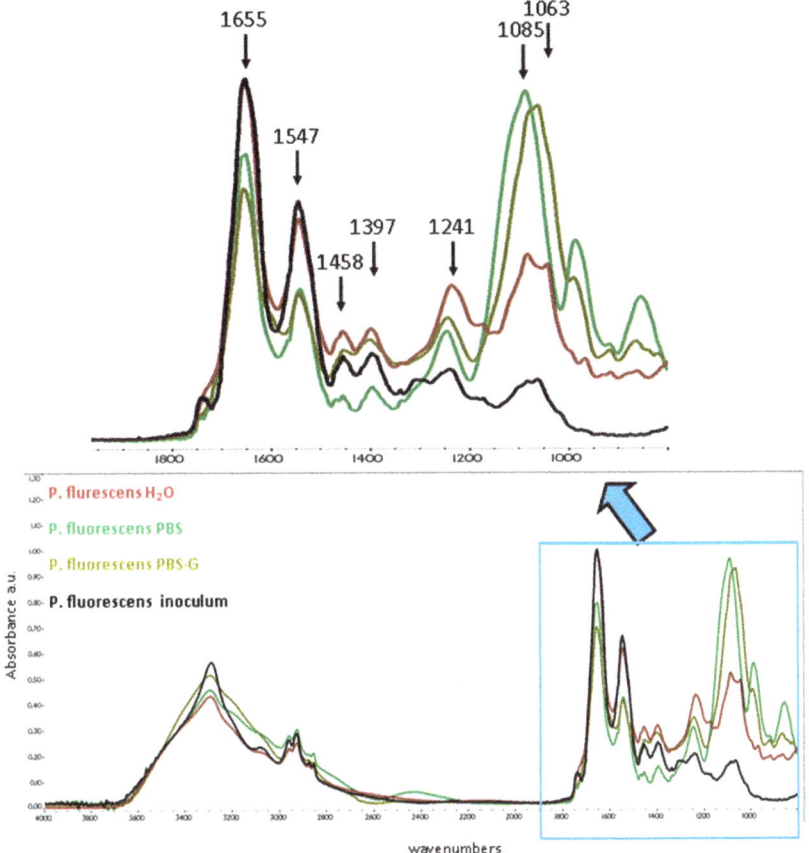

Figure 6. FTIR spectra of P. fluorescens after 180 min in H_2O, PBS, and PBS-G compared with inoculum.

The unexpected change of the band at 1235 cm^{-1}, mainly related to the bacteria wall, can be due to the amount of cells attached to the crystal surface.

3.6. Secondary-Structure Analysis

The differences in the IR spectra of SE and PF biofilms are corroborated and more clearly illustrated by their second derivatives calculated and presented in the most informative spectroscopic region (1800–1500 cm^{-1}) (Figures 7 and 8). The second derivative of FTIR spectra was used to distinguish the secondary structure of protein under different conditions, and the results have been summarized in Table 2. To eliminate the contribution of the water bands, the spectrum of water was subtracted from each sample spectrum until the baseline in the region above 1750–2000 cm^{-1} (where no absorption from the sample does occur) becomes a straight line. The analysis of derivative spectra demonstrated that the proteins of both bacterial species are minimally affected by external environment.

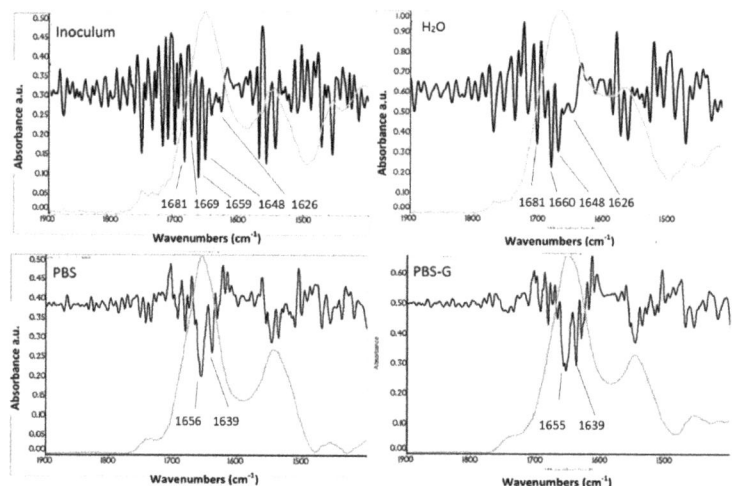

Figure 7. Second derivatives of the FTIR spectra of *S. epidermidis* (SE) biofilms at Inoculum (upper left), in H$_2$O (upper right), PBS (lower left) and PBS-G (lower right) (for related spectra see Figure 5).

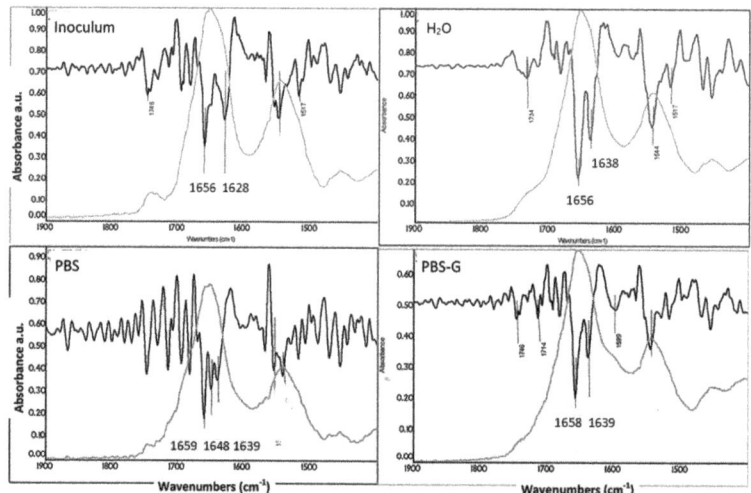

Figure 8. Second derivatives of the FTIR spectra of PF biofilms at Inoculum (upper left), in H$_2$O (upper right), in PBS (lower left) and PBS-G (lower right) (for related spectra see Figure 6).

Table 2. *P. fluorescens* (PF) and SE Second derivative Amide I band frequencies and assignment to proteins secondary structure in Inoculum, H_2O, PBS, and PBS-G media.

Inoculum	H_2O	PBS	PBS-G	Assignment
	Frequencies (cm^{-1})			
	Pseudomonas Fluorescent			
1656	1656	1659	1658	α-helix
-	-	1648	-	Random coils
1628	1638	1639	1639	β-sheet
	Staphylococcus Epidermidis			
1681, 1669	1681	-	-	β-sheet
1659	1660	1656	1655	α-helix
1648	1648	-	-	Random coils
1637, 1626	1626	1639	1639	β-sheet

Peaks in the second derivatives spectra, directed downwards, correspond both to peaks and to poorly resolved shoulders (i.e., spectral bands due to overlapped stronger neighboring absorptions).

Apart from differences in the carboxyl stretching regions (weak signals at 1727 and 1740 cm^{-1}), related to C=O in phospholipids and lipopolysaccharides typical for bacteria [46–49], it is noticeable that the secondary derivative plots in the Amide I (1600–1700 cm^{-1}) region demonstrated that proteins are predominantly in random coils and helices form as witnessed by the bands at 1659 and 1648 cm^{-1}. Proteins containing β-sheet, characterized by peaks around 1630 and 1680–1690 cm^{-1}, are present only in a very low amount in the inoculum and in H_2O samples and disappear completely in PBS and PBS-G samples [50].

In PF (Figure 8) the proteins are mainly in α-helix and β-sheet conformation and the ratio between these two structures is constant for all the investigated media. Only in the PBS medium part of the proteins are in random coils conformation. For SE the proteins in inoculum and in H_2O have quite similar conformations as α-helix, random coils and β-sheet. The random coil conformation could not be distinguished in PBS and PBS-G medium.

3.7. Bacteria Staining Test

Figure 9 shows the surface of the ATR crystal, after washing treatment, stained with safranin dye. In all the experiments pink colored surfaces are evident, thus highlighting the presence of bacteria attached to the surface. Both the bacteria under all the selected conditions tested positive on safranin test after 180 min of contact.

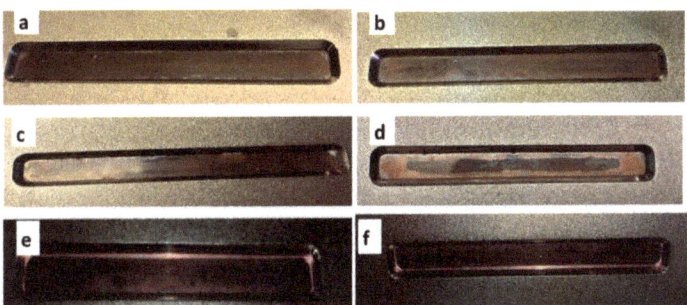

Figure 9. Attenuated total reflectance (ATR) crystal, after the experiment, stained with safranin dye (**a**) *P. fluorescens* in H_2O, (**b**) *S. epidermidis* in H_2O, (**c**) *P. fluorescens* in PBS, (**d**) *S. epidermidis* in PBS, (**e**) *P. fluorescens* in PBS-G, (**f**) *S. epidermidis* in PBS-G.

4. Conclusions

This study highlights the influence of environmental condition on the chemistry of bacteria cell surfaces and biofilm, during the early stage of bacteria attachment to the surface.

The amount of bacterial EPS in the different media was monitored by crystal violet assay for the same time interval as spectroscopy experiment. In addition, the presence of bacterial EPS on the ATR crystal was confirmed by O-safranin staining at the end of each experiment.

Dominant functional groups of bacterial EPS were, as expected, carboxyl, amide, phosphate, hydroxyl, and carbohydrate related moieties. Both the bacterial species, at planktonic phase, show all the IR bands expected, but with an evident difference in proteins/polysaccharides ratio. SE shows a strong band in Zone III due to polysaccharides vibrational absorption, while, conversely, the PF is characterized by a strong band in Zone I related to protein vibrational absorption, together with a weak polysaccharide absorption. The phospholipids and nucleic acid bands show similar intensities in both bacterial species, as expected, since the initial number of bacteria cells are almost the same and cell division and proliferation occur generally in a longer time with respect to the experiment timing.

Concerning the biofilm evolution over the time, both the bacterial species, are detectable on the surface within 45 and 75 min after the inoculation. In water both SE and PF start to produce plenty of proteins in a short time probably because they need to firmly attach to the surface, due to stressful lack of nutrients. The osmotic shock appears to be the driving force forcing the bacteria to attach to the surface. In fact, the presence of PBS (physiological ionic strength but no nutrients) and glucose mitigate this process. An exception was PF where a low amount of proteins was produced when glucose is added.

Independently on the environmental conditions (H_2O, PBS, PBS-G) both SE and PF start to increase their polysaccharide content, even if with different kinetics and amounts, probably because of biofilm needs to counteract the lack of nutrients. After 180 min all spectra of SE and PF show an increase of polysaccharide bands intensity with respect to the inoculum with PBS and PBS-G samples.

The prevalent structure of the proteins has been highlighted by the second derivative study. PF proteins are mainly in α-helix and β-sheet conformation and their ratio is constant and independent on the media. Only in PBS medium the random coils component is visible. SE protein conformation at inoculum and in H_2O are similar and composed of α-helix, random coils and β-sheet. The random coil conformation seems to decrease in PBS and PBS-G.

Author Contributions: Conceptualization, Validation, Formal analysis, Investigation, Writing—original draft, Writing—review and editing, M.C. and K.J.; Investigation and Validation, G.L. and A.P.; Conceptualization, Resources, Supervising, Coordination and Funding acquisition, C.R., E.R., K.W., A.B. and A.M. All authors have read and agreed to the published version of the manuscript.

Funding: This work was supported by the European Union's Horizon 2020 research and innovation programme under grant agreement No. 722871 and CO-FUND–MANUNET III-NON-ACT-project ref. MNET17/NMAT-0061.

Conflicts of Interest: The authors declare no conflict of interest.

References

1. Flemming, H.-C. The perfect slime. *Colloids Surf. B Biointerfaces* **2011**, *86*, 251–259. [CrossRef]
2. Pamp, S.J.; Gjermansen, M.; Tolker-Nielsen, T. The Biofilm Matrix: A Sticky Framework. In *Bacterial Biofilm Formation and Adaptation*; Givskov, M.C., Ed.; Horizon BioScience: Norfolk, UK, 2007; Volume 34, pp. 37–69.
3. Watnick, P.; Kolter, R. Biofilm, City of Microbes. *J. Bacteriol.* **2000**, *182*, 2675–2679. [CrossRef] [PubMed]
4. Flemming, H.-C.; Wingender, J.; Szewzyk, U.; Steinberg, P.; Rice, S.A.; Kjelleberg, S. Biofilms: An emergent form of bacterial life. *Nat. Rev. Microbiol.* **2016**, *14*, 563–575. [CrossRef] [PubMed]
5. Costerton, J.W. Bacterial Biofilms: A Common Cause of Persistent Infections. *Science* **1999**, *284*, 1318–1322. [CrossRef] [PubMed]
6. Bahn, Y.-S.; Xue, C.; Idnurm, A.; Rutherford, J.C.; Heitman, J.; Cardenas, M.E. Sensing the environment: Lessons from fungi. *Nat. Rev. Microbiol.* **2007**, *5*, 57–69. [CrossRef] [PubMed]

7. Yin, W.; Wang, Y.; Liu, L.; He, J. Biofilms: The Microbial "Protective Clothing" in Extreme Environments. *Int. J. Mol. Sci.* **2019**, *20*, 3423. [CrossRef]
8. Harrison, J.J.; Ceri, H.; Turner, R.J. Multimetal resistance and tolerance in microbial biofilms. *Nat. Rev. Microbiol.* **2007**, *5*, 928–938. [CrossRef]
9. Fulaz, S.; Vitale, S.; Quinn, L.; Casey, E. Nanoparticle–Biofilm Interactions: The Role of the EPS Matrix. *Trends Microbiol.* **2019**, *27*, 915–926. [CrossRef]
10. Whitchurch, C.B. Extracellular DNA Required for Bacterial Biofilm Formation. *Science* **2002**, *295*, 1487. [CrossRef]
11. Tolker-Nielsen, T. Biofilm Development. *Microbiol. Spectr.* **2015**, *3*. [CrossRef]
12. Davies, D.G. The Involvement of Cell-to-Cell Signals in the Development of a Bacterial Biofilm. *Science* **1998**, *280*, 295–298. [CrossRef] [PubMed]
13. De Kievit, T.R.; Gillis, R.; Marx, S.; Brown, C.; Iglewski, B.H. Quorum-Sensing Genes in *Pseudomonas aeruginosa* Biofilms: Their Role and Expression Patterns. *Appl. Environ. Microbiol.* **2001**, *67*, 1865–1873. [CrossRef]
14. Carniello, V.; Peterson, B.W.; van der Mei, H.C.; Busscher, H.J. Physico-chemistry from initial bacterial adhesion to surface-programmed biofilm growth. *Adv. Colloid Interface Sci.* **2018**, *261*, 1–14. [CrossRef] [PubMed]
15. Serra, D.O.; Richter, A.M.; Hengge, R. Cellulose as an Architectural Element in Spatially Structured *Escherichia coli* Biofilms. *J. Bacteriol.* **2013**, *195*, 5540–5554. [CrossRef] [PubMed]
16. Rohde, H.; Frankenberger, S.; Zähringer, U.; Mack, D. Structure, function and contribution of polysaccharide intercellular adhesin (PIA) to Staphylococcus epidermidis biofilm formation and pathogenesis of biomaterial-associated infections. *Eur. J. Cell Biol.* **2010**, *89*, 103–111. [CrossRef] [PubMed]
17. McDougald, D.; Rice, S.A.; Barraud, N.; Steinberg, P.D.; Kjelleberg, S. Should we stay or should we go: Mechanisms and ecological consequences for biofilm dispersal. *Nat. Rev. Microbiol.* **2012**, *10*, 39–50. [CrossRef]
18. Franklin, M.J.; Nivens, D.E.; Weadge, J.T.; Howell, P.L. Biosynthesis of the *Pseudomonas aeruginosa* Extracellular Polysaccharides, Alginate, Pel, and Psl. *Front. Microbiol.* **2011**, *2*, 167. [CrossRef]
19. Colvin, K.M.; Irie, Y.; Tart, C.S.; Urbano, R.; Whitney, J.C.; Ryder, C.; Howell, P.L.; Wozniak, D.J.; Parsek, M.R. The Pel and Psl polysaccharides provide *Pseudomonas aeruginosa* structural redundancy within the biofilm matrix: Polysaccharides of the *P. aeruginosa* biofilm matrix. *Environ. Microbiol.* **2012**, *14*, 1913–1928. [CrossRef]
20. Conrad, A.; Kontro, M.; Keinänen, M.M.; Cadoret, A.; Faure, P.; Mansuy-Huault, L.; Block, J.-C. Fatty acids of lipid fractions in extracellular polymeric substances of activated sludge flocs. *Lipids* **2003**, *38*, 1093–1105. [CrossRef]
21. Romero, D.; Vlamakis, H.; Losick, R.; Kolter, R. An accessory protein required for anchoring and assembly of amyloid fibres in *B. subtilis* biofilms: *B. subtilis* amyloid fibre accessory protein. *Mol. Microbiol.* **2011**, *80*, 1155–1168. [CrossRef]
22. Grunert, T.; Monahan, A.; Lassnig, C.; Vogl, C.; Müller, M.; Ehling-Schulz, M. Deciphering Host Genotype-Specific Impacts on the Metabolic Fingerprint of *Listeria monocytogenes* by FTIR Spectroscopy. *PLoS ONE* **2014**, *9*, e115959. [CrossRef] [PubMed]
23. Sutherland, I. The biofilm matrix—An immobilized but dynamic microbial environment. *Trends Microbiol.* **2001**, *9*, 222–227. [CrossRef]
24. Klausen, M.; Heydorn, A.; Ragas, P.; Lambertsen, L.; Aaes-Jørgensen, A.; Molin, S.; Tolker-Nielsen, T. Biofilm formation by *Pseudomonas aeruginosa* wild type, flagella and type IV pili mutants: Roles of bacterial motility in the formation of the flat *P. aeruginosa* biofilm. *Mol. Microbiol.* **2003**, *48*, 1511–1524. [CrossRef]
25. Chen, R.; Guo, C.; Chu, W.; Jiang, N.; Li, H. ATR-FTIR study of *Bacillus* sp. and *Escherichia coli* settlements on the bare and Al_2O_3 coated ZnSe internal reflection element. *Chin. Chem. Lett.* **2019**, *30*, 115–119. [CrossRef]
26. Genkawa, T.; Ahamed, T.; Noguchi, R.; Takigawa, T.; Ozaki, Y. Simple and rapid determination of free fatty acids in brown rice by FTIR spectroscopy in conjunction with a second-derivative treatment. *Food Chem.* **2016**, *191*, 7–11. [CrossRef] [PubMed]
27. Pink, J.; Smith-Palmer, T.; Chisholm, D.; Beveridge, T.J.; Pink, D.A. An FTIR study of *Pseudomonas aeruginosa* PAO1 biofilm development: Interpretation of ATR–FTIR data in the 1500–1180 cm^{-1} region. *Biofilms* **2005**, *2*, 165–175. [CrossRef]
28. Pink, J.; Smith-Palmer, T.; Beveridge, T.J.; Pink, D.A. An FTIR study of *Pseudomonas aeruginosa* PAO1 biofilm growth and dispersion. An improved ATR method for studying biofilms: The C–H stretch spectral region. *Biofilms* **2004**, *1*, 157–163. [CrossRef]

29. Cwalina, B.; Dec, W.; Michalska, J.K.; Jaworska-Kik, M.; Student, S. Initial stage of the biofilm formation on the NiTi and Ti6Al4V surface by the sulphur-oxidizing bacteria and sulphate-reducing bacteria. *J. Mater. Sci. Mater. Med.* **2017**, *28*, 173. [CrossRef]
30. Boualam, M.; Quilès, F.; Mathieu, L.; Block, J.-C. Monitoring the Effect of Organic Matter on Biofilm Growth in Low Nutritive Waters by ATR/FT-IR Spectroscopy. *Biofouling* **2002**, *18*, 73–81. [CrossRef]
31. Van der Kooij, D. The occurrence of *Pseudomonas* spp. in surface water and in tap water as determined on citrate media. *Antonie Leeuwenhoek* **1977**, *43*, 187–197. [CrossRef]
32. Van der Kooij, D.; Visser, A.; Oranje, J.P. Multiplication of fluorescent pseudomonads at low substrate concentrations in tap water. *Antonie Leeuwenhoek* **1982**, *48*, 229–243. [CrossRef] [PubMed]
33. Pop, C.; Apostu, S.; Rotar, A.M.; Semeniuc, C.A.; Sindic, M.; Mabon, N. FTIR spectroscopic characterization of a new biofilm obtained from kefiran. *J. Agroaliment. Process. Technol.* **2013**, *19*, 157–159.
34. Karadenizli, A.; Kolayli, F.; Ergen, K. A novel application of Fourier-transformed infrared spectroscopy: Classification of slime from staphylococci. *Biofouling* **2007**, *23*, 63–71. [CrossRef] [PubMed]
35. Cappelli, A.; Razzano, V.; Paolino, M.; Grisci, G.; Giuliani, G.; Donati, A.; Mendichi, R.; Samperi, F.; Battiato, S.; Boccia, A.C.; et al. Bithiophene-based polybenzofulvene derivatives with high stacking and hole mobility. *Polym. Chem.* **2015**, *6*, 7377–7388. [CrossRef]
36. Leone, G.; Consumi, M.; Franzi, C.; Tamasi, G.; Lamponi, S.; Donati, A.; Magnani, A.; Rossi, C.; Bonechi, C. Development of liposomal formulations to potentiate natural lovastatin inhibitory activity towards 3-hydroxy-3-methyl-glutaryl coenzyme A (HMG-CoA) reductase. *J. Drug Deliv. Sci. Technol.* **2018**, *43*, 107–112. [CrossRef]
37. Genevaux, P.; Muller, S.; Bauda, P. A rapid screening procedure to identify mini-Tn10 insertion mutants of *Escherichia coli* K-12 with altered adhesion properties. *FEMS Microbiol. Lett.* **1996**, *142*, 27–30. [CrossRef] [PubMed]
38. Quilès, F.; Humbert, F.; Delille, A. Analysis of changes in attenuated total reflection FTIR fingerprints of *Pseudomonas fluorescens* from planktonic state to nascent biofilm state. *Spectrochim. Acta Part A Mol. Biomol. Spectrosc.* **2010**, *75*, 610–616. [CrossRef]
39. Fasasi, Y.A.; Mirjankar, N.; Fasasi, A. Fourier Transform Infrared Spectroscopic Analysis of Protein Secondary Structures Found in Egusi. *Am. J. Appl. Ind. Chem.* **2015**, *1*, 1–4.
40. Bonechi, C.; Donati, A.; Tamasi, G.; Pardini, A.; Rostom, H.; Leone, G.; Lamponi, S.; Consumi, M.; Magnani, A.; Rossi, C. Chemical characterization of liposomes containing nutraceutical compounds: Tyrosol, hydroxytyrosol and oleuropein. *Biophys. Chem.* **2019**, *246*, 25–34. [CrossRef]
41. Pietralik, Z.; Mucha-Kruczynska, I.; Kozak, M. FTIR analysis of protein secondary structure in solid and solution states. *Synchrotron Radiation News* **2012**, *11*, 52.
42. Leone, G.; Consumi, M.; Pepi, S.; Lamponi, S.; Bonechi, C.; Tamasi, G.; Donati, A.; Rossi, C.; Magnani, A. Alginate–gelatin formulation to modify lovastatin release profile from red yeast rice for hypercholesterolemia therapy. *Ther. Deliv.* **2017**, *8*, 843–854. [CrossRef]
43. Leone, G.; Consumi, M.; Lamponi, S.; Bonechi, C.; Tamasi, G.; Donati, A.; Rossi, C.; Magnani, A. Thixotropic PVA hydrogel enclosing a hydrophilic PVP core as nucleus pulposus substitute. *Mater. Sci. Eng. C* **2019**, *98*, 696–704. [CrossRef] [PubMed]
44. Delille, A.; Quilès, F.; Humbert, F. In Situ Monitoring of the Nascent *Pseudomonas fluorescens* Biofilm Response to Variations in the Dissolved Organic Carbon Level in Low-Nutrient Water by Attenuated Total Reflectance-Fourier Transform Infrared Spectroscopy. *AEM* **2007**, *73*, 5782–5788. [CrossRef] [PubMed]
45. Jiang, W.; Saxena, A.; Song, B.; Ward, B.B.; Beveridge, T.J.; Myneni, S.C.B. Elucidation of Functional Groups on Gram-Positive and Gram-Negative Bacterial Surfaces Using Infrared Spectroscopy. *Langmuir* **2004**, *20*, 11433–11442. [CrossRef]
46. Jubeen, F.; Liaqat, A.; Amjad, F.; Sultan, M.; Iqbal, S.Z.; Sajid, I.; Khan Niazi, M.B.; Sher, F. Synthesis of 5-Fluorouracil Cocrystals with Novel Organic Acids as Coformers and Anticancer Evaluation against HCT-116 Colorectal Cell Lines. *Cryst. Growth Des.* **2020**, *20*, 2406–2414. [CrossRef]
47. Jubeen, F.; Liaqat, A.; Sultan, M.; Zafar Iqbal, S.; Sajid, I.; Sher, F. Green synthesis and biological evaluation of novel 5-fluorouracil derivatives as potent anticancer agents. *Saudi Pharm. J.* **2019**, *27*, 1164–1173. [CrossRef] [PubMed]

48. McWhirter, M.J.; Bremer, P.J.; McQuillan, A.J. Direct Infrared Spectroscopic Evidence of pH- and Ionic Strength-Induced Changes in Distance of Attached Pseudomonas a eruginosa from ZnSe Surfaces. *Langmuir* **2002**, *18*, 1904–1907. [CrossRef]
49. Lasch, P. Spectral pre-processing for biomedical vibrational spectroscopy and microspectroscopic imaging. *Chemom. Intell. Lab. Syst.* **2012**, *117*, 100–114. [CrossRef]
50. Bunaciu, A.A.; Fleschin, Ş.; Aboul-Enein, H.Y. Evaluation of the Protein Secondary Structures Using Fourier Transform Infrared Spectroscopy. *Gazi Univ. J. Sci.* **2014**, *27*, 637–644.

 © 2020 by the authors. Licensee MDPI, Basel, Switzerland. This article is an open access article distributed under the terms and conditions of the Creative Commons Attribution (CC BY) license (http://creativecommons.org/licenses/by/4.0/).

Article

Physico-Chemical Properties and In Vitro Antifungal Evaluation of Samarium Doped Hydroxyapatite Coatings

Steluta Carmen Ciobanu [1], Simona Liliana Iconaru [1], Daniela Predoi [1], Alina Mihaela Prodan [2,3] and Mihai Valentin Predoi [4,*]

1. National Institute of Materials Physics, Atomistilor Street, No. 405A, P.O. Box MG 07, 077125 Magurele, Romania; carmen.ciobanu@infim.ro or ciobanucs@gmail.com (S.C.C.); simona.iconaru@infim.ro or simonaiconaru@gmail.com (S.L.I.); dpredoi@infim.ro or dpredoi@gmail.com (D.P.)
2. Department of General Surgery, Carol Davila University of Medicine and Pharmacy, 8 Eroii Sanitari, Sector 5, 050474 Bucharest, Romania; prodan1084@gmail.com
3. Emergency Hospital Floreasca Bucharest, 8 Calea Floreasca, 014461 Bucharest, Romania
4. Department of Mechanics, University Politehnica of Bucharest, BN 002, 313 Splaiul Independentei, Sector 6, 060042 Bucharest, Romania
* Correspondence: mihai.predoi@upb.ro or predoi@gmail.com

Received: 29 July 2020; Accepted: 25 August 2020; Published: 27 August 2020

Abstract: Hydroxyapatite (HAp) and samarium doped hydroxyapatite, $Ca_{10-x}Sm_x(PO_4)_6(OH)_2$, $x_{Sm} = 0.05$, (5SmHAp), coatings were prepared by sol-gel process using the dip coating method. The stability of 5SmHAp suspension was evaluated by ultrasound measurements. Fourier transform infrared spectroscopy (FTIR) was used to examine the optical characteristics of HAp and 5SmHAp nanoparticles in suspension and coatings. The FTIR analysis revealed the presence of the functional groups specific to the structure of hydroxyapatite in the 5SmHAp suspensions and coatings. The morphology of 5SmHAp nanoparticles in suspension was evaluated by transmission electron microscopy (TEM). Moreover, scanning electron microscope (SEM) was used to evaluate the morphology of nanoparticle in suspension and the morphology of the surface on the coating. The SEM and TEM studies on 5SmHAp nanoparticles in suspension showed that our samples consist of nanometric particles with elongated morphology. The SEM micrographs of HAp and 5SmHAp coatings pointed out that the coatings are continuous and homogeneous. The surface morphology of the 5SmHAp coatings was also assessed by Atomic Force Microscopy (AFM) studies. The AFM results emphasized that the coatings presented the morphology of a uniformly deposited layer with no cracks and fissures. The crystal structure of 5SmHAp coating was characterized by X-ray diffraction (XRD). The surface composition of 5SmHAp coating was analyzed by X-ray photoelectron spectroscopy (XPS). The XRD and XPS analysis shown that the Sm^{3+} ions have been incorporated into the 5SmHAp synthesized material. The antifungal properties of the 5SmHAp suspensions and coatings were studied using *Candida albicans* ATCC 10231 (*C. albicans*) fungal strains. The quantitative results of the antifungal assay showed that colony forming unity development was inhibited from the early phase of adherence in the case of both suspensions and coatings. Furthermore, the adhesion, cell proliferation and biofilm formation of the *C. albicans* were also investigated by AFM, SEM and Confocal Laser Scanning Microscopy (CLSM) techniques. The results highlighted that the *C. albicans* adhesion and cell development was inhibited by the 5SmHAp coatings. Moreover, the data also revealed that the 5SmHAp coatings were effective in stopping the biofilm formation on their surface. The toxicity of the 5SmHap was also investigated in vitro using HeLa cell line.

Keywords: samarium; hydroxyapatite; dip coating method; antifungal activity

1. Introduction

Currently, the use of nanoparticles and coatings in the medical field is on an upward trend due to their remarkable properties that make them effective in many ways, starting from the fight against pathogenic microorganisms and ending with their antitumoral activity [1]. One of the most used biomaterials in the biomedical field is hydroxyapatite (HAp, $Ca_{10}(PO_4)_6(OH)_2$) which is the main mineral component of bone tissue. Hydroxyapatite, due to his unique biological properties, such as biocompatibility, low citotoxicity, osteoconductivity, bioactivity and so forth [2–4], is used in various applications in the medical field like implantology (metal prosthesis coating material, etc.) [5], dentistry [5], drug and gene delivery systems [6,7], as an antimicrobial agent [8], in bioimaging [9] and so forth.

In addition, due to its hexagonal structure, HAp on the one hand has the capacity to incorporate in its structure various ions (Eu^{3+}, Ag^+, Mg^{2+}, Zn^{2+}, and Sm^{3+}, etc.) [10–12] and on the other hand the HAp surface can be modified with various biopolymers/drugs [13], which leads to an improvement of the biological and physico-chemical properties, thus making HAp more efficient and useful for the medical sphere [2]. Hydroxyapatite can be obtained in various forms, from powder/gel and reaching to thin layers/coatings by various synthesis methods, among which we mention the sol-gel method, coprecipitation (these are often used to obtain powders/gels), dip coating, spin coating, magnetron sputtering and so forth (the latter being used to obtain thin HAp coatings) [14]. These methods, by controlling the synthesis parameters, allow the obtaining of both powders with nanometric/micrometric dimensions and the desired morphology and uniform and homogeneous layers [14].

In the current context of increasing the resistance of microorganisms to antibiotic treatment, recent studies have shown that doping hydroxyapatite with antimicrobial ions and the use of these new materials may be a viable alternative to conventional antibiotic treatment after implant surgery [2,5,10]. Recent studies reported in the literature have shown that the dopping of hydroxyapatite with lanthanide ions, especially Samarium (Sm^{3+}), leads to enhanced biological and antimicrobial properties [5,10,15,16]. According to these studies, Sm^{3+} doped HAp exihibits great antibacterial activity against bacterial strain such as *Staphylococcus aureus* (ATCC 25923), *Escherichia coli* (ATCC 25922), *Staphylococcus epidermidis* (ATCC 35984/RP62A), *Enterococcus faecalis* (ATCC 29212) and *Pseudomonas aeruginosa*.

Recently, it has been observed that the interest of researchers in the use of lanthanides in biomedical applications is increasing. Therefore, in their studies, Nakayama, et al. [17] showed that the smarium doped TiO_2 nanoparticles could improve the radiosensitising effects and could be used as theranostic agents in radiation therapy. In addition, Zhang et al. [18] showed that mesoporous bioactive glass microspheres doped with small amounts of samarium of could be used as a delivery system for doxorubicin in the treatment of bone cancer.

Also, in their recent work, Kannan et al. [19] highlighted that the presence of a samarium oxide coating on a Mg implant could prevent on the one hand the recurrence of bone tumors and metastases and on the other hand the appearance of post-implant infections.

The aim of the present research was to obtain a homogenous coating with antimicrobial properties by sol-gel process using the dip coating method on Si substrate. The evaluation of the stability of nanoparticles in suspension was conducted by ultrasound measurements. To the best of our knowledge, the samarium doped hydroxyapatite coatings with antimicrobial properties realized using the dip coating method was very little studied. The antimicrobial properties investigations of the 5SmHAp coating presented in this study revealed a very good behavior of these coatings.

2. Materials and Methods

2.1. Materials

The synthesis of hydroxyapatite and samarium doped hydroxyapatite (x_{Sm} = 0.05), $Ca_{10-x}Sm_x(PO_4)_6(OH)_2$ was effectuated using calcium nitrate tetrahydrate, $Ca(NO_3)_2 \cdot 4H_2O$ (≥99.0%), samarium nitrate hexahydrate, $Sm(NO_3)_3 \cdot 6H_2O$ (99.97% purity), ammonium hydrogen phosphate,

$(NH_4)_2HPO_4$ (≥99.0%), ethanol absolute, C_2H_5OH (≥99.8%) and double distilled water. The reagents, such as $Ca(NO_3)_2·4H_2O$, $(NH_4)_2HPO_4$ and C_2H_5OH, were processed by Sigma Aldrich, St. Louis, MO, USA while $Sm(NO_3)_3·6H_2O$ was processed by Alpha Aesar, Kandel, Germany.

2.2. Hydroxyapatite (HAp) and Samarium Doped Hydroxyapatite (SmHAp)

Samarium doped hydroxyapatite nanoparticles with the chemical formula $Ca_{10-x}Sm_x(PO_4)_6(OH)_2$, x_{Sm} = 0.05 were synthesized by the adapted chemical method [20] with (Ca + Sm)/P fixed to 1.67 [21–23]. The $Ca(NO_3)_2·4H_2O$ was dissolved in 100 mL C_2H_5OH and $(NH_4)HPO_4$ was dissolved in 25 mL deionized water (DI) were mixed in a 500 mL beaker and stirred together for 1 h at room temperature. $Sm(NO_3)_3·6H_2O$ was dissolved in 25 mL of DI water at room temperature and added drop by drop into the mixture under continuous stirring at a temperature of 100 °C. The resulting solution was stirred continuously for another 2 h at 100 °C. After that, the resulting solution was centrifuged, redispersed in ethanol absolute and stirred at 100 °C for 24 h. The hydroxyapatite nanoparticles were obtained as previously described by Ciobanu et al. [24]. The final suspension was analyzed and used to prepare the coatings.

2.3. Preparation of HAp and 5SmHAp Coatings

The HAp and 5SmHAp coatings have been deposited on the Si wafer by the sol-gel process using the dip coating method in agreement with previous studies [25]. The Si substrate was washed with ethanol absolute before coating. The 5SmHAp layer was dried at 100 °C for 4 h and heat treated at 500 °C for 2 h. Cooling was done at a rate of 5 °C/min.

2.4. Characterization Methods

Ultrasonic measurements were performed on 100 mL of concentrated suspension of 5SmHAp [22,26]. The digitalized ultrasonic signals were recorded on the digital oscilloscope (General-Electric, Krautkramer, Germany) at a very precise interval of 5.00 s. In order to have an accurate evaluation of the stability of the 5SmHAp suspension, the double distilled water (the most stable suspension) was chosen as the reference fluid, under the same experimental conditions.

The morphology of the 5SmHAp suspensions was analyzed using transmission electron microscopy (TEM) with a CM 20 (Philips-FEI, Hillsboro, OR, USA) transmission electron microscope having a Lab6 filament (Agar Scientific Ltd., Stansted, UK), which operates at 200 kV. The particle size distributions from the TEM micrographs was performed by measuring approximately 700 particles from different regions of the sample.

The morphology of 5SmHAp suspensions and coatings were evaluated by scanning electron microscopy (SEM) using a HITACHI S4500 microscope (Hitachi, Ltd., Tokyo, Japan). The 5SmHAp suspensions were prepared on a conductive carbon tape with double adhesion, dried and introduced in the microscope. The microscope was equipped with energy-dispersive X-ray spectroscopy (EDX) (Ametek EDAX Inc., Mahwah, NJ, USA) attachment operating at 20 kV. The particle size distributions from the SEM micrographs was performed by measuring approximately 200 particles from different regions of the sample. The surface morphology of the coatings was analyzed using atomic force microscopy (AFM), in a non-contact mode. The measurements were performed using a NT-MDT NTEGRA Probe NanoLaboratory system (NT-MDT, Moscow, Russia). The data was recorded with a silicon NT-MDT NSG01 cantilever coated with a 35 nm gold layer having a tetrahedral tip. The AFM micrographs were acquired on surface areas of 3 × 3 μm^2. The data analysis of the 2D surface topographies as well as the 3D representation of the AFM images were performed with the aid of Gwyddion 2.55 software [27].

The 5SmHAp coating was examined by the X-ray diffraction (XRD) (Bruker D8 Advance diffractometer, Billerica, MA, USA) with nickel filtered Cu Kα (Å) radiation.

The functional groups present in the prepared gel and coating were identified by Fourier-transform infrared spectroscopy (FTIR) analysis using a SpectrumBX spectrometer in the case of the prepared gel and by Fourier-transform infrared spectroscopy -Attenuated total reflection (FTIR-ATR) spectroscopy using a Perkin Elmer SP-100 spectrometer (Waltham, MS, USA), for the coatings. The spectra were taken in the spectral range of 400 to 4000 cm^{-1} with a resolution of 4 cm^{-1}. The curve fitting analysis of the spectra was achieved using a nonlinear least-squares data-fitting algorithm [28].

X-ray Photoelectron Spectroscopy (XPS) analysis was performed using a VG ESCA 3 MK II XPS installation ($E_{k\alpha}$ = 1486.7 eV). The vacuum analysis chamber pressure was P~3 × 10^{-8} torr. The XPS register spectrum involved an energy window w = 20 eV with the resolution R = 50 eV and 256 recording channels. The XPS spectra were analyzed using Spectral Data Processor v 2.3 (SDP) software.

2.5. In Vitro Antifungal Activity

The effects of the 5SmHAp coatings against fungi cells were assessed using the reference fungal strain *Candida albicans* ATCC 10231 acquired from the American Type Culture Collection (ATCC, Manassas, VA, USA). The in vitro antifungal assays were performed with fungal suspensions of approximately 10^5–10^6 colony forming units (CFU)/mL obtained from 15 to 18 h bacterial cultures as previously reported [29]. The qualitative assay of the fungal biofilm development, after incubation at 24, 48, and 72 h, was assessed by the visualization of the adherent fungal cells on the substrates using atomic force microscopy (AFM), scanning electron microscopy (SEM) and confocal laser scanning microscopy (CLSM). For the AFM and SEM observation, the 5SmHAp coatings were removed from the *C. albicans* ATCC 10231 culture medium after 24, 48, and 72 h of incubation, then washed with sterile saline solution and fixed using cold methanol. For the CLSM visualization, the 5SmHAp coatings after 24, 48, and 72 h of incubations with *C. albicans* fungal culture, were washed with sterile saline buffer solution (PBS) to remove the unattached fungal cells. After that, the unattached cells were fixed with cold methanol and stained in the dark with propidium iodide (PI) for 10 min at room temperature. After the PI staining, the excess of label marker was removed using filter paper. The samples were visualized directly after staining. The CLSM studies were performed both in reflection and fluorescence modes using a Leica TCS-SP confocal microscope (Leica Camera AG, Wetzlar, Germany), equipped with a PL FLUOTAR (40_ NA 0.7) objective and an Ar ion laser with a laser line at 488 nm.

The quantitative assays of the antifungal activity of both 5SmHAp suspensions and coatings were done using an adapted method (E2149-10; ASTM International) [30,31], as previously described [32]. The experiments were performed in triplicate and the results of the results were expressed as mean ± SD. Moreover, the 3D representation of the SEM and CLSM images were obtained Image J software (ImageJ 1.51j8, National Institutes of Health, Bethesda, MD, USA) was used [33].

2.6. In Vitro Cytotoxic Assay

The toxicity of the HAp and 5SmHAp solutions and coatings was assessed using the HeLa cell line (Sigma-Aldrich Corp., St. Louis, MO, USA). The in vitro interaction of the HeLa cells with the samples was studied at three different time intervals (24, 48, and 72 h) using the methodology previously described in Predoi et al. [34]. A cell viability using live/dead cell double staining kit (purchased from Merck/Sigma-Aldrich) was also performed. The kit allowed the simultaneously staining of both viable and dead cells. The number of live and dead cells were numbered and represented as function of percentage from the number of total cells. The experiments were performed in triplicate and the data represented as mean ± SD.

3. Results

The quality of the obtained coatings by sol-gel process is influenced by many factors. The most significant element is the stability of the concentrated suspension from which the coatings are obtained. A more accurate way of evaluating the stability of the suspensions is represented by ultrasound measurements. This analysis allows evaluation of the stability of the concentrated solution as opposed to the other traditional methods in which the suspension needs to be diluted. Due to the dilution, important information regarding the stability of the suspension can be lost.

Figure 1 shows information regarding the stability of the concentrated suspension of 5SmHAp obtained by the ultrasound measurements. The digitalised ultrasonic signals are recorded on the digital oscilloscope at a very precise interval of 5.00 s (Figure 1a). The peak amplitudes evolve in time in an unusual way for this suspension, compared to many others studied previously. There is an initial slowly decreasing amplitude, followed by a rising amplitude after 2400 s and a hyperpolic decreasing amplitude up to the end of the experiment (5000 s). Time delays between the first echo that was recorded in the analyzed suspension and the equivalent echo in the reference fluid, allowed a precise determination of the velocity of ultrasounds through the sample, for each signal that was recorded. The result for the velocity in the reference fluid was $c_0 = 1478.27$ m/s while the velocity in the analyzed sample was $c = 1492.07$ m/s and the temperature at which the experiment was performed was 23.3 °C. The ultrasound velocity in the sample is very close to the velocity in pure water and has a negligible variation during the experiment. To characterize the evolution in time of the sample's stability the evolution of amplitude in time was also evaluated (Figure 1b). The amplitude of the first echo was the only one measured with sufficient accuracy. The parameter that gives quantitative information on the stability of the suspension is closely related to the slope of the amplitude of the first echo vs. time. The value of the stability parameter computed by the algorithm was $s = \frac{1}{A_m}\left|\frac{dA}{dt}\right| = 1.3 \times 10^{-4} \cdot s^{-1}$, in which A_m is the averaged amplitude of the signals. Ignoring the local maximum at $t = 2400$ s, the overall stability parameter can be estimated as $s = 9 \times 10^{-4} \cdot s^{-1}$. The value obtained for the stability parameter shows a very good stability of 5SmHAp concentrate suspension.

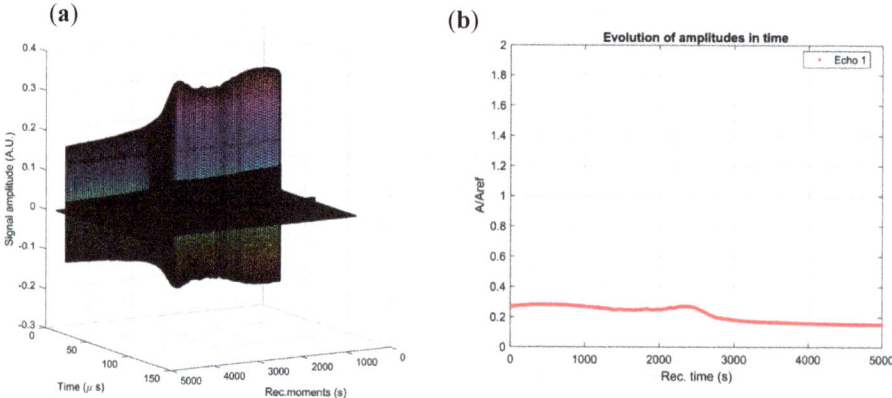

Figure 1. (a) Recorded signals at 5 s recording interval. A decreasing amplitude of the first echo is visible in colors, whereas the second is significantly weaker (in black); (b) Relative amplitudes evolution vs. the recording moments (a).

The relative spectral amplitude vs. time for the first echo and attenuation vs. time for the spectral components of echo 1 revealed the complex information concerning the stability of the analyzed concentrated suspension (Figure 2). Figure 2a shows the initial highest amplitude ratio of 0.55 for the frequency of 2 MHz and the lowest of 0.25 at 8 MHz. For accuracy, all values obtained are relative to the values of the reference fluid (double distilled water) measured under the same conditions. Before the peak at value of time $t = 2400$ s, the amplitudes at all selected frequencies are slowly decreasing. After the peak, the amplitudes decrease continuously, which is atypical for a dispersion after 5000 s. It can be expected that after a much longer period, the relative amplitudes will tend to 1, when most of the particles will settle at the bottom of the container. As shown in Figure 2b, the attenuation calculated for each spectral component depends on the moment during the experiment. The HApSm5 has extremely high attenuations from the first moment, between 20 nepper/m for the 2 MHz component, up to 69 nepper/m for the 8 MHz component of the ultrasonic signal. After the localized reduction of attenuation at $t = 2400$ s, follows a pronounced and continuous increase of the attenuation. The 2 MHz component reaches 28 nepper/m, whereas the 8 MHz component reaches 110 nepper/m. This increase of the attenuation is attributed to the increase of particles concentration during the sedimentation process. The progressive variation of the attenuation with frequency is normal for suspensions of particles which do not resonate at frequencies in the range of the selected transducer.

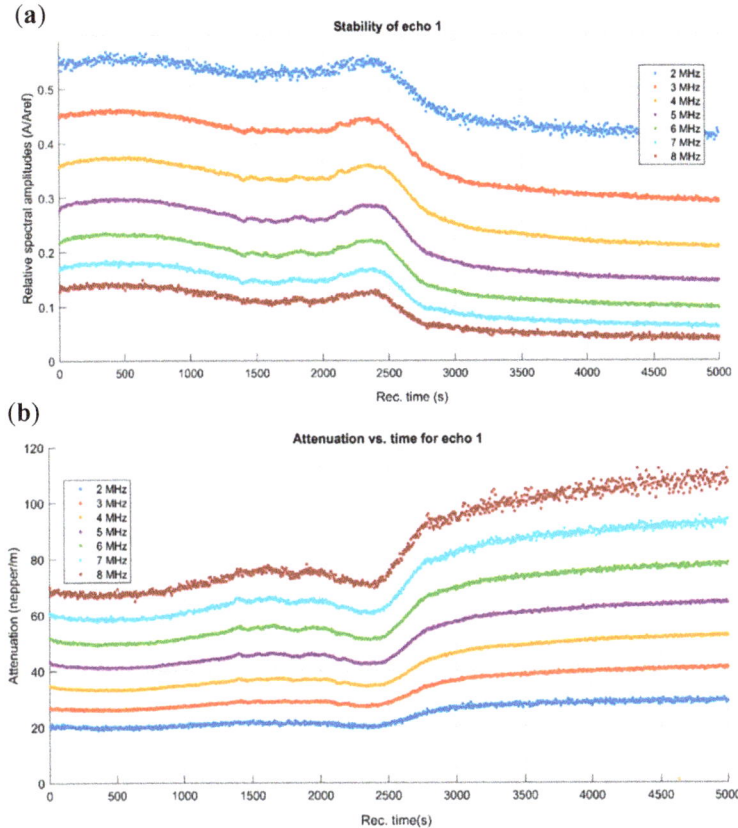

Figure 2. Spectral amplitudes relative variation vs. time, for the first echo (**a**); Attenuation vs. time for the spectral components of echo 1 (**b**).

The morphology of the HAp and 5SmHAp suspension was investigated through TEM and SEM studies. The results of these studies for the HAp suspension are depicted in Figure 3a,b. Also, in Figure 4a,b are presented the SEM and TEM micrographs obtained on the 5SmHAp suspension. In addition, the particle size distribution obtained from TEM and SEM measurements are presented in Figure 3c,d for HAp suspension and in Figure 4c,d for the 5SmHAp suspension.

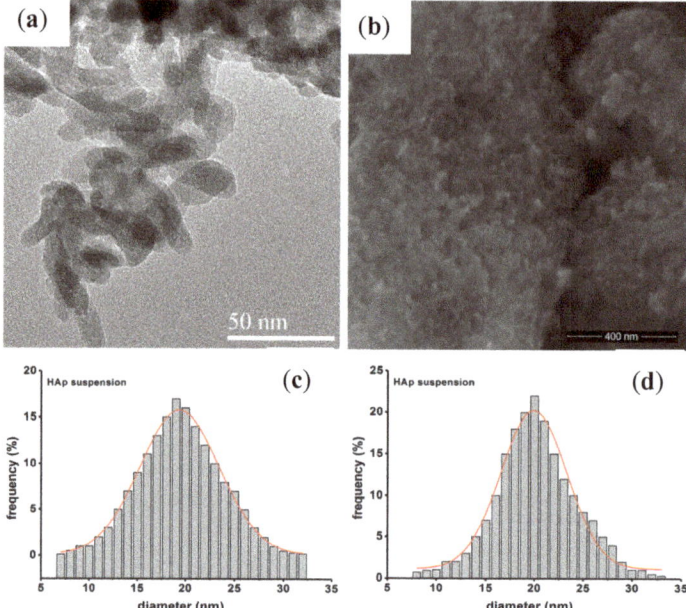

Figure 3. Transmission electron microscopy (TEM) (**a**) and scanning electron microscopy (SEM) (**b**) micrographs of HAp suspension. Particle size distribution of HAp suspension obtained by TEM (**c**) and SEM (**d**) studies.

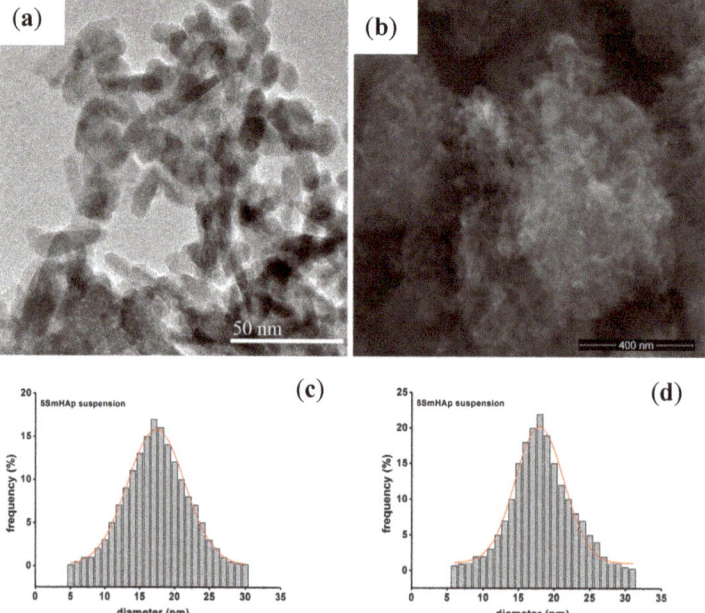

Figure 4. TEM (**a**) and SEM (**b**) micrographs of 5SmHAp suspension. Particle size distribution of 5SmHAp suspension obtained by TEM (**c**) and SEM (**d**) studies.

In the TEM micrographs (Figures 3a and 4a) it could be observed that the HAp and 5SmHAp suspensions consists of particles with nanometric dimensions and ellipsoidal morphology. The mean particle size estimated by TEM studies is around 19 nm in the case of 5SmHAp suspension and about 17 nm for the 5SmHAp suspension. Also, could be noticed that the presence of the samarium in the sample induces a slight decrease of particle size.

The results of the SEM investigations conducted on the HAp and 5SmHAp suspensions presented in Figures 3b–d and 4b–d revealed that for the both samples the nanoparticles tend to agglomerate and exhibit an elongated morphology. Furthermore, the mean particle size obtained by SEM studies was around 20 nm for the 5SmHAp suspension and about 18 nm for the 5SmHAp suspension. Therefore, it can be seen that the results of SEM studies are in good agreement with those obtained through TEM studies.

FTIR measurements were conducted in order to investigate the presence of the vibrational bands characteristic to the HAp structure in the samples. In Figures 5 and 6 are presented the general FTIR absorbance spectra of HAp and 5SmHAp coatings and suspension along with their FTIR deconvoluted spectra in the 400–700 cm^{-1} and 900–1200 cm^{-1} spectral regions.

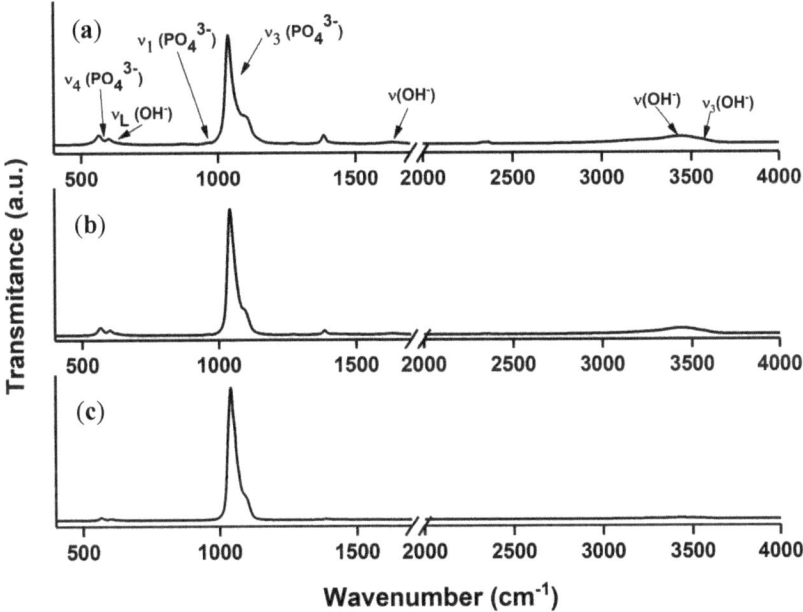

Figure 5. Fourier transform infrared (FTIR) absorbance spectra for HAp suspension (a) and 5SmHAp suspension (b) and coatings (c).

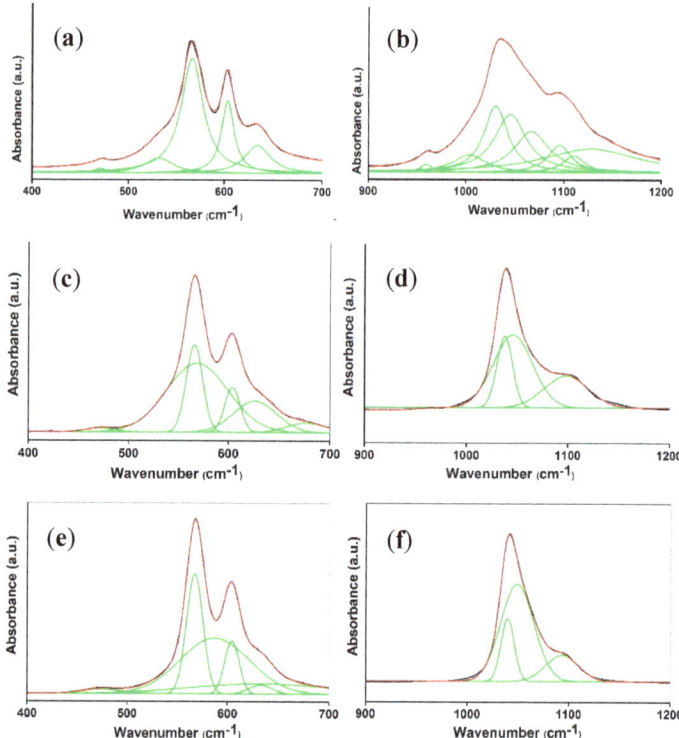

Figure 6. FTIR deconvoluted spectra in the 400–700 cm^{-1} and 900–1200 cm^{-1} spectral regions for HAp suspension (**a,b**) and 5SmHAp suspension (**c,d**) and coatings (**e,f**).

In the general FTIR absorbance spectra (Figure 5) of the studied samples have been identified the peaks belonging to the main vibrational modes of the adsorbed water (H–O–H), phosphate (PO$_4^{3-}$) and hydroxyl (OH$^-$) groups characteristic of the hydroxyapatite structure. The presence of the peak at about 635 cm^{-1} corresponds to the stretching librational mode (ν_L) of (OH$^-$) group [35,36]. At the same time, the presence of the peak at around 3570 cm^{-1} is due to stretching mode of (OH$^-$) groups. The presence of wide peaks in the 1600–1700 cm^{-1} and 3200–3600 cm^{-1} spectral regions are assigned to H–O–H bands of lattice water [35–37]. The presence of the peaks at around 1045 and 1093 cm^{-1} denote the presence of ν_3 of (PO$_4^{3-}$) group [35,36]. Furthermore, the peaks at around 604 and 568 cm^{-1} belong to ν_4 of (PO$_4^{3-}$) group. In agreement with the studies conducted by Iconaru et al. [35] a well crystalized hydroxyapatite structure is highlighted by the presence of the vibrational bands at around 635 cm^{-1} and at about 3570 cm^{-1} (which correspond to the OH$^-$ groups from the structure of hydroxyapatite). On the other hand, it can be seen that the vibration bands are more intense in the case of 5SmHAp coatings compared to HAp and 5SmHAp suspensions. This behaviour is atributed to the the fact that the 5SmHAp coatings are more crystaline comparative to the other two studied samples. Also, in the case of 5SmHAp coatings a slight displacement of the position of the peaks was observed.

Figure 6 depicts the deconvoluted FTIR spectra of the HAp suspension 5SmHAp coatings and suspension (green lines represents the individual subbands) in the 400–700 cm^{-1} and 900–1200 cm^{-1} spectral regions. In Figure 6a,b it can be observed that in order to obtain a good fit of the experimental data obtained oh HAp suspension, five subbands are needed in the spectral region 450–700 cm^{-1}, while for the spectral domain between 900 and 1200 cm^{-1}, eight subbands are needed. Also, it can be seen that in the case of 5SmHAp coatings in order to obtain a good fitting in the 400–700 cm^{-1} spectral region are needed six main components while for the 900–1200 cm^{-1} spectral region are needed four main components. In addition, our studies revealed that to have a good fit for the 5SmHAp suspension are needed six main components for the 400–700 cm^{-1} spectral region and four components for the 900–1200 cm^{-1} spectral region. Furthermore, an increase in the intensity of the subbands from the both deconvoluted regions specific to the 5SmHAp coatings could be noticed. This behavior is due to the superior crystallinity of the deposited coatings compared to that of the nanoparticles in suspensions.

In order to analyze the surface chemical elemental composition in the 5SmHAp coatings XPS measurements were conducted. The XPS general spectrum of the coating with $x_{Sm} = 0$ (HAp) and $x_{Sm} = 0.05$ (5SmHAp) was presented in Figure 7. The results indicated that elements such as C, O, Ca, and P were observed in the scan of HAp and 5SmHAp general spectrum (Figure 7a,b). In the scan of 5SmHAp XPS general spectra Sm was observed, too.

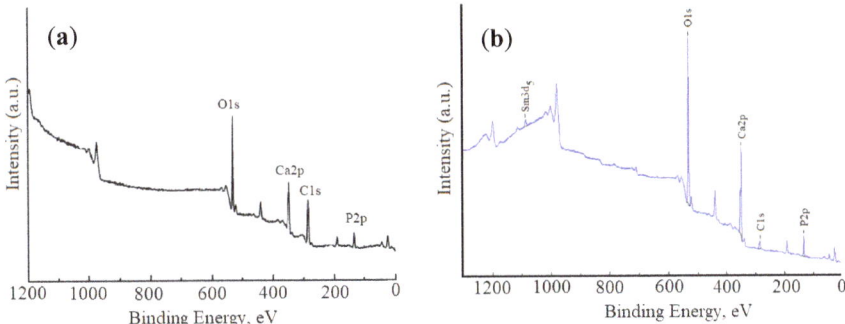

Figure 7. X-ray photoelectron spectroscopy (XPS) general spectrum of the coating with (**a**) $x_{Sm} = 0$ (HAp) and (**b**) $x_{Sm} = 0.05$ (5SmHAp).

The high resolution of C 1s, Ca 2p, O 1s, P 2p and Sm 3d are exhibited in Figure 8. The high-resolution C 1s spectra for the coating with $x_{Sm} = 0$ (HAp) and $x_{Sm} = 0.05$ (5SmHAp) are presented in Figure 8a,b. The peak in the C 1s region for HAp coating presented one component of the binding energy (BE) of 284.8 eV attributed to C–C bonds was used as a reference (Figure 8a). Three peaks in the binding energy (BE) region of C 1s were observed (Figure 8b). The peak of C 1s at 284.8 eV attributed to C–C bonds was used as a reference. Peak of C 1s at 286.74 eV can be assigned to C–O bonds while peak at 289.63 eV can be assigned to C=O bonds. In Figure 8c,d the high-resolution Ca 2p spectra for HAp and 5SmHap coatings is presented. The peaks in the BE region of Ca 2p exhibits a well-defined doublet with two components (Ca 2p3/2 and Ca 2p1/2). For the 5SmHAp coatings the BE shift slightly from 347.2 eV (for the HAp coatings) to 347.8 eV for the 5SmHAp coatings. The peak located at about 347.2 eV highlights that the calcium atoms are bound to a phosphate group (PO$_4^{3-}$). Kaciulis et al. showed [38] that the shift of the maximum to a higher value of the binding energy in the case of hydroxyapatite doped with different ions shows that the obtained sample is well crystallized. The high resolution XPS spectra of oxygen O 1s for pure HAp ($x_{Sm} = 0$) coating was shown in the Figure 8e. The peak at BE of 531.4 eV was assigned to hydroxyl groups that are the result of water or oxygen chemisorption [39]. The O 1s photoelectron peak was deconvoluted into two components (Figure 6f). The component located at 531.4 eV is characteristic to O 1s peak in HAp structure while the peak at 534.1 eV was attributed to the O 1s in adsorbed water [40,41]. The second component of O 1s was observed in the coatings obtained by sol-gel method [38,41]. In previous studies, Gaggiotti et al. [42] reported that the EB = 531.8 eV correspond to a position of the hydroxyl ion (OH$^-$). Kawabe et al. [43] appreciate that the two oxygen species (O– and OH), may be attributed to the peak at binding energy of 531.2 eV. Moreover, according to previous studies [44–46] the peak position of chemisorbed oxygen species O– could be attributed to a binding energy between EB = 531.0–531.5 eV. The P 2p photoelectron peak of pure HAp coating is shown in Figure 8g. The peak associated to P 2p revealed one component after the deconvolution data processing that was located at around EB = 133.1eV. The peak pf P 2p of 5SmHAp coating (Figure 8h) was deconvoluted into two components at 133.01 and 134.04 eV BE in agreement with previous studies [47].

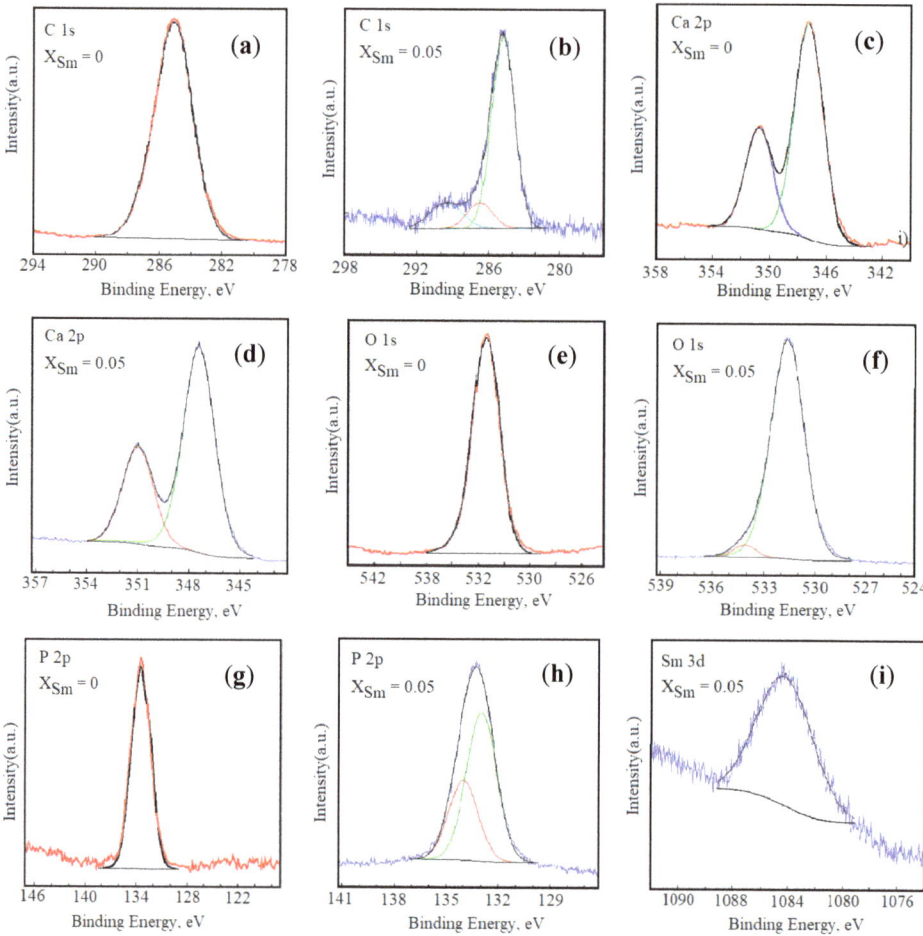

Figure 8. Deconvolution of XPS peaks of C 1s, O 1s, Ca 2p, P 2p and Sm 3d for the pure HAp (**a,c,e,g**) and 5SmHAp (**b,d,f,h,i**) coatings.

Previous studies on hydroxyapatite layers [48] revealed that the P 2p photoelectron line consists of a single component assigned to a peak at binding energy of 133.4 eV. On the other hand, the precedent XPS analysis [49] indicated that the binding energy of the photoelectron peaks for P and Ca are characteristic to their full oxidation states (P^{5+} and Ca^{2+}) for hydroxyapatite. The high resolution spectra of the Sm 2p3 is shown in Figure 8i. The peak corresponding to the binding energy of 1082.93 eV can be assigned to Sm 3d5/2 [50]. In agreement with previous studies [51], the binding energy of Sm 3d5/2 as revealed in Figure 8i suggested that the chemical state of Sm is +3. The XPS results provided obvious information for the successful doping of HAp with samarium.

Figure 9 shows the XRD diffractogram and SEM micrograph of HAp (a,c) and 5SmHAp coatings (b,d). The elemental mapping analysis of the chemical constituents of 5SmHAp coatings (Ca (a), O (b), P (c), Sm (d)) is also presented in Figure 10. In the SEM micrograph (Figure 9d) obtained on the 5SmHAp coatings could be noticed that the surface of the coatings is uniform and continuous with a granular morphology (granules formed by agglomeration of nanoparticles).

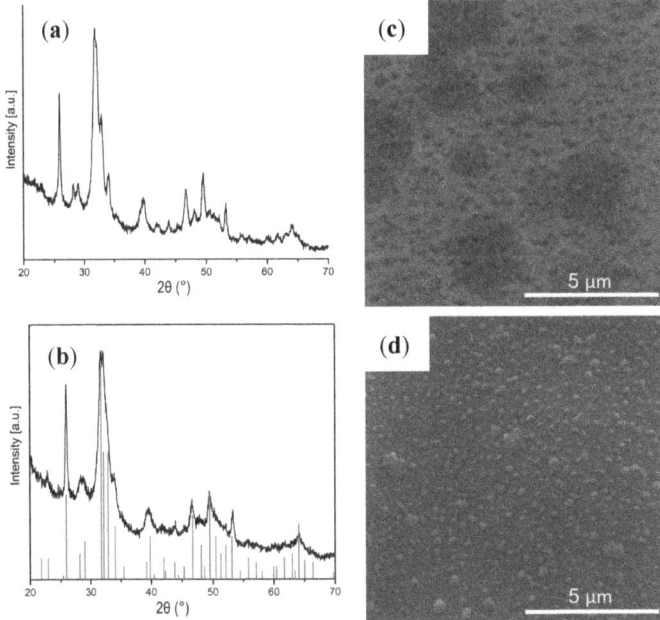

Figure 9. X-ray diffraction (XRD) diffractogram and SEM micrograph of HAp (**a**,**c**) and 5SmHAp coatings (**b**,**d**).

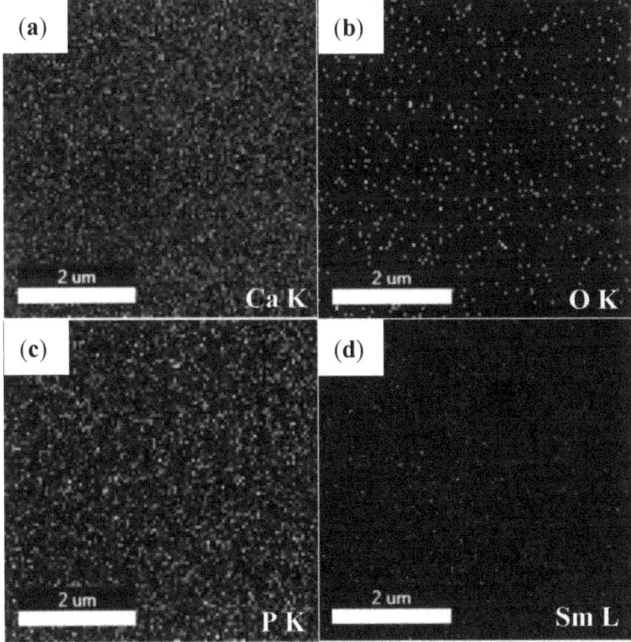

Figure 10. Elemental mapping analysis of the chemical constituent of 5SmHAp coating (Ca (**a**), O (**b**), P (**c**), Sm (**d**)).

In order to determine the crystal structure phase of HAp and 5SmHAp coating, the X-ray diffraction (XRD) was performed using Cu Kα radiation (λ = 1.5406 Å). The XRD patterns of the HAp coating were presented in Figure 9a. Figure 9b shows the XRD patterns of the 5SmHAp coating with $x_{Sm} = 0.05$. The XRD peaks were indexed by the hexagonal phase of pure hydroxyapatite (JCPDS no. 09-0432). It should be noted that no other secondary phases due to Sm^{3+} ions doped hydroxyapatite were observed after the analysis. This result highlights that Sm^{3+} ions substituted Ca^{2+} ions of 5SmHAp without changing the crystal structure. The diffraction peaks of HAp and 5SmHAp coating correspond to (002), (210), (211), (310), (202), (310), (311), (113), (222), (213), and (004) crystal planes of the hexagonal phase of pure HAp. The average crystallite sizes of the HAp and 5SmHAp coating calculated using Scherrer's formula was around 18 and 15 nm, respectively

On the other hand, in the SEM image of HAp and 5SmHAp coatings (Figure 9c,d) the presence of cracks or fissures on the surface of the coatings could not be observed. The results of the elemental mapping analysis of the chemical constituents of 5SmHAp coatings are presented in the (Figure 10a–d). In the Figure 10a–d can be seen that the main chemical constituents of the 5SmHAp coatings were Ca, O, P, and Sm. Moreover, our studies revealed that all the chemical constituents were evenly distributed in coatings.

The morphology of the HAp and 5SmHAp coatings surface topography has been investigated using atomic force microscopy. The results of the AFM investigations regarding the surface morphology of the HAp and 5SmHAp coatings are depicted in Figure 11a–d.

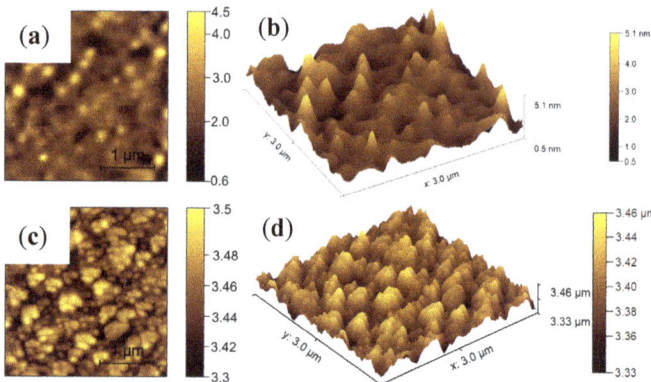

Figure 11. Characteristic 2D atomic force microscopy (AFM) image of HAp (**a**) and 5SmHAp coatings (**c**) and 3D representation of the HAp (**b**) and 5SmHAp (**d**) coatings surface collected on an area of 3 × 3 μm².

Figure 11 presents the AFM 2D micrograph and the 3D representation of the HAp and 5SmHAp surface topography of the HAp and 5SmHAp coatings. The results of the AFM studies emphasized that all the investigated coatings have the morphology of a uniformly deposited layer. Furthermore, the 2D representation of the surface highlighted that there is no visible evidence of the existence of cracks or fissures and that the deposited coatings both in the case of HAp and 5SmHAp samples consist of uniformly distributed nanoaggregates. The AFM results showed that the surface topography of the 5SmHAp coatings was also homogenous having a roughness (R_{RMS}) value of 21.45 nm. In addition, the roughness (R_{RMS}) value for the HAp coatings resulted from the AFM studies was 12.55 nm. The results obtained by AFM studies are in good agreement with the SEM studies, which also revealed that the coatings are uniform and homogenous.

The adhesion of the *C. albicans* cells on the surface of 5SmHAp coatings was also investigated by AFM studies. For this purpose, the surface of HAp and 5SmHAp coatings incubated with *C. albicans* fungal cells at three different time intervals was studied using AFM topography. The results of the AFM surface topography of the HAp and 5SmHAp coatings incubated with *C. albicans* fungal cells for 24, 48, and 72 h are presented in Figure 12. The 2D AFM images of the 5SmHAp coatings surface highlighted that the *C. albicans* cell development was inhibited by the coatings and that the inhibition rates were correlated with the incubation time. The AFM images collected on an area of 20 × 20 μm² evidenced that the *C. albicans* development was inhibited starting from 24 h and the inhibition continued and accentuated after 48 and 72 h. In the case of the coatings incubated for 72 h, the *C. albicans* cells were almost completely eradicated from the surface of the 5SmHAp coatings, demonstrating that the incubation time strongly affected the antifungal properties of the 5SmHAp coatings. Furthermore, the AFM images of the HAp coatings incubated with *C. albicans* fungal cells emphasized that the HAp coatings promoted the development of the fungal cells and provided a good adhesive surface for the *C. albicans* cells and allowed the development of fungal biofilm.

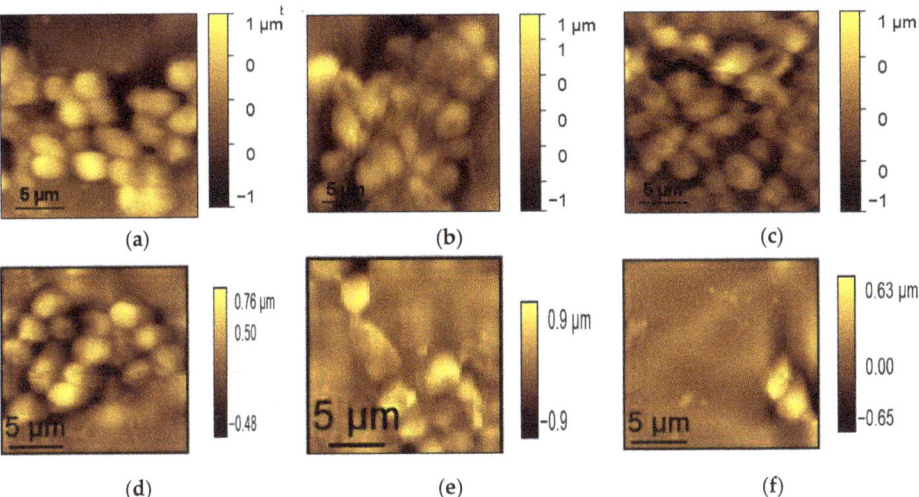

Figure 12. AFM surface topography of *Candida albicans* ATCC 10231 cell development on HAp (**a–c**) and 5SmHAp coatings (**d–f**) at different time intervals of incubation collected on an area of 20 × 20 μm².

The effects of the 5SmHAp coatings on the adhesion and proliferation of *C. albicans* were furthermore studied using SEM and CLSM investigations. The images of *C. albicans* cell development on the surfaces of 5SmHAp coatings after 24, 48, and 72 h of incubation with the *C. albicans* microbial cells, resulted from SEM observations were presented in Figure 13. The SEM visualization of the 5SmHAp coatings incubated with the fungal cell at different time intervals highlighted that the morphology of the fungal cells was typical to that of the *C. albicans* fungal strain, having round and oval shapes. Moreover, the SEM images emphasized that the size of the *C. albicans* fungal cells were in the range of 2.378–4.419 μm. In addition, the SEM analysis revealed that the fungal cells adhesion and development of biofilms on the 5SmHAp coatings was inhibited only after 24 h of incubation. Moreover, the SEM observations highlighted the inhibitory effects of the 5SmHAp coatings against *C. albicans* fungal strain at all tested time intervals. Furthermore, the results of the SEM visualization evidenced that the inhibitory effect of the 5SMHAp coatings was correlated with the incubation time.

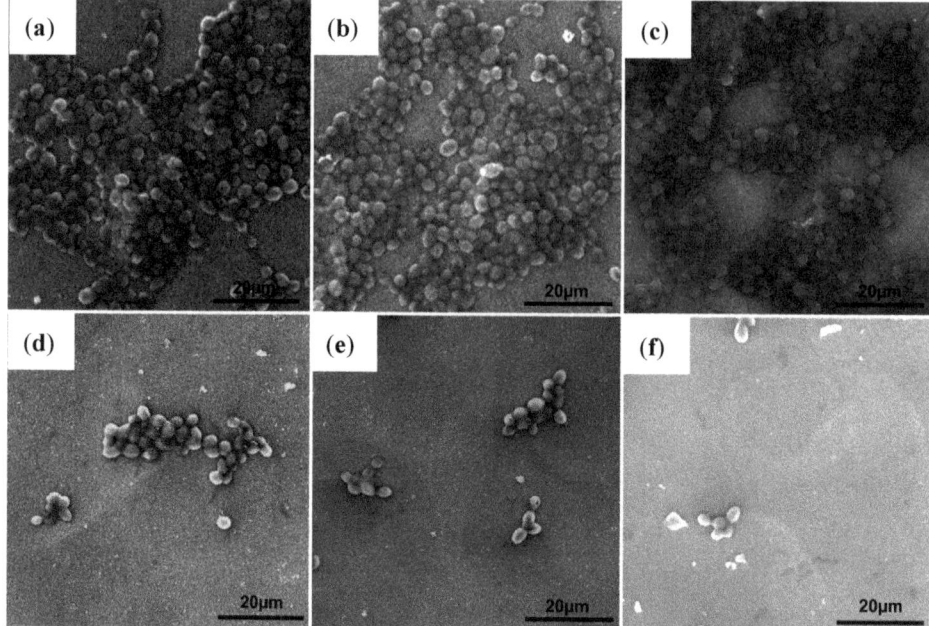

Figure 13. SEM images of *Candida albicans* ATCC 10231 cell development on HAp (**a–c**) and 5SmHAp coatings (**d–f**) at different time intervals of incubation 24 h (**a,d**) 48 h (**b,e**) and 72 h (**c,f**).

In addition, the *C. albicans* development on the HAp coatings was also assessed by SEM visualization at three different time intervals. The results emphasized that the HAp coatings promoted the development of the fungal cells and aided their proliferation and also allowed the formation of a fungal biofilm on their surface. These studies are in good agreement with previous conducted studies regarding the antifungal properties of HAp. The SEM images have revealed that the number of *C. albicans* fungal cells was considerably diminished after 48 and 72 h of incubation with the 5SmHAp coatings. Furthermore, the 3D representation of the SEM images with the *C. albicans* attached fungal cells was also performed using Image J software (Image J 1.51j8) [33]. The 3D representation of the SEM images of the 5SmHAp coatings incubated at three different time intervals with *C. albicans* fungal cells are presented in Figure 14.

The development and adhesion of *C. albicans* cells on the surface of 5SmHAp coatings after different time incubation intervals were also examined by CLSM. The images acquired using CLSM proved much more clearly, that the *C. albicans* cells development was inhibited by the 5SmHAp coatings and that the coatings did not allowed the fungal cells to proliferate and develop biofilms. These results are in good agreement with previous reported studies regarding the fungal properties of other biocomposite layers deposited on Si and Ti substrates [12,29,52,53]. CLSM is an optical imaging technique and a useful tool for imaging, qualitative analysis and quantification of cells [54,55]. Figure 15 shows the CLSM images of *C. albicans* cell growth on 5SmHAp coatings incubated with the fungal cells at different time intervals. The CLSM visualization was done using propidium iodide to label the fungal cells. The CLSM visualization evidenced that the surviving *C. albicans* cells were intact with round morphology and a smooth surface for all tested time intervals of incubation on the surface of HAp and 5SmHAp coatings. The results are depicted in Figure 15a–f. Furthermore, the CLSM examination confirmed the results obtained by the qualitative antimicrobial assay and was in good agreement with the results obtained from the SEM analysis. CLSM images highlighted that 5SmHAp coatings after 24 h of incubation exhibited a strong antifungal activity against *Candida albicans* ATCC

10231 and inhibited the biofilm formation. In addition, the results obtained by CLSM studies have emphasized that the time of incubation influenced the antifungal activity of the 5SmHAp coatings. The antifungal assays performed revealed that the 5SmHAp coatings could have a great potential to be used in medical application as antifungal coatings for medical devices.

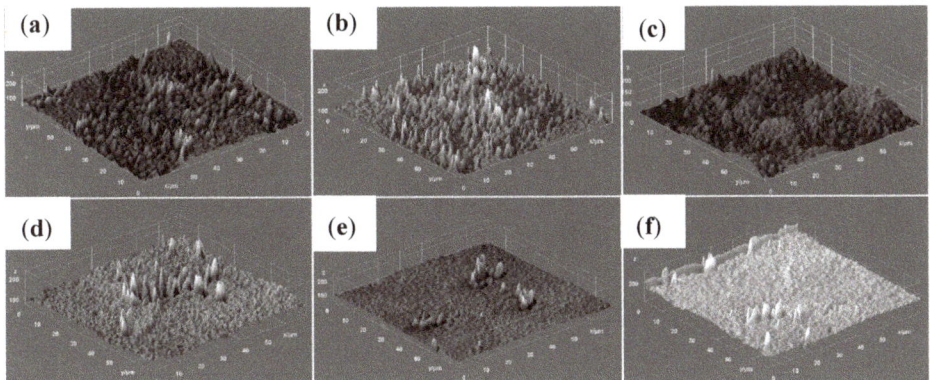

Figure 14. 3D representation of the SEM images of *Candida albicans* ATCC 10231 cell development on HAp (**a**–**c**) and 5SmHAp coatings (**d**–**f**) at different time intervals of incubation 24 h (**a,d**) 48 h (**b,e**) and 72 h (**c,f**).

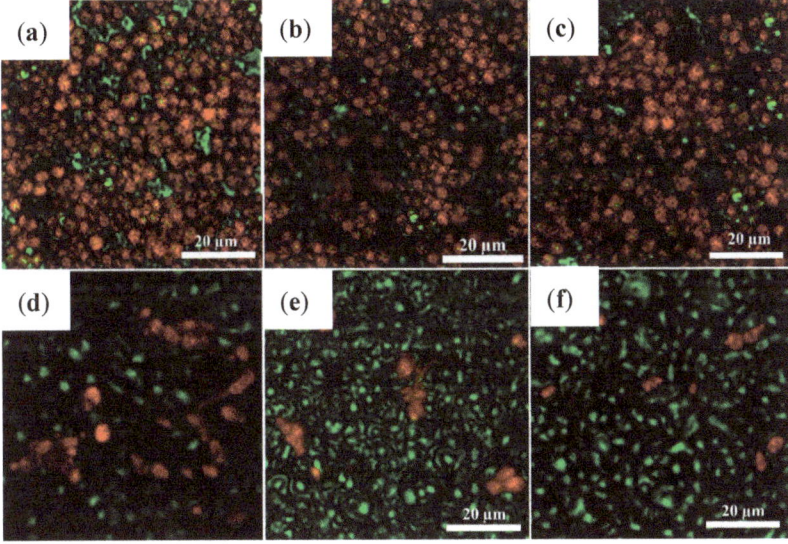

Figure 15. 2D confocal Laser Scanning Microscopy (CLSM) images of *Candida albicans* ATCC 10231 adhesion on HAp (**a**–**c**) and 5SmHAp (**d**–**f**) coatings deposited on Si after 24 h (**a,d**), 48 h (**b,e**) and 72 h (**c,f**) of incubation.

Furthermore, the 3D representation of the CLSM images of *C. albicans* fungal cells adhered on the surface of the 5SmHAp coatings after being incubated at three different time intervals analyzed using Image J software [33] are presented in Figure 16. The 3D representation display the structure and spatial distribution of *C. albicans* surviving cells on the 5SmHAp coatings after 24, 48, and 72 h of incubation with the fungal cells. The images exhibited in Figure 15 revealed the spatial distribution of *C. albicans* surviving cells (red color) along horizontal (coverage) and the vertical (thickness) distributions on the 5SmHAp surfaces. The CLSM images presented in Figure 16 demonstrated that the survival of *C. albicans* cells have been significantly reduced in the presence of 5SmHAp coatings and that their inhibition was strongly correlated with the incubation time.

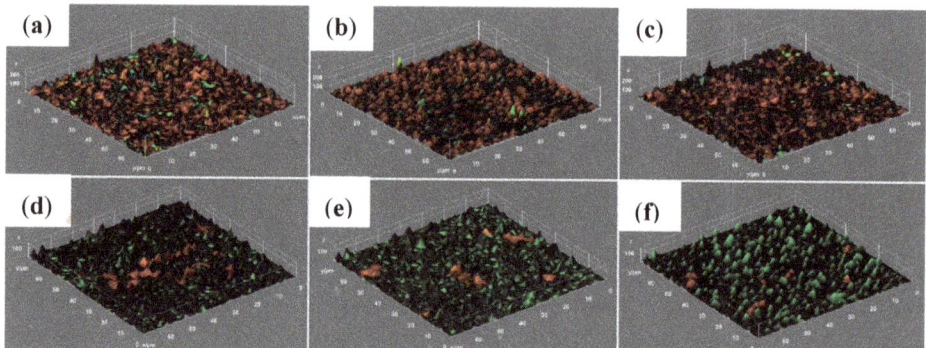

Figure 16. 3D representation of the CLSM images of *Candida albicans* ATCC 10231 adhesion on HAp (**a**–**c**) and 5SmHAp (**d**–**f**) coatings deposited on Si after 24 h (**a**,**d**), 48 h (**b**,**e**) and 72 h (**c**,**f**) of incubation.

The quantitative antifungal properties of the HAp and 5SmHAp coatings deposited on silicium substrate were also assessed using *C. albicans* ATCC 10231 fungal strain. The graphical representation of *C. albicans* ATCC 10231 colony-forming units on 5SmHAp suspensions and coatings at three different time intervals (24, 48, and 72 h) are depicted in Figure 17.

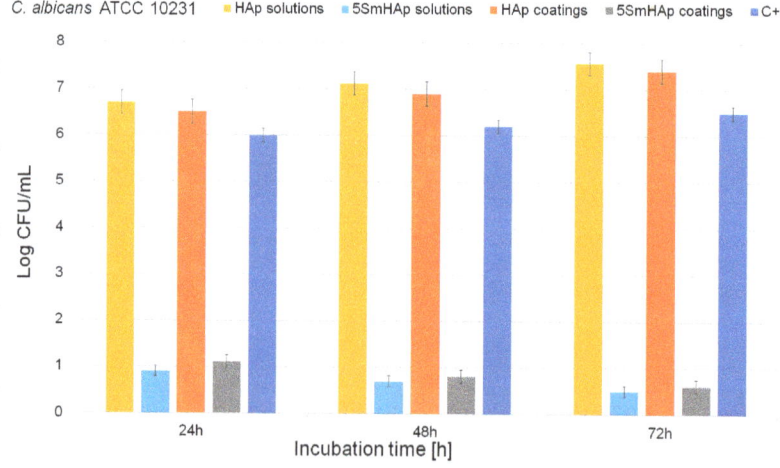

Figure 17. Graphical representation of the Log colony forming units (CFU)/mL of *C. albicans* as a function of time of exposure to the HAp and 5SmHAp suspensions and coatings.

The quantitative results emphasized that the fungal colony forming unity development was inhibited from the early phase of adherence in the case of 5SmHAp both suspensions and coatings. The colony forming unit count (CFUc) assay showed a significant decrease of the number of colonies in the case of both 5SmHAp suspensions and coatings compared to the number of colonies formed in the case of the control culture. Furthermore, the results emphasized that the antifungal activity of 5SmHAp suspensions was higher than that of 5SmHAp coatings for all tested intervals. The results obtained by the qualitative and quantitative studies conducted, evidenced that the 5SmHAp suspensions and coatings inhibited the *C. albicans* fungal strain development. The results are in good agreement with previous reported data [10,53,56–58] and emphasized that samarium ions exhibit antifungal properties.

The antifungal properties of HAp suspensions and HAp coatings were also investigated and a free *C. albicans* culture was used as positive control C+. The results demonstrated that both HAp suspensions and HAp coatings encouraged the proliferation of the fungal cells and influenced positive their growth for all tested time intervals. These results are in good agreement with previously reported studies regarding the lack of antimicrobial properties of hydroxyapatite [20,59,60]. Moreover, previous studies [15,56–64] reported that the antifungal effect of samarium could be explained by the fact that samarium ions have the ability to attach to the cell membrane leading to changes of its permeability. Furthermore, samarium ions that are released from the composite matrix could disrupt the bacterial membrane integrity thus affecting a multitude of cellular processes such like adhesion, ion conductivity and cell signaling [61].

The results of the qualitative and quantitative antifungal assays are in agreement with the previous data presented in the literature and highlighted that the Sm^{3+} ions present in the hydroxyapatite matrix of the 5SmHAp coatings is responsible for the antifungal activity of the coatings. Moreover, the data also suggested that the antifungal activity of both the 5SmHAp suspensions and coatings are strongly correlated with the incubation time. The qualitative studies conducted by AFM, SEM and CLSM suggested that the adhesion of *C. albicans* fungal cells was greatly reduced by the samarium doped composites. These results are similar with those reported in the literature, regarding the adherence of *S. aureus*, *S. epidermidis* and *P. aeruginosa* strains on composites doped with samarium [15]. The results reported by the authors showed that compared to glass-reinforced hydroxyapatite composites (GR-HA) composite the *S. aureus*, *S. epidermidis* and *P. aeruginosa* microbial cell development was considerably inhibited in the Sm doped composites and that the microbial cell reduction was correlated with the Sm content. The antibacterial effect of the composites were attributed also to the release of samarium ions from the composites. Thus, the development of 5SmHAp coatings with enhanced osteoblastic cell response given by the hydroxyapatite matrix and possessing antibacterial activity due to the presence of samarium ions could lead to a better outcome of bone graft implantation. The choice of Si as a substrate started from the intention to extend the applicability of these biocompatible materials with antimicrobial properties from dental field to devices that could be used in tissue engineering [62] or human prosthetic [63] in order to prevent postoperative infections. Ruffino and Torrisi [64] in their studies regarding the influence of interaction between film and substrate on the nanoscale film morphology showed that between Ag films deposited on SiO_2/Si and TiO_2/Ti substrates exist various differences. Also Jouanny et al. [65] reported that between the TiO_2 thin films deposited on the Si and Ti–6Al–4V substrate respectively, are differences from mechanical point of view. According to the studies reported by Barry et al. [66] the nature of the substrate could influence the surface roughness and coating thickness but there are no significant differences between the interface properties and the coating composition depending on the type of alloy. Moreover, Ferraris et al. [67] reported that the substrate also influences the biological properties of the samples.

Furthermore, due to the intended purpose of being used in biomedical applications, the toxicity of the HAp and 5SmHAp suspensions and coatings were also assessed using one of the most studied cell line. The toxicity of HAp and 5SmHAp suspensions and coatings was assessed against HeLa cells at three different time intervals (24, 48, and 72 h). The results of the cytotoxicity assay obtained by MTT 3-(4,5-dimethylthiazol-2-yl)-2,5-diphenyltetrazolium bromide (MTT) are presented in Figure 18. HeLa cell line is the oldest and the most commonly used human cell line and has been extensively used in scientific studies due to its remarkably durable and prolific properties. Since its discovery, the HeLa cells have been intensively and continually used for research into various fields of research such as cancer, AIDS, the effects of radiation and toxic substances, the toxicity of nanoparticles and other types of materials, gene mapping and countless other scientific pursuits.

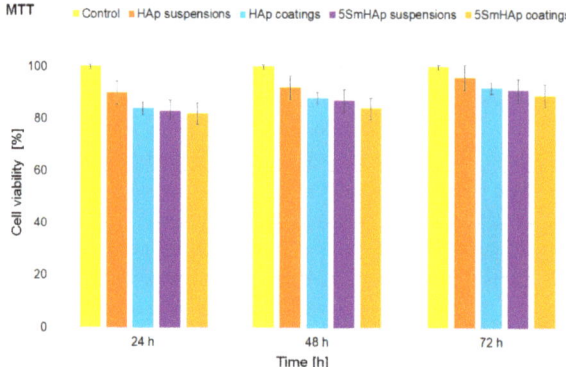

Figure 18. MTT assay for the viability of HeLa cells incubated with HAp and 5SmHAp suspensions and coatings at different time intervals.

The MTT studies revealed that after 24 h of incubation time, there were no representative differences between the cells viabilities of the HAp and 5SmHAp suspensions. The results have emphasized that after 24 h of incubation, the cell viability of HeLa cells was 90% in the case of the HAp suspensions and 84% in the case of the HAp coatings and also 83% in the case of 5SmHAp suspensions and 82% for the HeLa cells incubated with the 5SmHAp coatings. Moreover, the MTT assay results show that after 48 and 72 h of incubation a slight increase of the HeLa cell viability was observed for all the investigated samples.

The results highlighted that there is a correlation between the incubation time and the cell viability of the samples. Moreover, the MTT suggested that the HAp and 5SmHAp suspensions had better biocompatible properties than the HAp and 5SmHAp coatings. The results obtained in the present study are in good agreement with previous studies regarding the toxicity of hydroxyapatite suspensions and coatings [68–70]. In addition, a live/dead cell viability assay was performed in order to quantify the ratio of the live and dead cells after the incubation of the HeLa cells with 5SmHAp solutions and coatings at different time intervals. The results are depicted in Figure 19. The results are in agreement with the MTT assay and evidenced that both the 5SmHAp solutions and coatings did not present any toxicity against the tested cells for all tested time intervals. Furthermore, the data suggested that the percent of dead cells in the case of both samples and for all tested intervals was under 10% from the total number of counted cells.

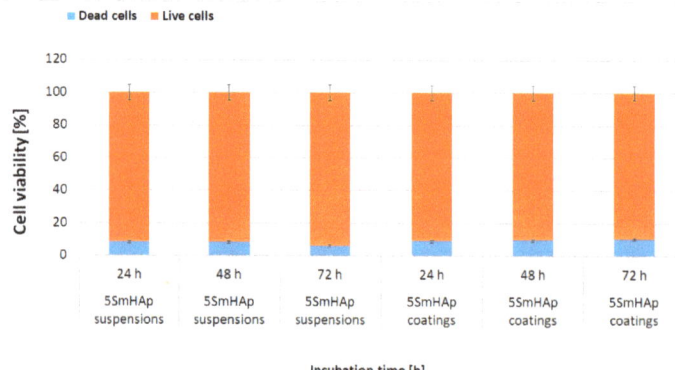

Figure 19. Percent of total counts of live cells and dead cells after incubation with 5SmHAp suspensions and coatings at different time intervals.

4. Conclusions

Hydroxyapatite and samarium doped hydroxyapatite (x_{Sm} = 0.05) coatings were synthesized by a simple sol-gel route using the dip coating method. Characterization studies of HAp and 5SmHAp nanoparticles in suspension and coatings using different techniques were performed. The stability of 5SmHAp suspension was highlighted by ultrasound measurements. Ultrasound measurements showed that the 5SmHAp suspension had a behavior similar to that of double-distilled water considered as a reference fluid. XPS and XRD studies on 5SmHAp coating confirmed that the Sm^{3+} ions have been incorporated into the 5SmHAp synthesized material by substituting the Ca^{2+} ions. The uniform distribution of constituent elements on 5SmHAp coating surface was confirmed by elemental mapping analysis. SEM and AFM investigations revealed the uniform and homogenous surface of 5SmHAp coating. In addition, FTIR studies confirmed the presence of the functional groups characteristic to the HAp structure in both samples. The AFM analysis of the surface coatings revealed the obtaining of homogenous and uniform coatings with no cracks or fissures. The antifungal activity of HAp and 5SmHAp coatings and suspensions was investigated using a *Candida albicans* ATCC 10231 fungal strain. The results of the quantitative assay of the antifungal activity revealed that both 5SmHAp coatings and suspensions inhibited the development of *C. albicans* fungal strain. The results obtained by the qualitative assays using AFM, SEM CLSM visualization of the *C. albicans* cell adherence on the surface coatings confirmed the quantitative results and evidenced that the antifungal properties of the coatings were influenced by the incubation time. Furthermore, the cytotoxic assay using HeLa cell line emphasized that both 5SmHAp suspensions and coatings did not present any toxicity against HeLa cells for all tested incubation intervals. Nowadays, following the occurrence of microbial infections affecting public health, the development of novel materials that could be used as a cheap alternative for the obtaining of coatings with high antimicrobial activity over time is of great interest.

Author Contributions: Conceptualization, D.P., S.C.C., M.V.P., S.L.I., and A.M.P.; methodology, D.P., S.C.C., M.V.P., S.L.I., and A.M.P.; software, M.V.P.; validation, D.P., S.C.C., M.V.P., S.L.I., and A.M.P.; formal analysis, D.P., S.C.C., M.V.P., S.L.I., and A.M.P.; investigation, D.P., S.C.C., M.V.P., S.L.I., and A.M.P.; resources, A.M.P., D.P., and M.V.P.; data curation, D.P., S.C.C., M.V.P., and S.L.I.; writing—original draft preparation, D.P., S.C.C., M.V.P., and S.L.I.; writing—review and editing, D.P., S.C.C., M.V.P., and S.L.I.; visualization, D.P., S.C.C., M.V.P., S.L.I., and A.M.P.; supervision, D.P., S.C.C., M.V.P., and S.L.I.; project administration, M.V.P.; funding acquisition, D.P. and A.M.P. All authors have read and agreed to the published version of the manuscript.

Funding: This research was funded by [Romanian Ministry of Research and Innovation] with the grant number PN-III-P1-1.2-PCCDI-2017-0629/contract No. 43PCCDI/2018, Core Program PN19-030101 (contract 21N/2019) and Scientific Research Contract Nr.1/4.06.2020] and The APC was funded by [Romanian Ministry of Research and Innovation] project number PN-III-P1-1.2-PCCDI-2017-0629/contract No. 43PCCDI/2018.

Acknowledgments: We would like to thank C.C. Negrila for his help in XPS data acquisition and to G. Stanciu and R. Hristu for their help in the CLSM data acquisition.

Conflicts of Interest: The authors declare no conflict of interest.

References

1. Nabeel, A.I. Samarium enriches antitumor activity of ZnO nanoparticles via downregulation of CXCR4 receptor and cytochrome P450. *Tumor Biol.* **2020**, *42*, 1–14. [CrossRef]
2. Alicka, M.; Sobierajska, P.; Kornicka, K.; Wiglusz, R.J.; Marycz, K. Lithium ions (Li Enhance expression of late osteogenic markers in adipose-derived stem cells. Potential theranostic application of nHAp doped with Li$^+$) and nanohydroxyapatite (nHAp) doped with Li$^+$ and co-doped with europium (III) and samarium (III) ions. *Mater. Sci. Eng. C* **2019**, *99*, 1257–1273. [CrossRef]
3. LeGeros, R.Z. Properties of osteoconductive biomaterials: Calcium phosphates. *Clin. Orthop.* **2002**, *395*, 81–98. [CrossRef] [PubMed]
4. Giannoudis, P.V.; Dinopoulos, H.; Tsiridis, E. Bone substitutes: An update. *Injury* **2005**, *36*, S20–S27. [CrossRef] [PubMed]

5. Sathishkumar, S.; Louis, K.; Shinyjoy, E.; Gopi, D. Tailoring the Sm/Gd-substituted hydroxyapatite coating on biomedical AISI 316L SS: Exploration of corrosion resistance, protein profiling, osteocompatibility, and osteogenic differentiation for orthopedic implant applications. *Ind. Eng. Chem. Res.* **2016**, *55*, 6331–6344. [CrossRef]
6. Barroug, A.; Glimcher, M.J. Hydroxyapatite crystals as a local delivery system for cisplatin: Adsorption and release of cisplatin in vitro. *J. Orthop. Res.* **2002**, *20*, 274–280. [CrossRef]
7. Uskoković, V.; Uskoković, D.P. Nanosized hydroxyapatite and other calcium phosphates: Chemistry of formation and application as drug and gene delivery agents. *J. Biomed. Mater. Res. B Appl. Biomater.* **2011**, *96*, 152–191. [CrossRef] [PubMed]
8. Raita, M.S.; Iconaru, S.L.; Groza, A.; Cimpeanu, C.; Predoi, G.; Ghegoiu, L.; Badea, M.L.; Chifiriuc, M.C.; Marutescu, L.; Trusca, R.; et al. Multifunctional hydroxyapatite coated with arthemisia absinthium composites. *Molecules* **2020**, *25*, 413. [CrossRef]
9. Zantyea, P.; Fernandesa, F.; Ramananb, S.R.; Kowshik, M. Rare earth doped hydroxyapatite nanoparticles for in vitro bioimaging applications. *Curr. Phys. Chem.* **2019**, *9*, 94–109. [CrossRef]
10. Turculet, C.S.; Prodan, A.M.; Negoi, I.; Teleanu, G.; Popa, M.; Andronescu, E.; Beuran, M.; Stanciu, G.A.; Hristu, R.; Badea, M.L.; et al. Preliminary evaluation of the antifungal activity of samarium doped hydroxyapatite thin films. *Rom. Biotechnol. Lett.* **2018**, *23*, 13928–13932.
11. Prodan, A.M.; Iconaru, S.L.; Predoi, M.V.; Predoi, D.; Motelica-Heino, M.; Turculet, C.S.; Beuran, M. Silver-doped hydroxyapatite thin layers obtained by sol-gel spin coating procedure. *Coatings* **2020**, *10*, 14. [CrossRef]
12. Predoi, D.; Iconaru, S.L.; Predoi, M.V. Bioceramic layers with antifungal properties. *Coatings* **2018**, *8*, 276. [CrossRef]
13. Predoi, D.; Iconaru, S.L.; Albu, M.; Petre, C.C.; Jiga, G. Physicochemical and antimicrobial properties of silver-doped hydroxyapatite collagen biocomposite. *Polym. Eng. Sci.* **2017**, *57*, 537–545. [CrossRef]
14. Šupová, M. Substituted hydroxyapatites for biomedical applications: A review. *Ceram. Int.* **2015**, *41*, 9203–9231. [CrossRef]
15. Morais, D.S.; Coelho, J.; Ferraz, M.P.; Gomes, P.S.; Fernandes, M.H.; Hussain, N.S.; Santos, J.D.; Lopes, M.A. Samarium doped glass-reinforced hydroxyapatite with enhanced osteoblastic performance and antibacterial properties for bone tissue regeneration. *J. Mater. Chem. B* **2014**, *2*, 5872–5881. [CrossRef] [PubMed]
16. Ciobanu, C.S.; Popa, C.L.; Predoi, D. Sm:HAp nanopowders present antibacterial activity against enterococcus faecalis. *J. Nanomater.* **2014**, *2014*, 780686. [CrossRef]
17. Nakayama, M.; Smith, C.L.; Feltis, B.N.; Piva, T.J.; Tabatabaie, F.; Harty, P.D.; Gagliardi, F.M.; Platts, K.; Otto, S.; Blencowe, A.; et al. Samarium doped titanium dioxide nanoparticles as theranostic agents in radiation therapy. *Phys. Med.* **2020**, *75*, 69–76. [CrossRef]
18. Zhang, Y.; Wang, X.; Su, Y.; Chen, D.; Zhong, W. A doxorubicin delivery system: Samarium/mesoporous bioactive glass/ alginate composite microspheres. *Mater. Sci. Eng. C Mater. Biol. Appl.* **2016**, *67*, 205–213. [CrossRef]
19. Kannan, S.; Nallaiyan, R. Anticancer activity of Samarium-coated magnesium implants for immunocompromised patients. *ACS Appl. Bio Mater.* **2020**, *3*, 4408–4416. [CrossRef]
20. Iconaru, S.L.; Predoi, M.V.; Stan, G.E.; Buton, N. Synthesis, characterization, and antimicrobial activity of Magnesium-doped hydroxyapatite suspensions. *Nanomaterials* **2019**, *9*, 1295. [CrossRef]
21. Ciobanu, C.S.; Iconaru, S.L.; Massuyeau, F.; Constantin, L.V.; Costescu, A.; Predoi, D. Synthesis, structure, and luminescent properties of europium-doped hydroxyapatite nanocrystalline powders. *J. Nanomater.* **2012**, *2012*, 942801. [CrossRef]
22. Predoi, D.; Iconaru, S.L.; Predoi, M.V.; Motelica-Heino, M.; Guegan, R.; Buton, N. Evaluation of antibacterial activity of zinc-doped hydroxyapatite colloids and dispersion stability using ultrasounds. *Nanomaterials* **2019**, *9*, 515. [CrossRef] [PubMed]
23. Ciobanu, C.S.; Iconaru, S.L.; Popa, C.L.; Motelica-Heino, M.; Predoi, D. Evaluation of samarium doped hydroxyapatite, ceramics for medical application: Antimicrobial activity. *J. Nanomater.* **2015**, *2015*, 849216. [CrossRef]
24. Ciobanu, C.S.; Constantin, L.V.; Predoi, D. Structural and physical properties of antibacterial Ag-doped nano-hydroxyapatite synthesized at 100 °C. *Nanoscale Res. Lett.* **2011**, *6*, 1–8. [CrossRef]
25. Rodriguez, L.; Matoušek, J. Preparation of TiO$_2$ sol-gel layers on glass. *Ceram. Silik.* **2003**, *47*, 28–31.

26. Predoi, D.; Iconaru, S.L.; Predoi, M.V. Dextran-coated zinc-doped hydroxyapatite for biomedical applications. *Polymers* **2019**, *11*, 886. [CrossRef]
27. Gwyddion. Available online: http://gwyddion.net/ (accessed on 20 January 2020).
28. Jastrzebski, W.; Sitarz, M.; Rokita, M.; Bułat, K. Infrared spectroscopy of different phosphates structures. *Spectrochim. Acta A Mol. Biomol. Spectrosc.* **2011**, *79*, 722–727. [CrossRef]
29. Iconaru, S.L.; Prodan, A.M.; Turculet, C.S.; Beuran, M.; Ghita, R.V.; Costescu, A.; Groza, A.; Chifiriuc, M.C.; Chapon, P.; Gaiaschi, S.; et al. Enamel based composite layers deposited on titanium substrate with antifungal activity. *J. Spectrosc.* **2016**, *2016*, 4361051. [CrossRef]
30. ASTM International. *ASTM E2149–13a Standard Test Method for Determining the Antimicrobial Activity of Antimicrobial Agents under Dynamic Contact Conditions*; ASTM International: West Conshohocken, PA, USA, 2013.
31. Fuchs, A.V.; Ritz, S.; Pütz, S.; Mailänder, V.; Landfester, K.; Ziener, U. Bioinspired phosphorylcholine containing polymer films with silver nanoparticles combining antifouling and antibacterial properties. *Biomater. Sci.* **2013**, *1*, 470–477. [CrossRef]
32. Predoi, D.; Iconaru, S.L.; Predoi, M.V.; Buton, N.; Motelica-Heino, M. Zinc doped hydroxyapatite thin films prepared by sol-gel spin coating procedure. *Coatings* **2019**, *9*, 156. [CrossRef]
33. ImageJ. Available online: http://imagej.nih.gov/ij (accessed on 10 January 2018).
34. Predoi, D.; Iconaru, S.L.; Predoi, M.V.; Buton, N.; Megier, C.; Motelica-Heino, M. Biocompatible layers obtained from functionalized iron oxide nanoparticles in suspension. *Coatings* **2019**, *9*, 773. [CrossRef]
35. Iconaru, S.L.; Motelica-Heino, M.; Predoi, D. Study on europium-doped hydroxyapatite nanoparticles by fourier transform infrared spectroscopy and their antimicrobial properties. *J. Spectrosc.* **2013**, *2013*, 284285. [CrossRef]
36. Ciobanu, C.S.; Iconaru, S.L.; Le Coustumer, P.; Constantin, L.V.; Predoi, D. Antibacterial activity of silver-doped hydroxyapatite nanoparticles against gram-positive and gram-negative bacteria. *Nanoscale Res. Lett.* **2012**, *7*, 324. [CrossRef] [PubMed]
37. Fowler, B.O. Infrared studies of apatites. I. Vibrational assignments for calcium, strontium, and barium hydroxyapatites utilizing isotopic substitution. *Inorg. Chem.* **1974**, *13*, 194–207. [CrossRef]
38. Kaciulis, S.; Mattogno, G.; Pandolfi, L.; Cavalli, M.; Gnappi, G.; Montenero, A. XPS study of apatite-based coatings prepared by sol-gel technique. *Appl. Surf. Sci.* **1999**, *151*, 1–5. [CrossRef]
39. Moulder, J.F.; Stickle, W.F.; Sobol, P.E.; Bomben, K.D. *Handbook of X-ray Photoelectron Spectroscopy*; Physical Electronics Inc.: Chanhassen, MN, USA, 1995.
40. Zhang, S. *Biological and Biomedical Coatings Handbook: Applications*, 1st ed.; CRC Press, Taylor & Francis Group: London, UK, 2016.
41. Massaro, C.; Baker, M.A.; Cosentino, F.; Ramires, P.A.; Klose, S.; Milella, E. Surface and biological evaluation of hydroxyapatite-based coatings on titanium deposited by different techniques. *J. Biomed. Mater. Res.* **2001**, *58*, 651–657. [CrossRef]
42. Gaggiotti, G.; Galdikas, A.; Kaciulis, S.; Mattongo, G.; Setkus, A. Surface chemistry of tin oxide based gas sensors. *J. Appl. Phys.* **1994**, *76*, 4467. [CrossRef]
43. Kawabe, T.; Shimomura, S.; Karasuda, T.; Tabata, K.; Suzuki, E.; Yamaguchi, Y. Photoemission study of dissociatively adsorbed methane on a pre-oxidized SnO_2 thin film. *Surf. Sci.* **2000**, *448*, 101–107. [CrossRef]
44. Hegde, M.S.; Ayyoob, M. O^{2-} and O^{1-} types of oxygen species on Ni and barium-dosed Ni and Cu surfaces. *Surf. Sci.* **1986**, *173*, L635–L640. [CrossRef]
45. Rao, C.N.R.; Vijayakrishnan, V.; Kulkarni, G.U.; Rajumon, M.K. A comparative study of the interaction of oxygen with clusters and single-crystal surfaces of nickel. *Appl. Surf. Sci.* **1995**, *84*, 285–289. [CrossRef]
46. Kulkarni, G.U.; Rao, C.N.R.; Roberts, M.W. Coadsorption of dioxygen and water on the Ni(110) surface: Role of O1-type species in the dissociation of water. *Langmuir* **1995**, *11*, 2572–2575. [CrossRef]
47. Mirzaee, M.; Vaezi, M.; Palizdar, Y. Synthesis and characterization of silver doped hydroxyapatite nanocomposite coatings and evaluation of their antibacterial and corrosion resistance properties in simulated body fluid. *Mater. Sci. Eng. C* **2016**, *69*, 675–684. [CrossRef] [PubMed]
48. Stoica, T.F.; Morosanu, C.; Slav, A.; Stoica, T.; Osiceanu, P.; Anastasescu, C.; Gartner, M.; Zaharescu, M. Hydroxyapatite films obtained by sol-gel and sputtering. *Thin Solid Films* **2008**, *516*, 8112–8116. [CrossRef]
49. Battistoni, C.; Casaletto, M.P.; Ingo, G.M.; Kaciulis, S.; Mattogno, G.; Pandolfi, L. Surface characterization of biocompatible hydroxyapatite coatings. *Surf. Interface Anal.* **2000**, *29*, 773–781. [CrossRef]

50. Jørgensen, B.; Christiansen, M. Samarium films on copper single crystals. *Surf. Sci.* **1991**, *251–252*, 519–523. [CrossRef]
51. Wei, L.; Yang, Y.; Xia, X.; Fan, R.; Su, T.; Shi, Y.; Yu, J.; Lia, L.; Jiang, Y. Band edge movement in dye sensitized Sm-doped TiO_2 solar cells: A study by variable temperature spectroelectrochemistry. *RSC Adv.* **2015**, *5*, 70512–70521. [CrossRef]
52. Iconaru, S.L.; Prodan, A.M.; Buton, N.; Predoi, D. Structural characterization and antifungal studies of Zinc-doped hydroxyapatite coatings. *Molecules* **2017**, *22*, 604. [CrossRef]
53. Iconaru, S.L.; Stanciu, G.A.; Hristu, R.; Ghita, R.V. Properties of Samarium doped hydroxyapatite thin films deposited by evaporation. *Rom. Rep. Phys.* **2017**, *69*, 508.
54. Patel, D.V.; McGhee, C.N. Contemporary in vivo confocal microscopy of the living human cornea using white light and laser scanning techniques: A major review. *Clin. Exp. Ophthalmol.* **2007**, *35*, 71–88. [CrossRef]
55. Pawley, J.B. *Handbook of Biological Confocal Microscopy*, 3rd ed.; Springer: Boston, MA, USA, 2006; pp. 20–42.
56. Bobbarala, V. *Antimicrobial Agents*; IntechOpen: London, UK, 2012.
57. Jankauskaitė, V.; Abzalbekuly, B.; Lisauskaitė, A.; Procyčevas, I.; Fataraitė, E.; Vitkauskienė, A.; Janakhmetov, U. Silicone rubber and microcrystalline cellulose composites with antimicrobial properties. *Mater. Sci.* **2014**, *20*, 42–49. [CrossRef]
58. Kramer, A.; Schwebke, I.; Kampf, G. How long do nosocomial pathogens persist on inanimate surfaces? A systematic review. *BMC Infect. Dis.* **2006**, *6*, 1–8. [CrossRef] [PubMed]
59. Predoi, D.; Iconaru, S.L.; Buton, N.; Badea, M.L.; Marutescu, L. Antimicrobial activity of new materials based on lavender and basil essential oils and hydroxyapatite. *Nanomaterials* **2018**, *8*, 291. [CrossRef] [PubMed]
60. Predoi, D.; Iconaru, S.L.; Predoi, M.V.; Groza, A.; Gaiaschi, S.; Rokosz, K.; Raaen, S.; Negrila, C.C.; Prodan, A.-M.; Costescu, A.; et al. Development of cerium-doped hydroxyapatite coatings with antimicrobial properties for biomedical applications. *Coatings* **2020**, *10*, 516. [CrossRef]
61. Morais, D.S.; Rodrigues, M.A.; Lopes, M.A.; Coelho, M.J.; Maurıcio, A.C.; Gomes, R.; Amorim, I.; Ferraz, M.P.; Santos, J.D.; Botelho, C.M. Biological evaluation of alginate-based hydrogels, with antimicrobial features by Ce(III) incorporation, as vehicles for a bone substitute. *J. Mater. Sci. Mater. Med.* **2013**, *24*, 2145–2155. [CrossRef] [PubMed]
62. Jouanny, I.; Labdi, S.; Aubert, P.; Buscema, C.; Maciejak, O.; Berger, M.-H.; Guipont, V.; Jeandin, M. Structural and mechanical properties of titanium oxide thin films for biomedical application. *Thin Solid Films* **2010**, *518*, 3212–3217. [CrossRef]
63. Dussan, A.; Bertel, S.D.; Melo, S.F.; Mesa, F. Synthesis and characterization of porous silicon as hydroxyapatite host matrix of biomedical applications. *PLoS ONE* **2017**, *2017*, e0173118. [CrossRef]
64. Xu, Y.; Hu, X.; Kundu, S.; Nag, A.; Afsarimanesh, N.; Sapra, S.; Mukhopadhyay, S.C.; Han, T. Silicon-based sensors for biomedical applications: A Review. *Sensors* **2019**, *19*, 2908. [CrossRef]
65. Ruffino, F.; Torrisi, V. Ag films deposited on Si and Ti: How the film-substrate interaction influences the nanoscale film morphology. *Superlattices Microstruct.* **2017**, *111*, 81–89. [CrossRef]
66. Barry, J.N.; Cowley, A.; McNally, P.J.; Dowling, D.P. Influence of substrate metal alloy type on the properties of hydroxyapatite coatings deposited using a novel ambient temperature deposition technique. *J. Biomed. Mater. Res. A* **2014**, *102*, 871–879. [CrossRef]
67. Ferraris, S.; Yamaguchi, S.; Barbani, N.; Cristallini, C.; Gautier di Confiengo, G.; Barberi, J.; Cazzola, M.; Miola, M.; Vernè, E.; Sprian, S. The mechanical and chemical stability of the interfaces in bioactive materials: The substrate-bioactive surface layer and hydroxyapatite bioactive surface layer interfaces. *Mater. Sci. Eng. C* **2020**, *116*, 111238. [CrossRef]
68. Zhao, S.F.; Jiang, Q.H.; Peel, S.; Wang, X.X.; He, F.M. Effects of magnesium-substituted nanohydroxyapatite coating on implant osseointegration. *Clin. Oral Implants Res.* **2013**, *24*, 34–41. [CrossRef] [PubMed]

69. Cai, Y.L.; Zhang, J.J.; Zhang, S.; Venkatraman, S.S.; Zeng, X.T.; Du, H.J.; Mondal, D. Osteoblastic cell response on fluoridated hydroxyapatite coatings: The effect of magnesium incorporation. *Biomed. Mater.* **2010**, *5*, 054114. [CrossRef] [PubMed]
70. Oliveira, A.L.; Mano, J.F.; Reis, R.L. Nature-inspired calcium phosphate coatings: Present status and novel advances in the science of mimicry. *Curr. Opin. Solid State Mater. Sci.* **2003**, *7*, 309–318. [CrossRef]

© 2020 by the authors. Licensee MDPI, Basel, Switzerland. This article is an open access article distributed under the terms and conditions of the Creative Commons Attribution (CC BY) license (http://creativecommons.org/licenses/by/4.0/).

Article

FTIR Characterization of the Development of Antimicrobial Catheter Coatings Loaded with Fluoroquinolones

Dorota Kowalczuk

Chair and Department of Medicinal Chemistry, Faculty of Pharmacy, The Medical University of Lublin, 20-090 Lublin, Poland; dorota.kowalczuk@umlub.pl; Tel.: +48-81-448-7380

Received: 3 August 2020; Accepted: 22 August 2020; Published: 24 August 2020

Abstract: The purpose of this paper was to present the development of antimicrobial coatings for different urinary catheters. Antimicrobial catheter coatings were prepared by immobilizing fluoroquinolones either with the use of linkers (covalent binding) or by activating the polymer matrix with iodine/bromine (noncovalent binding). The possibility of the deposition of antimicrobial agent(s) following bromine activation on latex, polyurethane, and silicone was evaluated. Fourier transform infrared spectroscopy (FTIR), used to monitor the changes in the catheter's molecular structure occurring over the course of its multi-stage modification, confirmed the presence of fluoroquinolones in the catheter matrix as well as site-specific reactions. The amounts of drugs embedded in the catheter matrix were determined by the HPLC method. Stability of the drug binding was checked by examining the drug release. The new antimicrobial coatings obtained with the participation of fluoroquinolone antibiotics have the potential to protect the patient against infections during catheterization.

Keywords: antimicrobial coatings; urinary catheters; catheter-associated urinary tract infections; FTIR; fluoroquinolones

---

1. Introduction

Scientists have long been searching for ideal biomaterials (i.e., those made up of biocompatible, antimicrobial, and antifouling materials) that would be capable of inhibiting bacterial adhesion, thereby preventing infections in the course of urinary catheterization.

The surface properties of materials such as morphology, functionality, surface charge density, interactions, hydrophilicity or hydrophobicity play a vital role in bacterial adhesion. With regard to the surface properties, the modifications are mainly accomplished by applying the following strategies: functionalization, coating, impregnation of the active molecule, and blending. The former two cause the development of bacteriostatic surfaces, whereas molecule impregnation and blending result in the development of bactericidal surfaces [1].

It appears that modification of the surface structure of the urinary catheter could reduce the incidence of urinary tract infections. Coating the catheter surfaces with a thin layer of a suitable material is apparently the most universal method aimed at developing antifouling and antibacterial properties. Thus far, thin films based on silicone, polytetrafluoroethylene, hydrogels, and other antifouling/antimicrobial materials, often containing silver or an alternative metal as well as other antimicrobials, have been proposed [1–3].

Scientific studies indicate that bacterial adhesion can be reduced by coating the catheters with titanium dioxide [4], a titanium oxide–silver composite [5], silver particles/silver nanoparticles [6], or impregnating the catheter with nitric oxide [7].

Sterility and prevention of bacterial adhesion (prevention of catheter-associated bacteriuria) can be achieved by incorporating into the catheter antiseptics such as triclosan, chlorhexidine [8,9], antibiotics/chemotherapeutic agents including fluoroquinolones [10–12], or other active substance/polymer [12,13].

Fourier transform infrared spectroscopy is the most commonly used technique to assess the surface changes of biomaterials that occur during their modification [14,15].

Since the research into the immobilization of active substances in solid biomaterials requires further investigation [16], this paper focused on investigating various ways of the immobilization of fluoroquinolone antibacterial substances and its aim was to present the development of antimicrobial coatings of different urinary catheters as well as the FTIR assessment of changes in the molecular structure of the catheter material during its multi-stage modification. This paper presents the formation and characterization of the new antimicrobial coatings, obtained with the participation of fluoroquinolone antibiotics in an uncomplicated way, which can protect urinary catheters against bacterial infections. As indicated by the literature data, modifications of the catheter are vital to ensure the patient's safety during urinary catheterization.

2. Materials and Methods

Sparfloxacin (SPA) and ciprofloxacin hydrochloride monohydrate (CIP) were purchased from Sigma-Aldrich. All reagents and chemicals used for the modification process of catheters were purchased from Sigma-Aldrich Co. (chitosan, CHT; glutaraldehyde GLU; adipic acid dihydrazide, ADH) or Avantor Performance Materials Poland S.A., Gliwice, Poland (iodine, bromine, methanol, acetonitrile). The Foley catheters were made of a natural latex; siliconized, silicone, polyurethane, and hydrogel-coated latex were obtained from a local medical supplier (Skamex, Łódź, Poland).

2.1. Preparation of Sparfloxacin-Treated Catheter Samples (SPA-M1 Method)

The siliconized latex catheters were subjected to the modification process using the selected patented procedures [17], adapted for this research.

Small samples of 0.5 cm-long catheters (average weight about 0.09 g) were coated with chitosan, then treated with glutaraldehyde and finally coupled with sparfloxacin in an organic medium. With regard to the biotoxic effect of glutaraldehyde, the non-crosslinked (trapped) glutaraldehyde was removed by repeatedly rinsing the crosslinked chitosan catheter with distilled water. Immobilization of SPA on the crosslinked chitosan-coated catheter surface was performed in the SPA solutions prepared in acetonitrile at the concentration of 0.1, 0.5, 1.0, and 2.0 mg/mL. The resulting catheter samples were rinsed with water, dried at 50 °C mand subjected to ATR (Attenuated Total Reflectance)-FTIR analysis.

2.2. Preparation of Sparfloxacin-Treated Catheter Samples (SPA-M2 and SPA-M3 Methods)

The hydrogel catheters (also polyurethane and silicone catheters in the case of the SPA-M3 method) were cut into 0.5 cm-long fragments (average weight about 0.11 g) and activated with the 1% methanol–$NaHCO_3$ iodine solution (SPA-M2 method) or 0.1–1% methanolic bromine solution (SPA-M3 method) for 1 h under gentle stirring at 40 °C. Next, the samples were removed, rinsed with distilled water, and dried. In the next stage, the samples were placed in Eppendorf tubes, immersed in 1 mL of SPA solution (0.1, 0.5, 1.0, and 2.0 mg/mL) and incubated in closed tubes for 12 h at 50 °C with stirring at 150 rpm. Then, the catheters were removed, rinsed with water, and left to dry.

2.3. Preparation of Ciprofloxacin-Treated Catheter Samples (CIP-M1 Method)

The siliconized latex catheters were subjected to the modification process using the selected patented procedures [17], adapted for this research.

The small samples of 0.5 cm-long catheters (average weight about 0.09 g) were coated with chitosan, then, sequentially treated with the 5% solution of glutaraldehyde and 5% solution of adipic acid dihydrazide and finally coupled with ciprofloxacin in an organic medium. With regard to the biotoxic effect of glutaraldehyde, the trapped glutaraldehyde was removed by repeatedly rinsing the crosslinked chitosan catheter with distilled water. Immobilization of CIP on the activated, chitosan-coated catheter surface was performed in the CIP solutions prepared in methanol at the concentrations of 0.1, 0.5, 1.0, and 2.5 mg/mL. The resulting catheter samples were rinsed with water, dried at 50 °C, and subjected to ATR-FTIR analysis.

2.4. FTIR Characterization

FTIR in transmittance mode was used to characterize the presence of specific chemical groups.

In the tested catheter samples. FTIR measurements were carried out on a Nicolet 6700 spectrometer (Thermo Fisher Scientific Inc., Warsaw, Poland) equipped with a deuterated triglycine sulfate detector (DTGS/KBr) and a versatile attenuated total reflectance (ATR) sampling accessory with a diamond crystal plate. The spectra were recorded in the spectral range of 4000–600 cm^{-1} at 4 cm^{-1} spectral resolution and 32 sample/background scans using OMNIC 8.1 computer software (Thermo Fisher Scientific Inc., Waltham, MA USA). The modified catheters were recorded and compared with the unmodified ones.

3. Results

The presented FTIR-ATR assessments show the effects of the functionalization of various types of urological catheters using selected chemotherapeutics (i.e., sparfloxacin and ciprofloxacin). With the use of the FTIR method, sparfloxacin was proven to be a particularly good choice to demonstrate the presence of this drug in the modified polymer matrix due to the unique absorption bands of the amine group in the range of 4000–3000 cm^{-1}. In addition, fluoroquinoline antibiotics showed a broad spectrum of antibacterial activity and effectively protected modified matrices against biofilm formation, which ensures the patient's safety during catherization.

3.1. Analysis of the SPA Modified Catheter with the Use of the SPA-M1 Method (CHT-Linker-SPA)

The FTIR spectrum of the siliconized latex catheter (Figure 1A) shows the absorption bands for silicone and latex. The peaks in the range of 1020 to 1260 cm^{-1} belong to Si–O–Si bonds of silicone with the peak at about 1100 cm^{-1} belonging to the C–O–C bonds of the latex, probably the polyacrylate latex, which also showed absorption peaks at about 3300 cm^{-1} as well as at 3000–2850 cm^{-1} and 1735 cm^{-1}, that are the result of the stretching vibrations of O–H, C–H, and C=O bonds, respectively. Furthermore, the absorption bands at 1480–1380 cm^{-1} can be attributed to the bending vibrations of the O–H and C–H bonds [18].

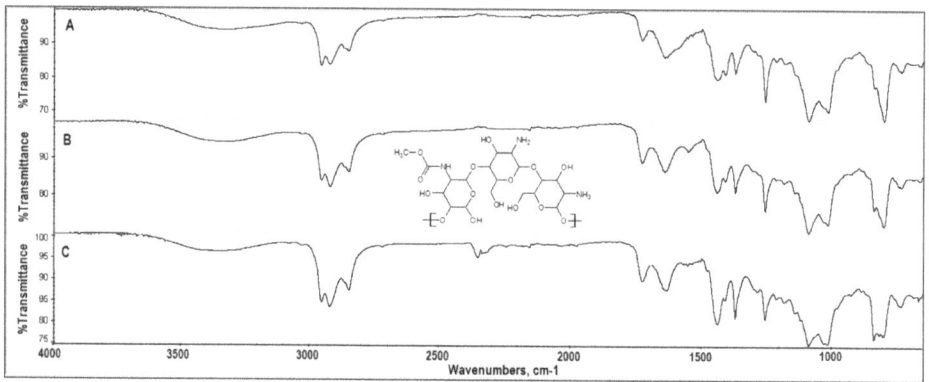

Figure 1. FTIR spectra of the siliconized latex catheter recorded during a three-stage modification using SPA-M1 method: the untreated catheter (**A**), CHT-treated and GLU-activated catheter (**B**), SPA-treated catheter (**C**).

Comparing curve B with curve A, at 3380 cm^{-1} there was no clear band related to the free amine group in the molecule of natural chitosan. A comparison of these curves showed an increased intensity in the peaks of the C–H stretching at 3000–2850 cm^{-1} and the carbonyl stretching (C=O) at 1735 cm^{-1}. Furthermore, two new absorption bands at 1644 cm^{-1} and at about 1551 cm^{-1} were observed in the modified chitosan (CHT-treated and GLU-activated catheter). These effects can be related to the interactions between chitosan and glutaraldehyde as a linker. The clear peak at 1644 cm^{-1} can be associated with the presence of the stretching vibrations of N=C bonds. The second peak at 1551 cm^{-1} can be attributed to the stretching vibrations of C=C bonds. The literature [19] provided evidence for the formation of the double imine bond (N=C) and the ethylenic double bond (C=C) in the chitosan–glutaraldehyde interactions, which depend on the pH value of the reaction medium and the GLU concentration [20]. In the process, crosslinking chains are formed [20,21].

It is apparent that curve C (spectrum of CHT-linker-SPA) was more intensified than curve B. The shape changes of the bands and the new peaks, typical of the SPA molecule [16] at about 1630 cm^{-1}, 1520–1550 cm^{-1}, 1430 cm^{-1}, 1290 cm^{-1}, and 850–800 cm^{-1} and attributed to the stretching and bending vibrations, may be indicative of the presence of SPA in the modified catheter. The absence of peaks for the N–H stretching bonds, a change in the shape of the imine peak, and a shift of this band to a lower frequency (from 1644 to 1630 cm^{-1}) may also suggest the formation of the C=N bond between SPA and the cross-linked chitosan (Table 1).

Table 1. Summary of FTIR analysis of the SPA-modified catheter following SPA-M1 method.

Prominent Peaks of SPA	Prominent Peaks of SPA-Modified Catheter
ν(N–H): 3460 cm^{-1} and 3336 cm^{-1}	ν(N–H): absence
ν(C=O): 1639 cm^{-1}	ν(C=N):~1630 cm^{-1}
ν(C=C in Ar) with δ(N–H): 1585–1495 cm^{-1}	ν(C=C in Ar):~1520–1550 cm^{-1}
δ(C-H in –CH2–): 1435 cm^{-1}	δ(C-H in -CH2-):~1430 cm^{-1},
ν(C–O in COOH): 1290 cm^{-1}	ν(C–O in COOH):~1290 cm^{-1}

Moreover, the binding of SPA to the cross-linked chitosan due to the Michael-type reaction may also be taken into consideration [21]. Probable covalent SPA binding sites are shown in Scheme 1.

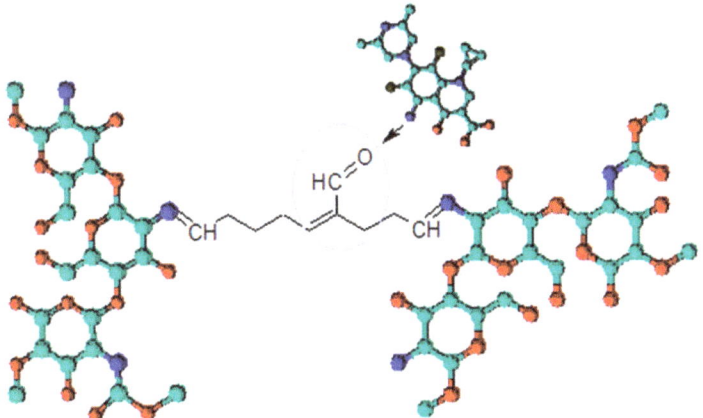

Scheme 1. Probable covalent SPA binding with the crosslinking chitosan.

3.2. Analysis of the SPA-Modified Catheter with the Use of SPA-M2 Method (Iodine–SPA)

A hydrogel, which is a cross-linked macromolecular polymer that absorbs relatively large volumes of a liquid, is a material frequently used for coating urinary catheters. The most commonly used hydrogel is made of poly(2-hydroxyethyl methacrylate) [22].

The FTIR spectrum of the hydrogel catheter (Figure 2A) showed a wide absorption band of O–H stretching vibrations at about 3300–3500 cm^{-1}, peaks corresponding to the C–H stretching vibrations at 3000–2850 cm^{-1}, the C=O stretching vibrations at 1710 cm^{-1}, the C–O–C stretching vibrations at about 1100 cm^{-1} as well as the bending vibrations of the C–H bonds in the region 1360–1480 cm^{-1} [23] The absorption bands indicate the presence of similar functional moieties, as in the case of the siliconized latex catheter (Figure 1A).

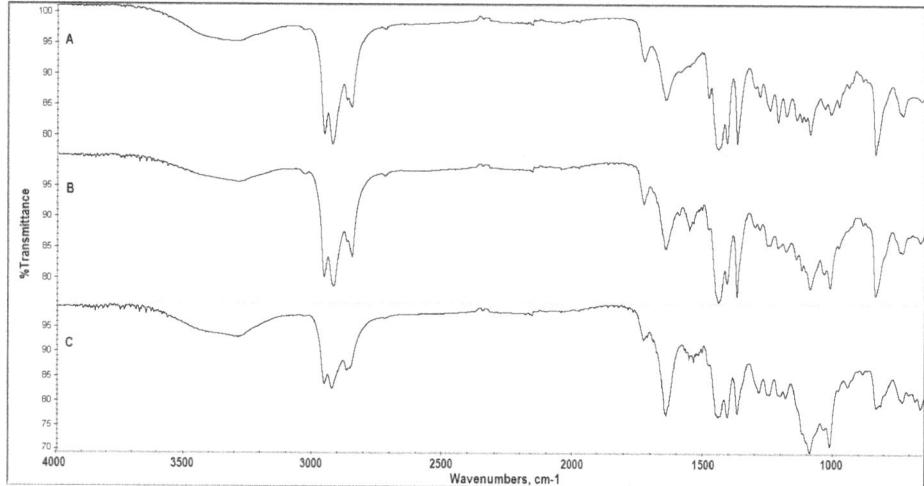

Figure 2. FTIR spectra of the hydrogel catheter recorded during a two-stage modification using the SPA-M2 method: the untreated hydrogel catheter (**A**), the iodine-activated catheter (**B**), SPA-treated catheter (**C**).

The spectra recorded after the second and third stage of modification (Figure 2A,B) were similar to those presented earlier [16]. However, the FTIR spectra demonstrated here were obtained by spectroscopic analysis where the latex-based catheter was coated with a hydrogel layer, treated with iodine in an organic-aqueous environment, and then with SPA as an antibacterial agent. The characteristic absorption bands appearing as new peaks in the FTIR spectrum of the iodine-activated and SPA-treated catheter (Figure 2C) clearly indicate the presence of SPA. The intensive absorption bands corresponding to the asymmetrical and symmetrical stretching vibrations of the N–H bonds of SPA overlapped with the O–H stretching occurred at 3500–3295 cm^{-1}. Comparing curve C with curve B, the absorption peak at 1639 cm^{-1} (the C=O stretching in the 4-quinolone ring of the SPA molecule) became more intensified and slightly shifted to a lower frequency (from 1647 to 1639 cm^{-1}). The peaks at 1585–1500 cm^{-1} resulted from the C=C stretching vibrations of the aromatic ring while the peak at 1291 cm^{-1} can be assigned to the C–O bond vibration of the carboxyl group (Table 2).

Table 2. Summary of the FTIR analysis of SPA-modified catheter following the SPA-M2 method.

Prominent Peaks of SPA	Prominent Peaks of SPA-Modified Catheter
ν(N–H): 3460 cm^{-1} and 3336 cm^{-1}	ν(N–H):~3500–3295 cm^{-1}
ν(C=O): 1639 cm^{-1}	ν(C=O):~1639 cm^{-1}
ν(C=C in Ar) with δ(N-H): 1585–1495 cm^{-1}	ν(C=C in Ar) with δ(N-H):~1585–1500 cm^{-1}
δ(C–H in –CH2–): 1435 cm^{-1}	δ(C–H in –CH2–):~1436 cm^{-1}
ν(C–O in COOH): 1290 cm^{-1}	ν(C–O in COOH):~1291 cm^{-1}

3.3. Analysis of the SPA Modified Catheter with the Use of SPA-M3 Method (Bromine-SPA)

In the case of functionalization of the hydrogel latex catheter with bromine as an activator and SPA as an antibacterial agent, the presence of SPA was even more pronounced (Figure 3).

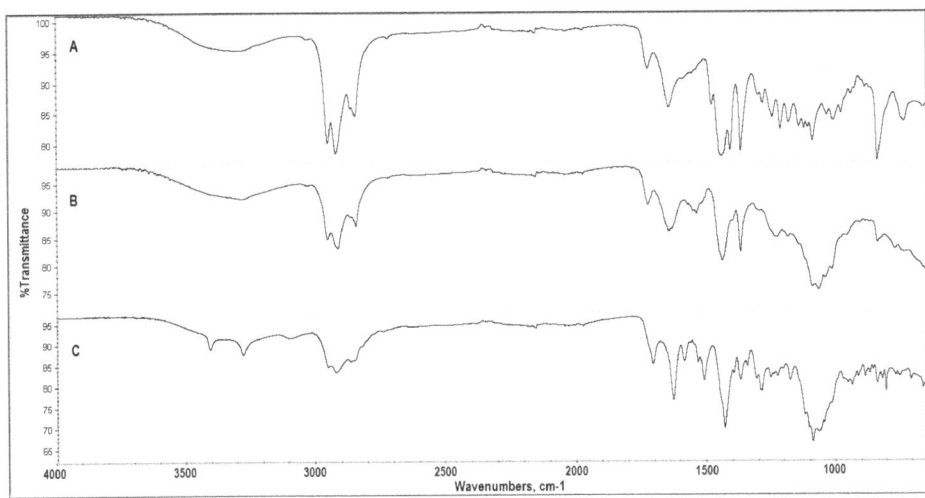

Figure 3. FTIR spectra of catheter surface recorded during a two-stage modification using the SPA-M3 method: the untreated hydrogel catheter (**A**), the bromine-activated catheter (**B**), and the SPA-treated catheter (**C**).

The spectrum of the modified catheter was more like the SPA molecule than the hydrogel latex matrix (Figure 3C). The absorption bands of SPA [16] appeared at 3411 cm^{-1} and 3285 cm^{-1} (the N–H asymmetrical and symmetrical stretching), 1636 cm^{-1} (the C=O stretching), 1536–1510 cm^{-1} (the C=C stretching of the aromatic ring), 1437 cm^{-1} (the C–H asymmetrical bending), and in the range of 1293–1095 cm^{-1} (the stretching vibrations of the C–N, Ar–F, C–O bonds) (Table 3).

Table 3. Summary of the FTIR analysis of the SPA-modified catheter following the SPA-M3 method.

Prominent Peaks of SPA	Prominent Peaks of SPA-Modified Catheter
ν(N–H): 3460 cm^{-1} and 3336 cm^{-1}	ν(N–H):~3411 cm^{-1} and 3285 cm^{-1}
ν(C=O in COOH): 1713 cm^{-1}	ν(C=O in COOH):~1709 cm^{-1}
ν(C=O): 1639 cm^{-1}	ν(C=O):~1636 cm^{-1}
ν(C=C in Ar) with δ(N–H): 1585–1495 cm^{-1}	ν(C=C in Ar) with δ(N–H):~1536–1510 cm^{-1}
δ(C–H in –CH2–): 1435 cm^{-1}	δ(C–H in –CH2–):~1437 cm^{-1}
ν(C–O in COOH): 1290 cm^{-1}	ν(C–O in COOH):~1293 cm^{-1}

3.4. Analysis of the Polyurethane and Silicone SPA-Modified Catheter Following the SPA-M3 Method

In order to show the universality of this procedure, the SPA-M3 method was also used for functionalizing the other catheter matrices (i.e., polyurethane or silicone). The immobilization of SPA in the polyurethane and silicone catheter was performed with satisfactory results.

Comparison of the polyurethane catheter before and after immobilization of SPA (Figures 3B and 4A) clearly demonstrated the changes in the catheter matrix. The FTIR spectrum of the modified catheter definitely resembled the spectrum of SPA [16].

The appearance of the absorption peaks at 3418 cm^{-1} and 3285 cm^{-1} (the N–H stretching), 1632 cm^{-1} (the C=O stretching), 1510 cm^{-1} (the C=C stretching of the aromatic ring), 1435 cm^{-1}

(the C-H asymmetrical bending), and in the range of 1300–1000 cm^{-1} (the stretching vibrations of the C–N, Ar–F, C–O bonds) confirmed the presence of SPA.

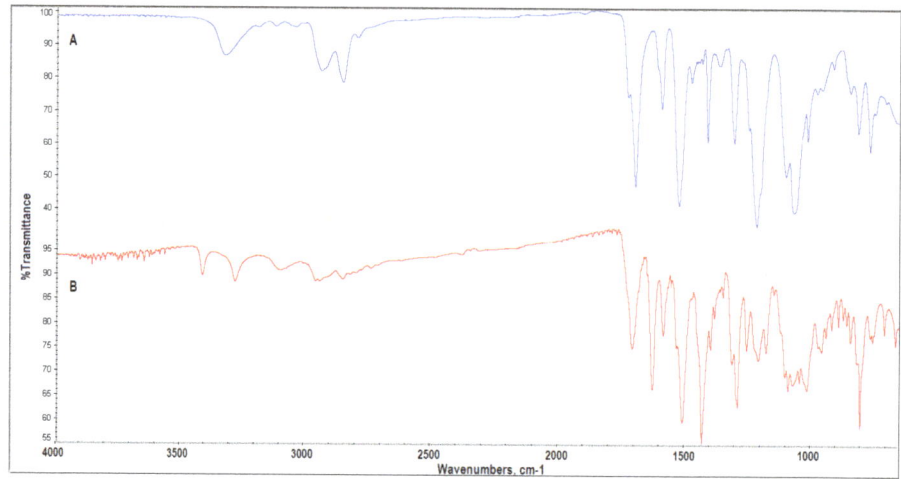

Figure 4. Comparison of the polyurethane catheter before (**A**) and after (**B**) immobilization of SPA.

Silicone has a relatively simple infrared spectrum, which is associated with its uncomplicated structure (organosilicon polymer). Silicone exhibits a typical distribution of the absorption bands for this type of material, located in the zone below 1300 cm^{-1}. The peaks in this region corresponded to the absorptions resulting from the antisymmetric and symmetric stretching vibrations of the oxygen atom in the Si–O–Si group [24] as well as the stretching vibrations of the carbon–silicon bond (C–Si) present in all organosilicon compounds. Another absorption band at 2954 cm^{-1} can be attributed to the C–H stretching vibrations of the methyl groups in the silicone structure.

The presented spectra (Figure 5) show changes of the silicone after immobilization of SPA. Numerous absorption bands characteristic of SPA, similar to the polyurethane matrix, confirmed incorporation of SPA to the silicone matrix. More intense bands were the effect of using a higher activator concentration (0.3% and 1.0% bromine solution). It was also found that with the increased concentration of bromine, the structure of the silicone catheter became more fragile.

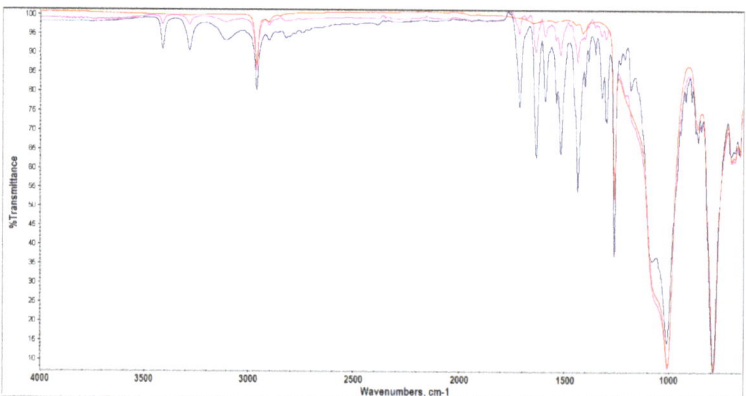

Figure 5. Comparison of the silicone catheter before (red curve) and after (pink and blue curves) immobilization of SPA; the activation with bromine at concentration 0.3% (pink curve) and 1% (blue curve).

3.5. Analysis of the CIP-Modified Catheter with the Use of the CIP-M1 Method

The FTIR spectrum of ciprofloxacin shows many absorption bands (Figure 6). The most characteristic bands can be assigned to the O–H stretching in COOH at 3530–3373 cm^{-1}, the N–H stretching of the ionized amine (>NH$_2^+$) at 2689–2463 cm^{-1}, the C=O stretching in COOH and 4-quinolone ring at 1703 cm^{-1} and 1623 cm^{-1}, respectively, the C=C stretching in aromatic ring at 1588–1495 cm^{-1}, the C–H bending at 1449 cm^{-1}, the C–N stretching at 1384 cm^{-1}, the C–O stretching in COOH at 1267 cm^{-1}, and the C–F stretching vibrations in the range of 1250–1100 cm^{-1}.

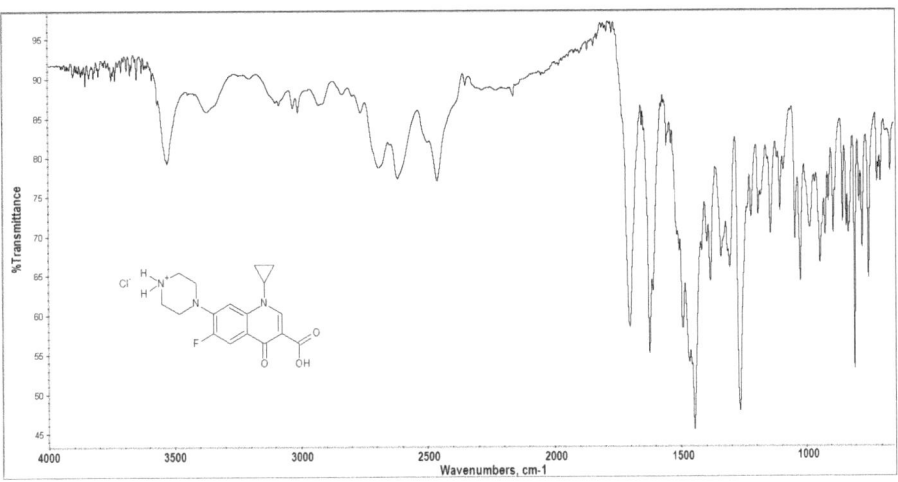

Figure 6. FTIR spectrum of ciprofloxacin hydrochloride monohydrate.

The changes occurring in the course of the first (CHT-treated catheter) and second (GLU-activated catheter) stage of the modification (Figure 7B) were the same as those observed during the formation of the crosslinking chitosan (Figure 1B): the double imine bond (N=C) at 1644 cm^{-1} and the ethylenic double bond (C=C) at 1551 cm^{-1}.

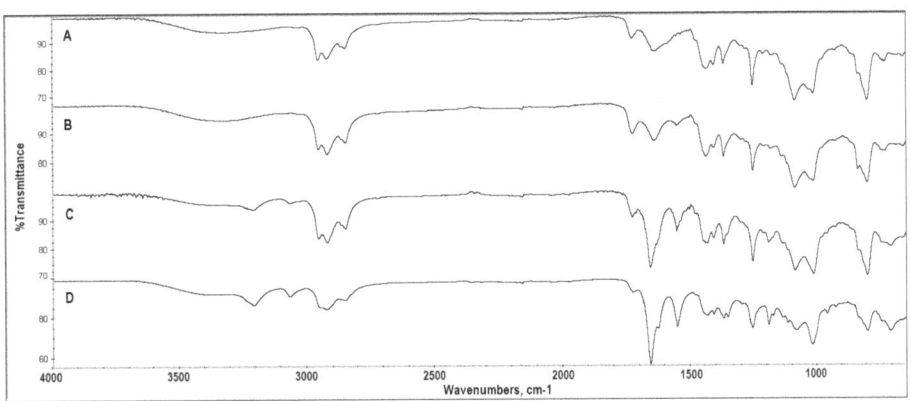

Figure 7. FTIR spectra of the siliconized latex catheter recorded during multi-stage modification using the CIP-M1 method: the untreated catheter (**A**), CHT-treated, GLU-activated catheter (**B**), CHT-treated, GLU-ADH activated catheter (**C**), and CIP-treated catheter (**D**).

In the next stage, the cross-linked chitosan was functionalized with adipic acid dihydrazide as the second linker, NH$_2$NH-CO-(CH$_2$)$_3$-CO-NHNH$_2$ [25].

The appearance of the strong peak (Figure 7C) at 1660 cm^{-1} belonging to the stretching vibrations of the N=C bond proves that ADH binds covalently with the cross-linked chitosan in order to introduce the amine groups whose peaks occur at 3215 cm^{-1} and 3068 cm^{-1}. Moreover, the new amido bonds' characteristic peaks at 1699 cm^{-1} and 1560 cm^{-1}, corresponding to the stretching vibrations of C=O and the bending vibrations of N–H, indicate the presence of ADH in the modified catheter.

In the last stage, the attachment of CIP to the GLU-ADH activated catheter was observed (Figure 7D). In the CIP-treated catheter spectrum, the C=N stretching band was intensified in comparison to the GLU-ADH activated catheter spectrum, which was indicative of the interaction between CIP and ADH. Moreover, the stronger peaks at 3300–3000 cm^{-1} and the broader band at 3000–2800 cm^{-1} can suggest that there are more N–H and CH$_2$ groups in the CIP-treated catheter than in the GLU-ADH activated catheter, which may indicate the introduction of CIP. Furthermore, the shape change of the bands in the region of 1600–1000 cm^{-1} and the appearance of new peaks may also be suggestive of the presence of CIP (Table 4).

Table 4. Summary of FTIR analysis of the CIP-modified catheter following the CIP-M1 method.

Prominent Peaks of CIP	Prominent Peaks of CIP-Modified Catheter
ν(N–H in >NH$_2^+$): 2689–2463 cm^{-1}	ν(N–H in >NH$_2^+$): absence
ν(C=O): 1623 cm^{-1}	ν(C=N):~1660 cm^{-1}
ν(C=C in Ar): 1588–1495 cm^{-1}	ν(C=C in Ar):~1555 cm^{-1}
δ(C–H in –CH2–): 1449 cm^{-1}	δ(C–H in –CH2–): 1445 cm^{-1}
ν(C–O in COOH): 1267 cm^{-1}	ν(C–O in COOH):~1267 cm^{-1}

A qualitative FTIR study of catheters treated with increasing concentrations of drugs during modification with use of all immobilization methods (SPA-M1, M2, M3, and CIP-M1) was performed to better assess the drug-specific absorption bands. The changes in the shape of the characteristic absorption bands in a series of the studied samples confirmed the presence of SPA and CIP in the catheter matrix.

The representative FTIR spectra in Figure 8 demonstrate the increasing intensity of SPA-characteristic bands as a consequence of the increasing amounts of SPA embedded in the catheter matrix. For this assessment, the N–H stretching of the SPA amine group at 3500–3295 cm^{-1}, and the C=O stretching vibrations at 1639 cm^{-1} derived from COOH group are particularly important. The presented FTIR spectra illustrate the successful binding of the drug with the catheter matrix.

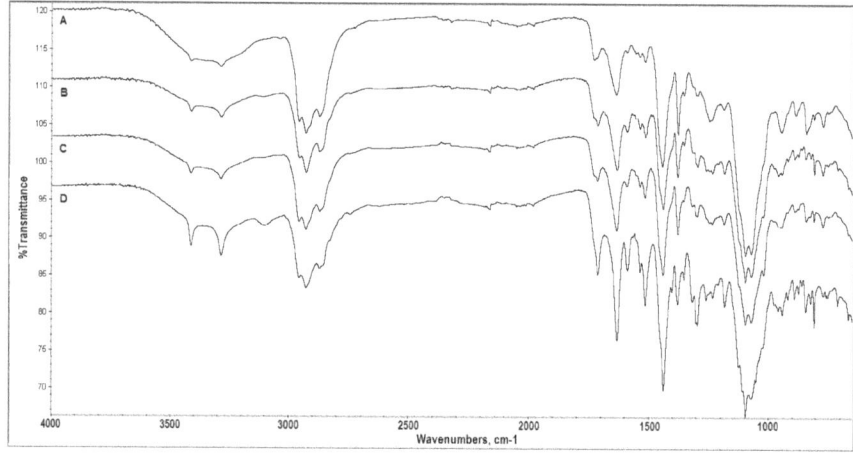

Figure 8. Typical FTIR spectra of the catheter with SPA immobilized from solutions with increasing SPA concentrations: 0.1 (**A**), 0.5 (**B**), 1.0 (**C**), and 2.0 (**D**) mg/mL (SPA-M3 method).

4. Discussion

Urinary tract infections are the most commonly occurring nosocomial infections in intensive care units. Most of these infections are associated with the presence of catheters in the urinary tract [1–3]. Two main problems that affect urinary catheters and make it more difficult to treat catheter-associated urinary tract infections are biofilm formation and encrustation. The biofilm bacteria can move to the kidneys, causing life-threatening complications. On the other hand, the incrustation/mineralization of the catheter with calcium and phosphorus salts promotes bacterial adhesion [26–28]. A common treatment for urinary infections is the use of antimicrobials in urinary catheters, which can reduce or eliminate colonization of bacteria and stop the biofilm formation.

In order to achieve an antifouling and antimicrobial coating, polymers are mostly impregnated with many types of antimicrobials with the participation of various methods of immobilization, which include: a procedure in which the active agents are covalently bound to the polymer (grafting); physical adsorption in which the active agents are non-covalently coupled to the polymer by hydrogen, ionic, or steric interactions; surface-initiated immobilization in which the synthesis of the antimicrobial on the polymer surface occurs due to the use of a covalently bonded initiator; and immobilization in which the synthesis of the active agent occurs within the substrate when it is formed [2].

Coating urinary catheters with the polymer hydrogel, often impregnated with antimicrobials, is a popular proposition in order to reduce diminishing the adhesion of microbes. In such a preventive strategy, polymers such as polyvinylpyrrolidone, chitosan [29,30], curdlan [31], heparin [32], mucin—the main component of the mucosa with anti-adhesive properties [33]—have been recommended. It was shown that the biomaterial coated with polymer was colonized by bacteria to a lesser extent than an unaltered one and was often less prone to incrustation, which is especially important for the biomaterials used in urology [28,32].

The above-mentioned studies were the motivation for coating urological catheters with the chitosan hydrogel and conducting further modifications with an intention to achieve effective, multidirectional, and long-term antibacterial protection. The following process of functionalization of the chitosan hydrogel consisted in its activation via crosslinking with the use of glutaraldehyde (Figure 1B) or glutaraldehyde and adipic acid dihydrazide (Figure 7B,C). Finally, fluoroquinolone—the main component of the antibacterial protection—was bound with the activated catheter, the first strategy of antibacterial modification (Figures 1C and 7D). It is known that glutaraldehyde is one of the crosslinking agents that is potentially cytotoxic. However, the reactivity of glutaraldehyde with amine and other functional groups of proteins and enzymes has led to its widespread use as a crosslinking agent for proteins and other compounds. The studies suggest that the toxic effect of glutaraldehyde can be quenched by glutamic acid [34] or other cell-friendly amine compounds. Our previous study indicates that the chitosan coat crosslinked with glutaraldehyde and then treated with tosufloxacin provides good protection against the cytotoxic effect of the silicone latex material; it probably limits leaching of harmful agents [35]. Moreover, recent studies have revealed that the glutaraldehyde cross-linked chitosan exhibits an antibacterial activity. This property appears to be desirable in the case of creating antibacterial coatings [36].

The second strategy of the antibacterial modification consisted in the functionalization of the ready-made hydrogel catheters by the hydrogel activation with iodine (Figure 2B) or bromine (Figure 3B), followed by incorporation of the antimicrobials into the catheter matrix (Figures 2C and 3C, respectively).

Modification of the surface with the participation of sparfloxacin was conducted according to the first and second strategy by applying three different immobilization methods marked as SPA-M1, SPA-M2, and SPA-M3. In the case of the SPA-M1 method, the catheter's surface was coated with the chitosan hydrogel, activated with GLU (crosslinking), and treated with SPA, according to the patented procedure [17]. In the case of the SPA-M2 and SPA-M3 methods, the ready-made hydrogel catheters were activated with iodine or bromine, respectively, and then treated with SPA. The SPA-M3 method was also used for functionalizing the polyurethane and silicone catheters in order to show the universality of this procedure. The immobilization of SPA in the polyurethane and silicone catheter produced satisfactory results (Figures 4 and 5, respectively). However, the structure of the silicone catheter underwent gradual destruction under the influence of bromine. Therefore, this way of modification cannot be proposed in the case of silicone materials.

Modifications with the application of ciprofloxacin were performed according to the first strategy (Figure 7) and with the use of the immobilization method marked as CIP-M1. The catheter surface was coated with the chitosan hydrogel, activated with GLU and adipic acid dihydrazide, and finally treated with CIP according to the patented procedure [17].

The manner of binding CIP and SPA with the catheter matrix was assessed on the basis of the recorded FTIR spectra after each stage of modification. Frequent recording of the spectrum of the tested material after each subsequent stage of modification enables finding even the smallest changes. It is not necessary to use the spectrum library, because during the spectrum analysis, it is easy to deduce which functional group in the drug molecule or polymer took part in the interaction [16,18]. Analysis of the FTIR spectra of pure substances CIP (Figure 6) and SPA [17] was used for identification of the most characteristic absorption bands corresponding to the vibrations of the respective functional groups in the molecules of both compounds. It was proven that CIP and SPA immobilized, according to the first method (CIP-M1, SPA-M1), bind with the catheter matrix by covalent bonds, probably in a mixed covalent–noncovalent manner. The increase in the intensity of absorption bands at 1630 cm^{-1} (SPA) and 1660 cm^{-1} (CIP), which corresponds to the stretching vibrations of the imine bond (C=N) formed between the amine or carbonyl groups of the molecules of drugs and the modified chitosan hydrogel (Figures 1C and 7D) may be evidence of covalent interactions. The mixed binding of the drugs can confirm the inhibition zones of bacterial growth under and around the catheter in the microbiological test: the zone inhibition test against the selected Gram-positive and Gram-negative bacteria. Similar effects were observed previously (i.e., during immobilization of SPA on the catheter coated with heparin [37]).

In the case of the immobilization of SPA according to the second and third method (SPA-M2, SPA-M3), the recorded FTIR spectra showed distinct bands deriving from the moieties that are characteristic of the SPA molecule, especially the bands at 3500–3280, corresponding to the SPA primary amine group. The presence of a free amine group indicates that the drug's bond is noncovalent (Figures 2C and 3C).

On the basis of the intensification of the absorption bands belonging to the vibrations of the bonds of the functional groups, it was concluded that the increase in the drug concentration in the solutions used for immobilization resulted in the increase in the amounts of drugs found in the catheter matrix (Figure 8).

In the present study, the HPLC method [38] was adopted for assessing the degree of CIP/SPA binding with the modified latex catheter. The drugs were determined in the solution before and after immobilization, and their concentrations were calculated from the equations of regression: $y = 131{,}196.3\,x + 21{,}192.7$ for CIP and $y = 96{,}999.7\,x + 10{,}571.1$ for SPA.

The results show that CIP bonded in 15–20% (1.5–2.3 mg per 1 g catheter), whereas SPA in 26–47% (2.9–5.2 mg per 1 g catheter) when the CIP-M1 and SPA-M1 methods were used, respectively. When the SPA-M2 and SPA-M3 methods were in use, 25–32% of SPA (2.8–3.6 mg per 1 g catheter) and 48–60% of SPA (5.4–6.7 mg per g catheter), respectively, were immobilized in the catheter matrix.

In order to assess the durability of the drug binding, the release profiles of both drugs from the modified latex catheter were determined. It was found that the drugs immobilized with the use of all methods were released gradually in small doses, providing long-term antibacterial protection of the catheter as well as the environment around it. Amounts of about 0.6% of CIP and 0.7% of SPA for the M1 method, 1.9% of SPA for the M2 method, and 1.3% of SPA for the M3 method were released by shaking the SPA/CIP loaded catheter samples in 10 mL of phosphate buffer at pH 5.5 for 30 min daily at 37 °C for a month. In the case of the M1 method, the drug release below 1% within one month confirmed a considerable stability of the C=N bonds and thus the stability of coatings based on chitosan crosslinked with glutaraldehyde.

5. Conclusions

Since the ideal antibacterial material has not yet been found, there is still a need to look for new ways of the prevention of catheter-associated urinary tract infections. New antibacterial coatings and materials must be thoroughly tested using reliable characterization methods to achieve the highest antimicrobial efficacy and safety.

The findings presented in this paper allow for the conclusion that the new antimicrobial coatings, obtained with the participation of fluoroquinolone antibiotics in an uncomplicated way, are able to protect the urinary catheters against bacterial infections during catheterization. The formation methods of antimicrobial catheter coatings through the covalent and noncovalent immobilization of fluoroquinolones were developed. In the case of covalent immobilization, it is predicted that the deposited coatings present a crosslinked chitosan thin film with the bonded drug, which preserves the antibacterial properties. It is also estimated that in the case of a noncovalent binding, the drug becomes trapped in the micro pores of the polymer matrix formed while being treated with bromine/iodine. The fluoroquinolones were successfully incorporated into the catheter matrix, which was confirmed by the FTIR analysis. Incorporation of the drug following activation with bromine is a suitable method of modifying latex and polyurethane-based catheters, but in the case of silicone, its structure is damaged. A dissolution study confirmed the stability of the drug binding to the matrix. A relatively durable binding of the drug enables long-term protection of the catheter against infection. In contrast, the low dose release protects the environment around the catheter. However, in order to evaluate their potential application in practice, further research should focus on the confirmation of their mechanical properties, efficacy, and safety.

Funding: This research was funded by the Polish Ministry of Scientific Research and Information Technology, grant number N N405 385037.

Acknowledgments: The author thanks Renata Steliga and Jakub Krol, students at the Chair and Department of Medicinal Chemistry, the Medical University of Lublin, for their assistance in preparing the modified catheter samples.

Conflicts of Interest: The author declares no conflict of interest.

References

1. Anjum, S.; Singh, S.; Benedicte, L.; Roger, P.; Panigrahi, M.; Gupta, B. Biomodification Strategies for the Development of Antimicrobial Urinary Catheters: Overview and Advances. *Glob. Chall.* **2018**, *2*, 1700068. [CrossRef]
2. Singha, P.; Locklin, J.; Handa, H. A Review of the Recent Advances in Antimicrobial Coatings for Urinary Catheters. *Acta Biomater.* **2017**, *50*, 20–40. [CrossRef]
3. Andersen, M.J.; Flores-Mireles, A.L. Urinary Catheter Coating Modifications: The Race against Catheter-Associated Infections. *Coatings* **2020**, *10*, 23. [CrossRef]
4. Sekiguchi, Y.; Yao, Y.; Ohko, Y.; Tanaka, K.; Ishido, T.; Fujishima, A.; Kubota, Y. Self-sterilizing catheters with titanium dioxide photocatalyst thin films for clean intermittent catheterization: Basis and study of clinical use. *Int. J. Urol.* **2007**, *14*, 426–430. [CrossRef] [PubMed]
5. Yao, Y.; Ohko, Y.; Sekiguchi, Y.; Fujishima, A.; Kubota, Y. Self-sterilization using silicone catheters coated with Ag and TiO_2 nanocomposite thin film. *J. Biomed. Mater. Res. B Appl. Biomater.* **2008**, *85*, 453–460. [CrossRef] [PubMed]
6. Roe, D.; Karandikar, B.; Bonn-Savage, N.; Gibbins, B.; Roullet, J.B. Antimicrobial surface functionalization of plastic catheters by silver nanoparticles. *J. Antimicrob. Chemother.* **2008**, *61*, 869–876. [CrossRef]
7. Regev-Shoshani, G.; Ko, M.; Av-Gay, Y. Slow release of nitric oxide from charged catheters and its effect formation on biofilm formation by Escherichia coli. *Antimicrob. Agents Chemother.* **2010**, *54*, 273–279. [CrossRef]

8. Jones, G.L.; Muller, C.T.; O' Reilly, M.; Stickler, D.J. Effect of triclosan on the development of bacterial biofilms by urinary tract pathogens on urinary catheters. *J. Antimicrob. Chemother.* **2006**, *57*, 266–272. [CrossRef]
9. Gaonkar, T.A.; Caraos, L.; Modak, S. Efficacy of a silicone urinary catheter impregnated with chlorhexidine and triclosan against colonization with Proteus mirabilis and other uropathogenes. *Infect. Control Hosp. Epidemiol.* **2007**, *28*, 596–598. [CrossRef]
10. Cormio, L.; La Forgia, P.; La Forgia, D.; Siitonen, A.; Ruutu, M. Bacterial adhesion to urethral catheters: Role of coating materials and immersion in antibiotic solution. *Eur. Urol. Urol.* **2001**, *40*, 354–359. [CrossRef]
11. Park, J.H.; Cho, Y.W.; Cho, Y.; Choi, J.M.; Shin, J.S.; Bae, Y.H.; Chung, H.; Jeong, S.Y.; Kwon, I.C. Norfloxacin—Releasing urethral catheter for long-term catheterization. *J. Biomater. Sci. Polym. Edn* **2003**, *14*, 951–962. [CrossRef]
12. Johnson, J.R.; Johnston, B.; Kuskowski, M.A. In vitro comparison of nitrofurazone- and silver alloy-coated foley catheters for contact-dependent and diffusible inhibition of urinary tract infection-associated microorganisms. *Antimicrob. Agents Chemother.* **2012**, *56*, 4969–4972. [CrossRef] [PubMed]
13. Thompson, V.C.; Adamson, P.J.; Dilag, J.; Liyanage, D.B.U.; Srikantharajah, K.; Blok, A.; Ellis, A.V.; Gordonb, D.L.; Koper, I. Biocompatible anti-microbial coatings for urinary catheters. *RSC Adv.* **2016**, *6*, 53303–53309. [CrossRef]
14. Kondyurina, I.; Nechitailo, G.S.; Svistkov, A.L.; Kondyurin, A.; Bilek, M. Urinary catheter with polyurethane coating modified by ion implantation. *Nucl. Instrum. Methods Phys. Res. B* **2015**, *342*, 39–46. [CrossRef]
15. Floroian, L.; Ristoscu, C.; Mihailescu, N.; Negut, I.; Badea, M.; Ursutiu, D.; Chifiriuc, M.C.; Urzica, I.; Dyia, H.M.; Bleotu, C.; et al. Functionalized Antimicrobial Composite Thin Films Printing for Stainless Steel Implant Coatings. *Molecules* **2016**, *21*, 740. [CrossRef]
16. Kowalczuk, D.; Pitucha, M. Application of FTIR Method for the Assessment of Immobilization of Active Substances in the Matrix of Biomedical Materials. *Materials* **2019**, *12*, 2972. [CrossRef]
17. Kowalczuk, D.; Ginalska, G. The Method for Preparing the Antimicrobial Biomaterial via Immobilization of the Antibacterial Substance on its Surface. Medical University of Lublin, Poland. Patent PL 214742 B1, 30 September 2013.
18. Xiao, X.; Xu, R. Preparation and surface properties of core-shell polyacrylate latex containing fluorine and silicon in the shell. *J. Appl. Polym. Sci.* **2011**, *119*, 1576–1585. [CrossRef]
19. Monteiro, O.A.C.; Airoldi, C. Some studies of crosslinking chitosan–glutaraldehyde interaction in a homogeneous system. *Int. J. Biol. Macromol.* **1999**, *26*, 119–128. [CrossRef]
20. Kil'deeva, N.R.; Perminov, P.A.; Vladimirov, L.V.; Novikov, V.V.; Mikhailov, S.N. Mechanism of the reaction of glutaraldehyde with chitosan. *Russ. J. Bioorg. Chem.* **2009**, *35*, 397–407. [CrossRef]
21. Islam, N.; Taha, M.O.; Dmour, I. Degradability of chitosan micro/nanoparticles for pulmonary drug delivery. *Heliyon* **2019**, *5*, e01684. [CrossRef]
22. Lawrence, E.L.; Turner, I.G. Materials for urinary catheters: A review of their history and development in the UK. *Med. Eng. Phys.* **2005**, *27*, 443–453. [CrossRef]
23. Perova, T.S.; Vij, J.K.; Xu, H. Fourier transform infrared study of poly(2-hydroxyethyl methacrylate) PHEMA. *Colloid Polym. Sci.* **1997**, *275*, 323–332. [CrossRef]
24. Abenojar, J.; Martınez, M.A.; Encinas, N.; Velasco, F. Modification of glass surfaces adhesion properties by atmospheric pressure plasma torch. *Int. J. Adhes. Adhes.* **2013**, *44*, 1–8. [CrossRef]
25. Zheng, E.; Dang, Q.; Liu, C.; Fan, B.; Yan, J.; Yu, Z.; Zhang, H. Preparation and evaluation of adipic acid dihydrazide cross-linked carboxymethyl chitosan microspheres for copper ion adsorption. *Colloids Surf. A Physicochem. Eng. Asp.* **2016**, *502*, 34–43. [CrossRef]
26. Jacobsen, S.M.; Stickler, D.J.; Mobley, H.L.T.; Shirtliff, M.E. Complicated Catheter-Associated Urinary Tract Infections Due to Escherichia coli and Proteus mirabilis. *Clin. Microbiol. Rev.* **2008**, *21*, 26–59. [CrossRef] [PubMed]
27. Cicciu, M.; Fiorillo, L.; Herford, A.S.; Crimi, S.; Bianchi, A.; D'Amico, C.; Laino, L.; Cervino, G. Bioactive Titanium Surfaces: Interactions of Eukaryotic and Prokaryotic Cells of Nano Devices Applied to Dental Practice. *Biomedicines* **2019**, *7*, 12. [CrossRef] [PubMed]
28. Fiorillo, L.; Cervino, G.; Laino, L.; D'Amico, C.; Mauceri, R.; Tozum, T.F.; Gaeta, M.; Cicciù, M. Porphyromonas gingivalis, Periodontal and Systemic Implications: A Systematic Review. *Dent. J.* **2019**, *7*, 114. [CrossRef]

29. Borowska, M.; Glinka, M.; Filipowicz, N.; Terebieniec, A.; Szarlej, P.; Kot-Wasik, A.; Kucińska-Lipka, J. Polymer biodegradable coatings as active substance release systems for urological applications. *Monatsh. Chem.* **2019**, *150*, 1697–1702. [CrossRef]
30. Yang, S.H.; Lee, Y.S.; Lin, F.H.; Yang, J.M.; Chen, K.S. Chitosan/poly(vinyl alcohol) blending hydrogel coating improves the surface characteristics of segmented polyurethane urethral catheters. *J. Biomed. Mater. Res. B* **2007**, 304–313. [CrossRef]
31. Khandwekar, A.P.; Patil, D.P.; Khandwekar, V.; Shouche, Y.S.; Sawant, S.; Doble, M. TecoflexTM functionalization by curdlan and its effect on protein adsorption and bacterial and tissue cell adhesion. *J. Mater. Sci. Mater. Med.* **2009**, *20*, 1115–1129. [CrossRef]
32. Tenke, P.; Riedl, C.R.; Jones, G.L.; Williams, G.J.; Stickler, D.; Nagyd, E. Bacterial biofilm formation on urologic devices and heparin coating as preventive strategy. *Int. J. Antimicrob. Agents.* **2004**, *23*, S67–S74. [CrossRef] [PubMed]
33. Svensson, O.; Arnebrant, T. Mucin layers and multilayers—Physicochemical properties and applications. *Curr. Opin. Colloid Interface Sci.* **2010**, *15*, 395–405. [CrossRef]
34. Gough, J.E.; Scotchford, C.A.; Downes, S. Cytotoxicity of glutaraldehyde crosslinked collagen/poly(vinyl alcohol) films is by the mechanism of apoptosis. *J. Biomed. Mater. Res.* **2002**, *61*, 121–130. [CrossRef] [PubMed]
35. Kowalczuk, D.; Przekora, A.; Ginalska, G. Biological safety evaluation of the modified urinary catheter. *Mater. Sci. Eng. C* **2015**, *49*, 274–280. [CrossRef]
36. Li, B.; Shan, C.-L.; Zhou, Q.; Fang, Y.; Wang, Y.-L.; Xu, F.; Han, L.-R.; Ibrahim, M.; Guo, L.-B.; Xie, G.-L.; et al. Synthesis, Characterization, and Antibacterial Activity of Cross-Linked Chitosan-Glutaraldehyde. *Mar. Drugs* **2013**, *11*, 1534–1552. [CrossRef]
37. Kowalczuk, D.; Ginalska, G.; Golus, J. Characterization of the developed antimicrobial urological catheters. *Int. J. Pharm.* **2010**, *402*, 175–183. [CrossRef]
38. Kowalczuk, D.; Ginalska, G.; Gowin, E. Development and comparison of HPLC method with fluorescence and spectrophotometric detections for determination of sparfloxacin. *Ann. UMCS Sect. DDD* **2011**, *24*, 163–169.

© 2020 by the author. Licensee MDPI, Basel, Switzerland. This article is an open access article distributed under the terms and conditions of the Creative Commons Attribution (CC BY) license (http://creativecommons.org/licenses/by/4.0/).

Communication

Copper-Silver Alloy Coated Door Handles as a Potential Antibacterial Strategy in Clinical Settings

Nicole Ciacotich [1,2], Lasse Kvich [3], Nicholas Sanford [4], Joseph Wolcott [4], Thomas Bjarnsholt [3,5] and Lone Gram [2,*]

1. Elplatek A/S, Bybjergvej 7, DK-3060 Espergærde, Denmark; nc@elplatek.dk
2. Department of Biotechnology and Biomedicine, Technical University of Denmark, Søltofts Plads bldg. 221, DK-2800 Kgs Lyngby, Denmark
3. Department of Immunology and Microbiology, Costerton Biofilm Center, Faculty of Health and Medical Sciences, University of Copenhagen, Blegdamsvej 3B, DK-2200 Copenhagen N, Denmark; lkvich@sund.ku.dk (L.K.); tbjarnsholt@sund.ku.dk (T.B.)
4. Southwest Regional Wound Care Center, 2002 Oxford Ave, Lubbock, TX 79410, USA; nick@josephwolcott.com (N.S.); joe@josephwolcott.com (J.W.)
5. Department of Clinical Microbiology, Rigshospitalet, Juliane Maries vej 22, 2100 Copenhagen Ø, Denmark
* Correspondence: gram@bio.dtu.dk; Tel.: +45-2368-8295

Received: 28 July 2020; Accepted: 11 August 2020; Published: 14 August 2020

Abstract: Coating surfaces with a copper-silver alloy in clinical settings can be an alternative or complementary antibacterial strategy to other existing technologies and disinfection interventions. A newly developed copper-silver alloy coating has a high antibacterial efficacy against common pathogenic bacteria in laboratory setups, and the purpose of this study is to determine the antibacterial efficacy of this copper-silvery alloy in real-world clinical settings. Two field trials were carried out at a private clinic and a wound care center. Door handles coated with the copper-silver alloy had a lower total aerobic plate count (1.3 ± 0.4 Log CFU/cm^2 and 0.8 ± 0.3 Log CFU/cm^2, CFU stands for Colony Forming Units) than the reference uncoated material on-site (2.4 ± 0.4 Log CFU/cm^2 for the stainless steel and 1.7 ± 0.4 Log CFU/cm^2 for the satin brass). The copper-silver alloy did not selectively reduce specific bacterial species. This study points to the possibility of a successful long-term implementation of the copper-silver alloy coating as an antibacterial strategy.

Keywords: antibacterial coating; healthcare-associated infections; touch-surfaces

1. Introduction

Extensive laboratory evidence has demonstrated the antibacterial properties of copper alloys, which has led to several field trials aimed at providing a proof of concept, particularly in clinical settings [1–3]. A multihospital clinical trial of six US Environmental Protection Agency (EPA)-registered antimicrobial copper alloys found that the microbial burden of copper alloy surfaces was six times lower (465 CFU/100 cm^2) than that of conventional surfaces such as plastics, coated carbon steel, aluminum and stainless steel (2674 CFU/100 cm^2) [4]. The microbial burdens on both the copper alloy and conventional surfaces were above those proposed as harmless on a surface immediately after cleaning (250 CFU/100 cm^2) [5]. However, the results of the trial showed a reduction in the rate of infections of 58% in "copper" rooms, compared to the "non-copper" rooms.

In a Finnish study, door handles made of copper alloys (99.8 wt.% Cu) and brass (60.5 wt.% Cu, 36.5 wt.% Zn) were installed in a hospital, a kindergarten, a retirement home and an office building, and microbial levels were compared to reference chromed door handles [3]. In terms of total aerobic plate count, door handles made of copper alloys outperformed those made of brass, which, in turn, did not (on average) show significant differences with the chromed material [3]. Lower levels of both

Gram-negative bacteria and *Staphylococcus aureus* were found on copper alloy surfaces than on brass and reference surfaces.

Door handle surfaces have the highest levels of bacterial contamination in clinical environments [6]. Recently published studies have tested a newly developed copper-silver alloy coating and found high antibacterial efficacy against *S. aureus* MSSA and MRSA, *Pseudomonas aeruginosa*, *Escherichia coli* and *Enterobacter aerogenes* [7,8]. Based on these findings, the present study aims to investigate the antibacterial efficacy of this copper-silver alloy coating on door handles in a private clinic, FamilieLægerne Espergærde, and a wound care center, Southwest Regional (SWR) Wound Care Center. In addition, it isolates and identifies microorganisms from the surfaces of interest to determine any species-specific effects. Finally, it evaluates the durability of the copper-silver alloy coating under the conditions tested.

2. Materials and Methods

2.1. Manufacturing and Installation of Door Handles

The stainless-steel door handles (Ruko Assa Abloy, Stockholm, Sweden) were electroplated with a copper-silver alloy coating at Elplatek A/S [7]. In the private clinic FamilieLægerne Espergærde (Egeskovvej 20, 3490 Kvistgård, DK), copper-silver alloy coated door handles (hereafter referred to as "test door handles") were installed on the doors of two doctors' and two nurses' exam rooms. Stainless steel door handles of four other offices were used as a reference material and sampled for microorganisms. Weekly sampling was performed for six weeks. At the SWR Wound Care Center (2002 Oxford Ave, Lubbock, TX 79410, USA), test door handles were installed on the doors of seven exam rooms, one public restroom and two laboratory rooms. The original satin brass door handles of six other exam rooms, three public restrooms and one laboratory room were sampled as reference material. Weekly sampling was performed for six weeks. All reference and test door handles were disinfected with 70% ethanol when the field trials started. Samplings were performed by the same person between 7.45 am and 8.00 am on Thursdays. The routine cleaning of door handles at the private clinic and the wound care center was carried out by wiping the surfaces with a dry cloth on Mondays.

2.2. Microbiological Sampling and Matrix-Assisted Laser Desorption Ionization Time-of-Flight Mass Spectrometry (MALDI-TOF MS) Analysis

In the private clinic in Denmark, every week the door handle surfaces (100 cm^2) were swabbed thoroughly (horizontal and vertical sweeps) with a flocked, sterile swab applicator (BD™ ESwab Regular Collection Kit, Franklin Lakes, NJ, USA). The swab was inserted into a sampling tube, which contained 1 mL of Liquid Amies Medium, and samples were transported to the laboratory at the Technical University of Denmark within 1 h. The sampling tubes were sonicated for 2 min at 28 kHz (Delta 220; Deltasonic, Meaux, France) and vortexed for 15 s. Four-hundred µL of the sampling suspensions were plated in duplicates on 5% blood agar (BA) plates (BD™, Franklin Lakes, NJ, USA). The plates were incubated at 37 ± 1 °C, and a total aerobic plate count was performed after 48 h. Colony-forming units (CFU) per plate corresponded to CFU per door handle surface (100 cm^2). Average values of CFU/100 cm^2 for test and reference door handles were log-transformed and presented as total average values ± standard deviation (SD). All isolates collected from the test door handles at the third and last sampling were re-streaked on BA plates, and single colonies were stored for later identification at −80 °C in a freezing medium (Tryptone Soy Broth 30 g/L, Glucose 5 g/L, Skim milk powder 20 g/L, Glycerol 40 g/L in distilled water). Single colonies were also randomly selected from the reference door handles and stored in the same way. Identification of microbial species was performed using MALDI-TOF MS on a Microflex LT instrument (Bruker Daltonik GmbH, Bremen, Germany). Protein profiles were acquired with the FlexControl 3.3 software (Bruker Daltonik GmbH, Bremen, Germany) and analyzed with FlexAnalysis 3.3 (Bruker Daltonik GmbH, Bremen, Germany). The database used to match spectra was Bruker Taxonomy (7311 MSPs). MALDI-TOF MS scores $x > 2.0$ were used to identify isolates to species level, while scores of $x = 1.8$–2.0 were used to identify them to genus level. Scores of $x < 1.8$ were not considered in this study.

2.3. Microbiological Sampling and Analyses of Bacterial Load by Direct Sequence Analysis

Microbiological sampling and analyses were performed on-site at the medical laboratory at the SWR Wound Care Center, which is accredited by the College of American Pathologists (CAP). Every week, the door handle surfaces (100 cm^2) were swabbed thoroughly (horizontal and vertical sweeps) with a sterile cotton swab. Cotton swabs were inserted into 2 mL sterile screw cap microtubes, and 500 µL of phosphate-buffered saline solution (PBS; Dulbecco A; Oxoid) was added. Bacteria were detached from the cotton swabs by shaking at 20 Hz for 2 min using a Qiagen TissueLyser (Qiagen Inc., Valencia, CA, USA). In the first two weekly samplings (out of six), 500 µL was added to sterile screw cap microtubes and genomic DNA was extracted using the Roche High Pure PCR Template Preparation kit (Roche Life Sciences, Indianapolis, IN, USA) according to the manufacturer's specifications. Sample lysates for DNA extraction were prepared using the Qiagen TissueLyser and 0.5 mm zirconium oxide beads (Next Advance, Averill Park, NY, USA). Targeting the universal 16S rRNA gene sequence, a semi-quantitative determination of bacterial load was performed using TaqMan real-time PCR Assay with the LightCycler® 480 (Roche Life Sciences). Forward (5'-CCATGAAGTCGGAATCGCTAG-3') and reverse (5'-GCTTGACGGGCGGTGT-3') 16S rDNA primers (20 µM each) were used with a 16S rDNA probe (5'-TACAAGGCCCGGGAACGTATTCACCG-3') in Quanta PerfeCTa® qPCR ToughMix (Quanta Biosciences, Beverly, MA, USA). The template DNA (2.5 µL) was added to the master mix containing primers and probe (10 µL each), and the reaction was run with the following thermal cycling profile: 50 °C for 2 min, 95 °C for 10 min, 35 cycles at 95 °C for 15 s, 60 °C for 1 min, and 40 °C for 30 s. *E. coli* c600 (ATCC 23724, Manassas, VA, USA) genomic DNA was used as a positive 16S rDNA control, and molecular grade water (Phenix Research Products, Chandler, NC, USA) was used as a no-template control.

2.4. Microbiological Sampling and Analyses of Bacterial Load by Plating and Sequence Analysis

For the remaining four weekly samplings (out of six) at the SWR Wound Care Center, 500 µL from the swab collection tubes was plated on Tryptone Soy Agar (TSA) (Oxoid CM0131) plates. Plates were incubated at 37 ± 1 °C and counted after 48 h. For each sampling, the average values of CFU/100 cm^2 for test and reference door handles were log-transformed and presented as total average values ± SD. Plates were then washed using 1 mL PBS, and bacterial material was collected into sterile Eppendorf tubes.

2.5. Microbiological Sampling, Identification of Pathogens and Determination of Resistance Genes by Sequence Analysis

The possible presence of pathogenic bacteria or resistance genes, or both, was tested using the TaqMan real-time PCR Assay for *P. aeruginosa*, *Serratia marcescens*, *S. aureus*, *Streptococcus pyrogenes*, *Streptococcus agalactiae* and the *mecA* and *vanA* genes. This was done using primer sequences that were the property of the CAP-accredited medical laboratory at the SWR Wound Care Center. This analysis was performed on all bacterial DNA extracted directly from the swabs and on DNA extracted from bacterial colonies on the agar plates. In the third sampling, four DNA extracts from the test door handles had Ct (cycle threshold) values for the 16S rRNA gene below 30 (the cut-off value), and were selected for further screening to detect the possible presence of pathogenic bacteria or resistance genes, or both. In the same sampling, eight DNA extracts from the reference door handles (control) had Ct-values below 30, of which six samples were randomly chosen as representatives of the control group. For the remaining samplings, the same number of DNA extracts were selected (four for the test and six for the control), and they were randomly chosen from the test and control DNA extracts with Ct-values below 30. The remaining 500 µL of the 1 mL washing suspensions from the agar plates was spread on TSA plates and incubated at 37 ± 1 °C for 24 h, and single colonies were re-streaked on

Mannitol salt agar (selective for staphylococci and micrococcaceae) and Cetrimide agar plates (selective for *P. aeruginosa*). This was done to further verify the presence or absence of these species.

2.6. Energy Dispersive X-Ray Spectroscopy (EDS) Analysis on Copper-Silver Alloy Coated Door Handles

The chemical compositions of selected copper-silver alloy coated door handles were checked prior to and after field testing, using the Hitachi TM3030 Plus Tabletop Microscope (Hitachi, Krefeld, Germany) operated at 15 kV and equipped with the Oxford Inca software and the Bruker Quantax 70 EDS System. EDS analysis was performed on three different spots at the surface of each sample. The output values (normalized weight percentage) were averaged and re-calculated with respect to the total content of copper and silver, to evaluate the difference before and after installation of the door handles. The presence of other elements (C, O) was also reported in normalized weight percentage if greater than 5 wt.%, to evaluate possible changes in surface composition of the coating during usage.

2.7. Statistical Analysis

Average values of $CFU/100\ cm^2$ for test and reference door handles at each sampling were log-transformed. Values were tested for equal or unequal variance with the *F*-test, and statistical significance of the difference between test and reference door handles was verified using the *t*-test.

3. Results and Discussion

Both the reference stainless steel door handles and satin brass door handles had a microbial load that was approximately twice as large as the copper-silver alloy coated door handles. At FamilieLægerne Espergærde, the averaged total aerobic plate count from copper-silver alloy coated and uncoated reference stainless steel door handles were, respectively, 1.3 ± 0.4 and 2.4 ± 0.4 Log $CFU/100\ cm^2$ (*p*-value 0.0008). At the SWR Wound Care Center, these values were, respectively, 0.8 ± 0.3 and 1.7 ± 0.4 Log $CFU/100\ cm^2$ (*p*-value 0.0068) (Figure 1). All surfaces in the field tests, except for stainless steel, had a microbial load below 2.4 Log $CFU/100\ cm^2$ (which is the standard for acceptable microbial level on a surface immediately after terminal cleaning) [5]. The microbial load on the satin brass reference door handles (1.7 ± 0.4 Log CFU/cm^2) was lower than that of the stainless-steel reference door handles (2.4 ± 0.4 Log CFU/cm^2), probably due to the antibacterial activity of brass (Figure 1).

Interestingly, when DNA was directly extracted from the swabs, there was no difference in bacterial load between the coated and reference door handles (Ct-values of 24.5 and 24.6, respectively), as estimated by qPCR of the 16S rRNA gene (Table 1). This might be due to the fact that DNA was recovered from both living and dead bacterial cells on the surfaces. However, bacteria with fewer alleles of the 16S rRNA gene could also have been selectively targeted.

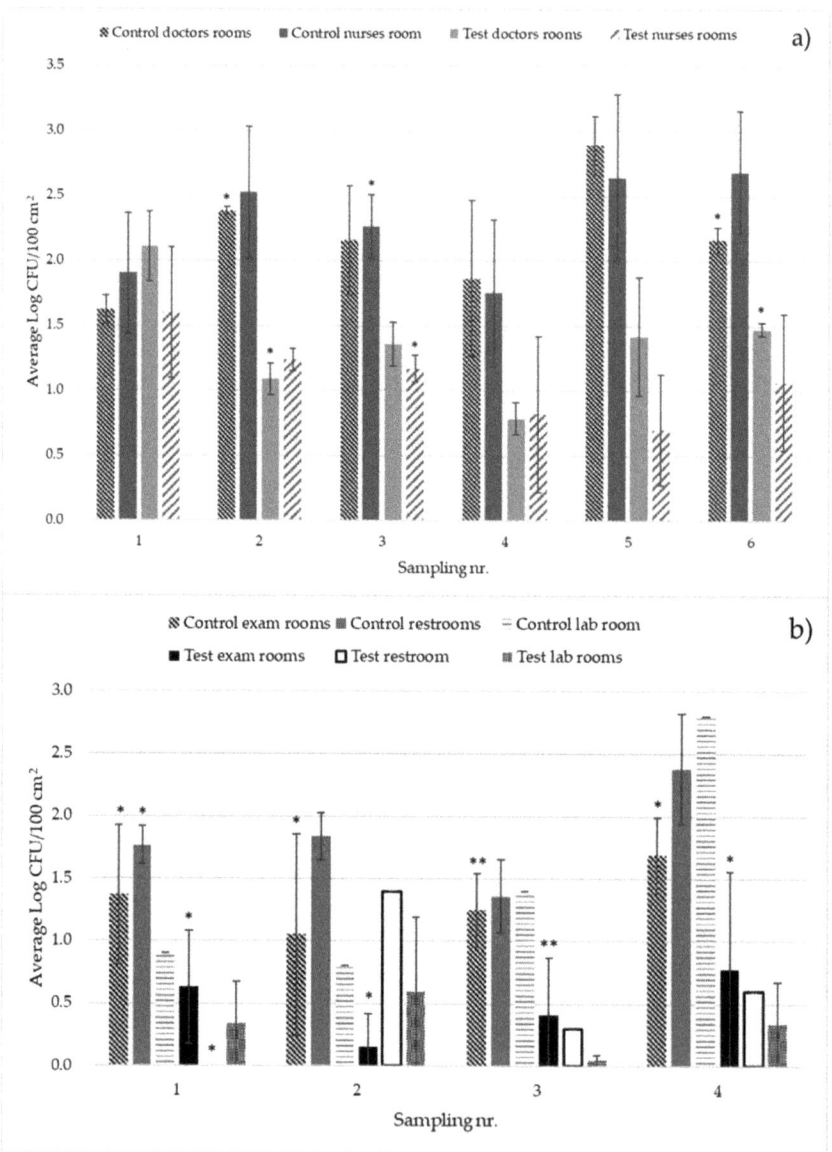

Figure 1. Averaged aerobic plate count presented as logarithmic transformed values (Log CFU/100 cm^2) from copper-silver alloy coated (test) and reference (control) door handles installed at doctors' and nurses' rooms at FamilieLægerne Espergærde (**a**), and from copper-silver alloy coated (test) and reference (control) door handles installed at exam rooms, laboratory rooms and restrooms at Southwest Regional (SWR) Wound Care Center (**b**). Error bars indicate standard deviations among the aerobic plate count values of test and control door handles groups for each sampling. The statistical difference among the different pair cases is indicated as * for $p < 0.05$ and ** for $p < 0.01$.

Table 1. Cycle threshold (Ct) values for 16S rDNA from test (copper-silver alloy coated door handles) and control (satin brass door handles) DNA extracts, when DNA extraction was performed directly from the swabs and after the growth step on agar, as described in MM. Occurrence (+/−) of S. aureus, mecA and vanA genes among the selected DNA extracts after the growth step on agar is also reported. NA = not analyzed.

		16s rDNA (Direct Extr.) Ct-Values			S. aureus (Agar Plates) +/−				mecA (Agar Plates) +/−				vanA (Agar Plates) +/−			
	Room	# Sampling			# Sampling				# Sampling				# Sampling			
		1	2	Average	3	4	5	6	3	4	5	6	3	4	5	6
Test	Exam 2	24.1	25.5		NA	NA	NA	−	+	NA	NA	+	−	NA	NA	−
	Exam 4	23.6	25.4		NA	NA	NA	NA	−	NA	NA	NA	−	NA	NA	NA
	Exam 6	24.1	25.2		−	NA	−	NA	−	NA	−	NA	−	NA	−	NA
	Exam 8	24.0	25.3		NA	NA	NA	NA	NA	−	NA	NA	NA	−	NA	NA
	Exam 10	24.1	25.0		−	−	+	−	NA	−	+	+	−	−	−	−
	Exam 12	23.9	24.0		−	+	NA	−	NA	NA	NA	NA	−	NA	NA	−
	Exam 14	25.0	24.4		NA	NA	NA	NA	NA	NA	NA	NA	NA	NA	NA	NA
	Men's West	25.1	23.8		NA	−	NA	−	NA	−	NA	−	NA	−	NA	−
	Lab Extraction	25.1	23.8		NA	NA	−	−	−	NA	−	−	NA	NA	−	−
	Lab Reaction	25.2	23.8		+	−	NA	NA	NA	−	NA	NA	NA	−	NA	NA
	Average	**24.4**	**24.6**	**24.5**												
Control	Exam 1	24.0	25.3		−	NA	NA	−	+	NA	NA	+	−	NA	NA	−
	Exam 3	23.9	25.2		−	NA	NA	NA	+	NA	NA	NA	−	NA	NA	NA
	Exam 5	24.0	25.5		−	NA	−	−	+	NA	+	+	−	NA	−	−
	Exam 7	23.9	24.8		NA	NA	NA	NA	NA	NA	NA	NA	NA	NA	NA	NA
	Exam 9	23.8	26.1		NA	−	−	−	NA	+	+	−	NA	−	−	−
	Exam 11	23.8	25.2		NA	−	−	NA	NA	+	+	NA	NA	−	−	NA
	Men's East	25.2	23.9		−	+	−	−	+	+	+	+	−	−	−	+
	Women's East	25.4	24.1		−	−	−	−	+	+	+	NA	−	−	−	NA
	Women's West	25.1	24.0		−	NA	NA	+	+	−	NA	+	−	−	NA	−
	Lab Detection	25.2	23.7		−	−	+	−	+	−	+	+	−	−	−	−
	Average	**24.4**	**24.8**	**24.6**												

In the Danish clinic, MALDI-TOF MS analysis was performed on randomly chosen isolates from both copper-silver alloy coated and uncoated stainless steel surfaces. There was no marked difference among the surfaces in terms of surviving bacterial species. The most abundant bacterial species were *Micrococcus luteus* and staphylococci (*S. hominis*, *S. epidermidis* and *S. capitis*) on both copper-silver alloy coated and uncoated door handles. *S. aureus* was found on the stainless steel but not on the copper-silver alloy coated door handles (Figure 2).

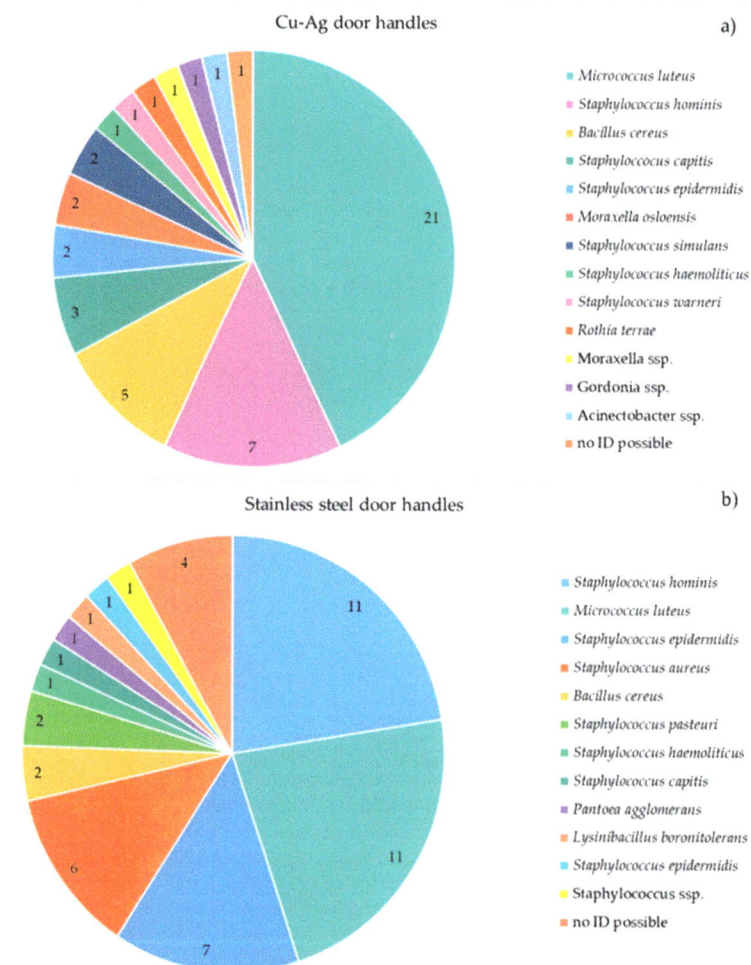

Figure 2. Species abundance of 98 microbial isolates recovered from copper-silver alloy coated (**a**) and uncoated stainless steel (**b**) door handles at FamilieLægerne Espergærde. Identification was performed with MALDI-TOF MS. The species are ordered according to their abundance in the column stacks and correspondingly in the legend. Identification at the species level was not possible for Bacillus ssp. and score values below 2.00, thus the genus is reported. It was not possible to recover and identify one isolate from stainless steel door handle due to lack of growth.

DNA extracted directly from the swabs was tested by qPCR for the presence of six pathogenic bacteria and two resistance genes, and none of the samples were positive. DNA from colonies on agar-plates from 16 of the test samples and from 24 control samples at the SWR Wound Care Center was tested for the presence of *S. aureus* by PCR. Three out of the sixteen test samples (19%) and 3 out of the 24 reference samples (13%) were positive (Table 1).

The presence of *S. aureus* was also confirmed by the yellow discoloration on selective (mannitol salt) agar plates. *P. aeruginosa*, *S. marcescens*, *S. pyrogenes* and *S. agalactiae* were not detected in any of the biomass-plate samples.

Resistance genes were detected in the biomass-plate samples, possibly because of their much higher bacterial load than in the amount extracted directly from the swabs. Seven out of the sixteen test samples (44%) and 22 out of the 24 reference samples (92%) were positive for the *mecA* gene (Table 1). In the last sampling, the *vanA* gene was detected in one biomass-plate sample from the control group. The greater occurrence of the *mecA* gene in the control group could be due to the larger bacterial counts on the plates and hence a larger amount of biomass. Selective antibacterial efficacy of copper surfaces against Gram-negative bacteria and *S. aureus*, as previously suggested, was not observed [3].

The EDS analysis on coated door handles prior to and after field testing at FamilieLægerne Espergærde revealed a 5 ± 1 wt.% relative difference in terms of copper and silver content, whereas there was basically no change in relative composition of copper and silver prior to and after installation at the SWR Wound Care Center. Carbon and oxygen could be detected on the surfaces after the field tests, but only the amount of carbon was above 5 wt.% (11.4 ± 2.6 wt.% and 12.3 ± 1.9 wt.%). No significant surface oxidation was observed on the surface prior to and after field testing. It is likely that door usage and other environmental affecting factors may have helped to reduce the copper content in the copper-silver alloy coated door handles at FamilieLægerne Espergærde, as compared to the ones at the Southwest Regional Wound Care Center. To our knowledge, the cleaning procedure should not have influenced the surface chemistry, since door handles were not subject to extensive disinfection in these environments. However, considering the reduction in copper content in the copper-silver alloy coated door handles after the field test in the Danish clinic, the lifetime (durability) of the coating could be safely estimated to 72 weeks. At that point, the door handles should be recoated to maintain a constant efficacy. During usage, a complementary cleaning procedure to remove dirt and filth (detected as presence of carbon at the surface) and to ensure direct contact between bacteria and the alloy coating would be recommended. Periodical cleaning would both increase efficacy and lifetime of the coating, and in turn the presence of the coating would reduce the amount and need for harsh cleaning chemicals and extensive disinfection interventions.

4. Conclusions

Copper-silver alloy coated door handles carried a lower bacterial load than stainless steel or satin brass door handles. Therefore, using copper-silver alloy coated door handles could also reduce healthcare-associated infections. The installation of coated door handles should not replace regular cleaning interventions, which in turn are recommended to ensure the efficacy of the active surface over time. Hence, this study provides a promising basis for a clinical trial where the infection rates are monitored pre and post installation of coated door handles (or similar items). This could pave the way for a future long-term usage of this coating as an antibacterial strategy.

Author Contributions: Conceptualization, N.C., T.B. and L.G.; methodology, N.C., L.K., N.S., T.B., and L.G.; software, N.C.; validation, N.C., L.K., N.S., T.B., and L.G.; formal analysis, N.C., L.K., and N.S.; investigation, N.C., L.K., and N.S; resources, T.B., J.W., and L.G.; data curation, N.C., L.K., N.S., and L.G.; writing—original draft preparation, N.C. and L.G.; writing—review and editing, N.C., L.K., N.S., T.B., and L.G.; visualization, N.C. and L.G.; supervision, T.B. and L.G.; project administration, N.C., T.B., N.S., J.W., and L.G.; funding acquisition, N.C, T.B., and L.G. All authors have read and agreed to the published version of the manuscript.

Funding: This research was funded by the Innovation Fund of Denmark, case grant no. 5189-00091B.

Acknowledgments: We would like to acknowledge FamilieLægerne Espergærde and the Southwest Regional Wound Care Center for making these field tests possible. We also thank the clinical staff for their help and support.

Conflicts of Interest: The authors declare no conflict of interest.

References

1. Michels, H.T.; Keevil, C.W.; Salgado, C.D.; Schmidt, M.G. From laboratory research to a clinical trial: Copper alloy surfaces kill bacteria and reduce hospital-acquired infections. *HERD* **2015**, *9*, 64–79. [CrossRef] [PubMed]
2. Hinsa-Leasure, S.M.; Nartey, Q.; Vaverka, J.; Schmidt, M.G. Copper alloy surfaces sustain terminal cleaning levels in a rural hospital. *Am. J. Infect. Control* **2016**, *44*, 195–203. [CrossRef] [PubMed]
3. Inkinen, J.; Mäkinen, R.; Keinänen-Toivola, M.M.; Nordström, K.; Ahonen, M. Copper as an antibacterial material in different facilities. *Lett. Appl. Microbiol.* **2017**, *64*, 19–26. [CrossRef] [PubMed]
4. Schmidt, M.G.; Attaway, H.H.; Sharpe, P.A.; John, J.; Sepkowitz, K.A.; Morgan, A.; Fairey, S.E.; Singh, S.; Steed, L.L.; Cantey, J.R.; et al. Sustained reduction of microbial burden on common hospital surfaces through introduction of copper. *J. Clin. Microbiol.* **2012**, *50*, 2217–2223. [CrossRef] [PubMed]
5. Dancer, S.J. How do we assess hospital cleaning? A proposal for microbiological standards for surface hygiene in hospitals. *J. Hosp. Infect.* **2004**, *56*, 10–15. [CrossRef] [PubMed]
6. Wojgani, H.; Kehsa, C.; Cloutman-green, E.; Gray, C.; Gant, V.; Klein, N. Hospital door handle design and their contamination with bacteria: A real life observational study. Are we pulling against closed doors? *PLoS ONE* **2012**, *7*, 1–6. [CrossRef] [PubMed]
7. Ciacotich, N.; Din, U.R.; Sloth, J.J.; Møller, P.; Gram, L. An electroplated copper–silver alloy as antibacterial coating on stainless steel. *Surf. Coat. Technol.* **2018**, *345*, 96–104. [CrossRef]
8. Ciacotich, N.; Kragh, N.K.; Lichtenberg, M.; Tesdorpf, E.J.; Bjarnsholt, T.; Gram, L. In situ monitoring of the antibacterial activity of a copper–silver alloy using confocal laser scanning microscopy and pH microsensors. *Glob. Chall.* **2019**, *3*, 1900044. [CrossRef] [PubMed]

© 2020 by the authors. Licensee MDPI, Basel, Switzerland. This article is an open access article distributed under the terms and conditions of the Creative Commons Attribution (CC BY) license (http://creativecommons.org/licenses/by/4.0/).

Article

Assessment of *Streptococcus Mutans* Adhesion to the Surface of Biomimetically-Modified Orthodontic Archwires

Santiago Arango-Santander [1,*], Carolina Gonzalez [2], Anizac Aguilar [2], Alejandro Cano [2], Sergio Castro [2], Juliana Sanchez-Garzon [2,3] and John Franco [2,4]

1. GIOM Group, Faculty of Dentistry, Universidad Cooperativa de Colombia, Envigado 055422, Colombia
2. Faculty of Dentistry, Universidad Cooperativa de Colombia, Envigado 055422, Colombia; carolina.gonzalezve@campusucc.edu.co (C.G.); anizac.aguilar@campusucc.edu.co (A.A.); alejandro.canom@campusucc.edu.co (A.C.); sergioa.castro@campusucc.edu.co (S.C.); juliana.sanchezga@campusucc.edu.co (J.S.-G.); john.francoa@campusucc.edu.co (J.F.)
3. CES University, Medellín 050021, Colombia
4. Universidad de Antioquia, Medellín 050010, Colombia
* Correspondence: santiago.arango@campusucc.edu.co; Tel.: +57-4-4446065

Received: 30 December 2019; Accepted: 3 February 2020; Published: 26 February 2020

Abstract: Bacterial adhesion and biofilm formation on the surfaces of dental and orthodontic biomaterials is primary responsible for oral diseases and biomaterial deterioration. A number of alternatives to reduce bacterial adhesion to biomaterials, including surface modification using a variety of techniques, has been proposed. Even though surface modification has demonstrated a reduction in bacterial adhesion, information on surface modification and biomimetics to reduce bacterial adhesion to a surface is scarce. Therefore, the main objective of this work was to assess bacterial adhesion to orthodontic archwires that were modified following a biomimetic approach. The sample consisted of 0.017 × 0.025, 10 mm-long 316L stainless steel and NiTi orthodontic archwire fragments. For soft lithography, a polydimethylsiloxane (PDMS) stamp was obtained after duplicating the surface of *Colocasia esculenta* (L) Schott leaves. Topography transfer to the archwires was performed using silica sol. Surface hydrophobicity was assessed by contact angle and surface roughness by atomic force microscopy. Bacterial adhesion was evaluated using *Streptococcus mutans*. The topography of the *Colocasia esculenta* (L) Schott leaf was successfully transferred to the surface of the archwires. Contact angle and roughness between modified and unmodified archwire surfaces was statistically significant. A statistically significant reduction in *Streptococcus mutans* adhesion to modified archwires was also observed.

Keywords: bacterial adhesion; soft lithography; biomimetics; surface modification

1. Introduction

Many elements are involved in the orthodontic treatment of dental malocclusions. Stainless steel and titanium alloys are essential to manufacture brackets and archwires. These alloys have been widely used due to their outstanding mechanical properties and high corrosion resistance [1,2]. In the oral cavity, alloys are exposed to saliva, biological fluids and bacteria, which ultimately may cause deterioration and corrosion [3]. The presence of bacteria is a fundamental event that leads to metal corrosion and degradation as bacteria first adhere to a surface and biofilm will eventually be formed [4]. Biofilm formation has been largely associated with treatment failure in the biomedical field [5], including orthodontics [6]. Corrosive degradation might occur in biometals when bacteria adhere to the surface and may absorb and metabolize the metal or they might increase the acidity level around the biometal by reducing the pH, which leads to metal corrosion [1]. In orthodontic alloys, this corrosion might affect the biocompatibility and performance of the appliance, which will lead to treatment failure or an increase in treatment time [3,6]. However, titanium alloys are among the most resistant to corrosion caused by oral microorganisms [7]. Furthermore, adhesion and colonization of bacteria on dental biomaterial surfaces might lead to the onset of different conditions, namely secondary caries or periodontal disease as the two most relevant [8].

Different alternatives have been proposed to reduce bacterial adhesion to the surface of biomaterials. Among such alternatives, physical surface modifications, including roughening of zirconia surfaces [9] or modification of the topography of materials to observe cellular [10,11] or bacterial behavior [12], have received attention in the last years.

An array of direct and indirect techniques to modify the surface of a material have been evaluated, including soft lithography, photolithography or dip-pen nanolithography [13–15]. Soft lithography is a set of easy-to-use and low-cost techniques that allows surface modification at nano and micro scales. This technique is based on the construction of a stamp, typically made of polydimethylsiloxane (PDMS), to duplicate the topography of one surface and then transfer it to another surface of interest [16]. Traditionally, photolithography has been used to fabricate the master model that is duplicated with soft lithographic techniques [17–21]. However, photolithography presents some disadvantages, including high costs, no control over the surface chemistry and impossibility to be applied on non-planar surfaces [22]. Therefore, alternatives have been investigated to fabricate the master model [12].

In this regard, biomimetics may be presented as an alternative to photolithography since nature has created master models on plant or animal surfaces over millions of years of evolution to resolve its own problems. Humans have been exploring such master models to use them in industrial and biomedical applications, among many others [23,24]. Chung et al. [25] used the topography of the shark skin as a master model to assess bacterial adhesion, while Bixler et al. [26] created a bio-inspired, anti-fouling surface based on the topography of rice leaves and butterfly wings. A well-known example of biomimetics is the lotus effect. In this effect, adhesion to the surface of the *Lotus* leaf is impaired by physical arrays and chemical processes that lead to the impossibility of dust or other agents to remain on the surface as they are washed off by water. This phenomenon is known as self-cleaning [24,27] and is related to the hydrophobicity of the surface.

Many plants and leaves that show the super hydrophobicity and self-cleaning effects are reported in the literature. Among them, *Colocasia esculenta* (L) Schott, a plant from tropical Asia but found in tropical and sub-tropical areas around the world [28], has been recognized as having super hydrophobic leaves. Surface hydrophobicity is measured using the contact angle (CA) measurement, which is the angle that forms between a drop of liquid and the solid surface. Depending on the CA, surfaces may be classified as super hydrophilic (CA<10°), hydrophilic (CA >10° but < 90°), hydrophobic (CA >90° but < 150°) and super hydrophobic (CA>150°) [24]. According to Neinhuis and Barthlott [29], *C. esculenta* shows a contact angle value of 164°, which places it in the super hydrophobic category. Hüger et al. [30] and Bhushan and Jung [31] demonstrated that the leaf of this plant presents a hierarchical structure that is responsible for the super hydrophobic behavior and optimal self-cleaning properties. In addition, these authors established that the leaf is also covered by wax, which in combination with the surface roughness, provides the super hydrophobic characteristic.

Current research in surface modification to reduce bacterial adhesion has produced interesting results. Chung et al. [25] modified the surface of PDMS and demonstrated a reduction in the adhesion of *Staphylococcus aureus*. Vasudevan et al. [32] compared *Staphylococcus cloacae* adhesion to modified vs. unmodified surfaces and found a higher rate of biofilm formation on the unmodified (50%) vs. the modified surface (<20%). Hochbaum y Aizenberg [33] concluded that physically modified surfaces reduced bacterial adhesion and Xu y Siedlecki [34] compared *Staphylococcus epidermidis* and *S. aureus* adhesion to polyurethane urea and a reduction in biofilm formation on the modified surface was observed. However, surface modification approaches are scarce in the fields of dentistry and orthodontics, particularly following a biomimetic approach. Moreover, to the best of our knowledge, the use of *C. esculenta* as a master model for surface modification in the biomedical field has not been investigated. Therefore, the objective of this work was to assess bacterial adhesion to the surface of modified stainless steel and NiTi orthodontic archwires following a biomimetic approach.

2. Materials and Methods

2.1. Substrate

0.017 × 0.025 in, 10 mm-long 316L stainless steel and NiTi orthodontic archwire fragments (ORMCO, Orange, CA, USA) were used. The archwire fragments of the aforementioned dimensions were sequentially cleaned, following a protocol established by our group [12,35], using surfactant, acetone (99.8% v/v, Merck Millipore, Burlington, MA, USA), distilled water (Protokimica, Medellin, Colombia), and absolute ethanol (99% v/v, Merck Millipore, Burlington, MA, USA) for 8 min each in an ultrasound bath and let dry in air.

2.2. Silica sol Synthesis

The one-stage sol-gel method was used to synthetize the silica sol [36]. Tetraethylorthosilicate (TEOS) and methyltriethoxysilane (MTES) (ABCR GmbH & Co., Karlsruhe, Germany) were used as silica precursors for the hybrid sol, 0.1N nitric acid (Merck Millipore, Burlington, MA, USA) and acetic acid (glacial, 100% v/v, Merck Millipore, Burlington, MA, USA) were used as catalysts, and absolute ethanol (99.9% v/v, Merck Millipore, Burlington, MA, USA) was used as solvent. The obtained final concentration of SiO_2 was 18 gL^{-1}.

2.3. Surface Modification

For the experimental groups, the *C. esculenta* leaf was cut in segments of 5.0 cm in diameter avoiding the central vein. Leaf fragments were placed at the bottom of a silicone container, the lamina of the leaf facing upward. Polydimethylsiloxane (PDMS) (Silastic T-2, Dow Corning Corporation, Midland, MI, USA) was used to duplicate the surface of the leaf. PDMS was prepared according to the manufacturer, poured to cover the leaf and cured for 24 h. Then, the leaf was carefully removed from the surface and the PDMS stamp was inspected to verify its integrity before completing polymerization at 80 °C for 3 h. The PDMS was used as a microstamp to transfer SiO_2 to one archwire surface. One of the 0.025 in surfaces was selected due to the higher area. 2 µL of the silica sol were deposited on the archwire surface, a microstamp was placed over the drop applying mild pressure, and the sol was allowed to gel for 4 h at RT. Then, the PDMS stamp was removed, and the archwire with the transferred pattern was heat treated in a furnace at 450 °C for 30 min [12]. Two experimental groups resulted: SS316L and NiTi modified wires (SS316L_e and NiTi_e). The control groups consisted of cleaned, unmodified 316L stainless steel and NiTi orthodontic archwire fragments of the same dimensions as the experimental groups (SS316L_c and NiTi_c).

2.4. Surface Characterization

The lamina of the *C. esculenta* Schott leaf and the archwire fragments from the experimental and control groups were characterized through atomic force microscopy (AFM) (Nanosurf Easyscan 2, Nanosurf AG, Liestal, Switzerland). For AFM acquisition, a NCLR tip (Nanosensors™, Neuchâtel, Switzerland) in tapping mode at a constant force of 48 N/m was used. Images were processed using software AxioVision (V 4.9.1.0, Carl Zeiss Microscopy GmbH, Jena, Germany), software Image J 1.51 J (Laboratory for Optical and Computational Instrumentation, University of Winsconsin, Madison, WI, USA) [37], and software WSxM 5.0 (Nanotec Electrónica and New Microscopy Laboratory, Madrid, Spain) [38]. 10 AFM images of 50 µm × 50 µm were used to measure surface roughness and the arithmetic average of the roughness profile (Ra) was calculated using software for AFM analysis (Gwyddion 2.34, Department of Nanometrology, Czech Metrology Institute, Brno, Czech Republic). To measure surface hydrophobicity, the sessile drop method was applied on 10 random fragments from each group. The contact angle images were attained using a camera (Canon EOS Rebel XS, Tokyo, Japan) and a macrolens (105mmF2.8 EX DG OS, Sigma, Ronkonkoma, NY, USA). Contact angle values were obtained using software AxioVision.

2.5. Bacterial Adhesion

Adhesion of *Streptococcus mutans* (ATCC 25175, Microbiologics, St. Cloud, MN, USA) to experimental and control surfaces was evaluated following a protocol previously published by our group [12,35]. *S. mutans* were grown in Heart-Brain Infusion (BHI) agar (Scharlab S.L., Barcelona, Spain) supplemented with 0.2 U/mL bacitracin (Sigma Fluka, St. Louis, MO, USA) for 24 h at 37° ± 1° C. Then, they were cultured in a solution of peptone water at 37° ± 1°C for 24 h. After centrifugation of the bacterial solution, the supernatant was discarded, followed by resuspension of the bacterial pellet in peptone water. The nephelometric turbidity unit (NTU)) (based on a calibration curve of NTU vs CFU/mL) was used to obtain a 10^{-7} CFUs/mL concentration. 40 archwire fragments from control and experimental groups (10 archwire fragments per group) were placed in wells of 24-well non-treated polystyrene plates (Costar, Corning Inc., NY, USA), and 500 µL of bacterial solution was added to each well. At the bottom of each well, a rubber matrix was placed and the respective modified archwire was embedded leaving only the modified surface exposed to the bacterial solution. In the case of unmodified archwires, the fragment was also embedded leaving only one 0.025 in surface exposed to the bacterial solution. This procedure was performed to ensure that bacteria could only adhere to the modified surface, or one surface in the case of unmodified archwires, and not to untreated surfaces.

Polystyrene plates were incubated at 37° ± 1° C for 2 and 6 h. After each incubation time, archwires were removed from the wells and washed three times with 500 µL of 0.9% saline solution (Corpaul, Medellin, Colombia) to remove non-adherent bacteria. Archwire fragments from control and experimental surfaces were then subjected to sonication at 50% power (Qsonica 125, Newtown, CT, USA) in 10 mL of 0.9% saline solution for 3 seconds. The sonicated solutions were serially diluted and 10 µL from each dilution were cultured in BHI agar in triplicate following the drop plate method [39]. Culture plates were incubated at 37° ± 1°C for 48 h, followed by counting the colony-forming units (CFUs). This process was repeated in triplicate during different periods.

2.6. Statistical Analysis

A univariate analysis to describe roughness, contact angle and bacterial adhesion by estimation of central tendency and dispersion was performed. Results are presented as the mean ± standard deviation (SD). The one-way ANOVA test with post hoc Tukey method was used for comparison among groups after previous verification of the assumption of normality and homogeneity of variances through the Shapiro Wilk and Levene tests, respectively. Values of $p < 0.05$ were considered statistically significant. Software SPSS (V. 25) (IBM Corp., Armonk, NY, USA) and GraphPad Prism 8.3.0 (GraphPad Software, San Diego, CA, USA) were used for statistical analyses.

3. Results

3.1. Surface Modification

Figure 1 shows 100 µm × 100 µm (a) and 50 µm × 50 µm (b) AFM images of the lamina of the *C. esculenta* leaf. Figure 2 shows 50 µm × 50 µm AFM images of control stainless steel 316L (a) and nickel-titanium (b) surfaces. In addition, successful transference of the *C. esculenta* topography to the surface of stainless steel (c) and NiTi (d) orthodontic archwires is shown.

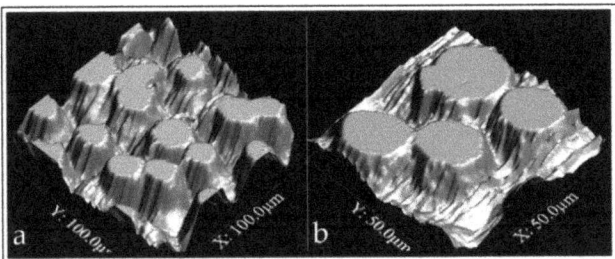

Figure 1. AFM image of the *C. esculenta* Schott leaf at 100 × 100 μm (**a**) and 50 × 50μm (**b**).

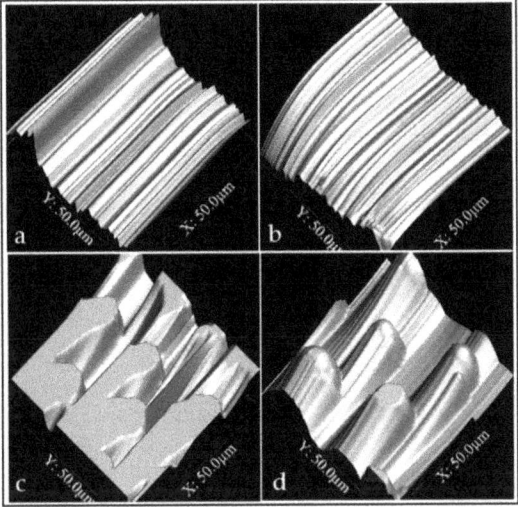

Figure 2. AFM images of control SS316L (**a**) and NiTi (**b**). Transference of the topography from *C. esculenta* to SS316L (**c**) and NiTi (**d**) were successful.

3.2. Surface Characterization

Contact angle measurements for control and experimental groups for both substrates are shown in Figure 3a. The most hydrophobic surface was the experimental NiTi (CA=129.4 ± 4.8°) and the least hydrophobic surface was the control 316L (CA=76.8 ± 6.9°). The difference between control and experimental surfaces for both substrates was statistically significant ($P<0.0001$), as well as the difference between both substrates ($P<0.0001$). The contact angle value of the *C. esculenta* Schott leaf was 131.2 ± 7.1°.

Regarding roughness measurements (Ra) for control and experimental groups for both substrates, results are shown in Figure 3b. The roughest surface was the experimental NiTi (Ra=142.3 ± 31.9 nm) and the smoothest was the control NiTi (Ra=34.1 ± 4.9 nm). The difference between SS316L_c and NiTi_c was statistically significant (P=0.0005), as well as the difference between SS316L_e and NiTi_c (P=<0.0001) and NiTi_c and NiTi_e ($P<0.0001$). The Ra of the *C. esculenta* Schott leaf was 270.6 ± 37.0 nm. Table 1 summarizes the results of contact angle and roughness values for the different groups.

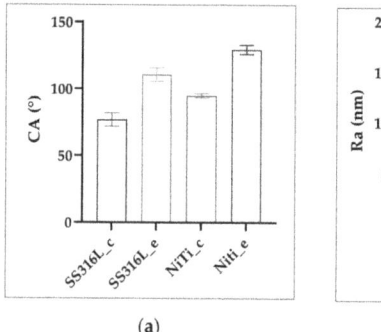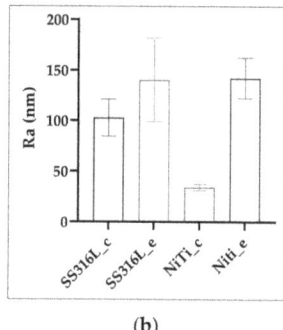

Figure 3. Contact angle (**a**) and Ra (**b**) measurements of experimental and control surfaces.

Table 1. Values for contact angle (CA) and roughness (Ra).

Group	CA (°)					Ra (nm)				
	Mean	SD	Min	Median	Max	Mean	SD	Min	Median	Max
SS316L_c	76.8	6.9	64.0	79.5	83.0	102.8	28.7	64.0	100.5	162.0
SS316L_e	110.8	7.1	96.0	113.5	118.0	140.4	65.2	54.0	134.0	251.0
NiTi_c	94.9	2.2	90.5	95.5	97.5	34.1	4.9	29.2	31.7	42.8
NiTi_e	129.4	4.8	120.2	130.8	135.4	142.3	31.9	119.9	125.1	204.5

3.3. Bacterial Adhesion

S. mutans was allowed to adhere to the four surfaces during the evaluated periods. The highest adhesion of *S. mutans* was to the SS316L_c surface, followed by adhesion to the NiTi_c surface. The difference in adhesion between both control surfaces was statistically significant ($P=0.0242$). The SS316L_e surface showed the lowest adhesion and no statistically significant difference was found at 2 and 6h. However, statistically significant differences were found between SS316L_e vs SS316L_c, SS316L_c vs NiTi_e, SS316L_e vs NiTi_c and NiTi_c vs NiTi_e ($P=0.0015$; 0.0024;0.0161 and 0.0369, respectively) The NiTi_e surface showed the second lowest adhesion and there was no statistically significant difference in time. The difference in adhesion to both experimental surfaces was not statistically significant ($P = 0.6477$, Figure 4). Table 2 summarizes the values of *S. mutans* adhesion to experimental and control surfaces at the evaluated times.

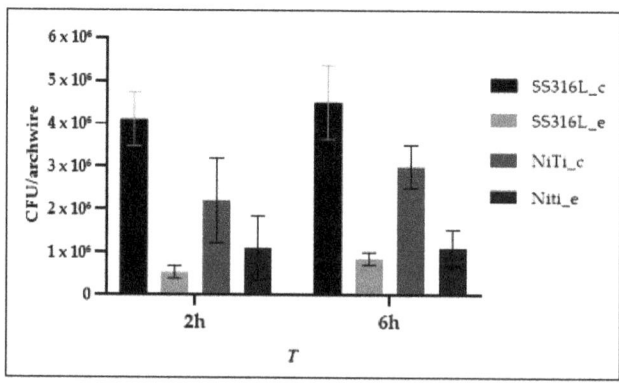

Figure 4. *S. mutans* adhesion to control and experimental surfaces at 2 and 6 hours.

Table 2. Values of *S. mutans* adhesion to control an experimental surfaces at the evaluated times.

Group	2h					6h				
	Mean	SD	Min	Median	Max	Mean	SD	Min	Median	Max
SS316L_c	4.1×10^6	6.3×10^5	3.4×10^6	4.2×10^6	5.0×10^6	4.5×10^6	8.6×10^5	3.2×10^6	4.6×10^6	5.6×10^6
SS316L_e	5.3×10^5	1.5×10^5	3.5×10^5	5.0×10^5	7.5×10^5	8.5×10^5	1.5×10^5	6.5×10^5	8.0×10^5	1.0×10^6
NiTi_c	2.2×10^6	9.9×10^5	1.4×10^6	1.9×10^6	3.8×10^6	3.0×10^6	5.0×10^5	2.3×10^6	3.0×10^6	3.6×10^6
NiTi_e	1.1×10^6	7.4×10^5	3.0×10^5	9.0×10^5	2.3×10^6	1.1×10^6	4.3×10^5	3.5×10^5	1.3×10^6	1.5×10^6

4. Discussion

Bacterial adhesion to a surface is a dynamic process and many variables related to the surface of the material and the bacterial species are involved. In orthodontics, bacterial adhesion and colonization have also been related to bracket presence and position [40], ligature method [41] and archwire material [42].

In the current work, surface characterization showed an increase in hydrophobicity on both substrates after surface modification was performed, which is in agreement with other results reported in the literature [12,43,44]. SS 316L modified surfaces showed an average of 110.8° in the contact angle measurement, while modified NiTi surfaces showed an average of 129.4° in the current work. Yang et al. [43] found contact angles in excess of 120° after using TEOS and MTES as silica precursors. Such angles depended on the morphology of the silica coatings and surface chemistry of the materials. Our previous work found an increase in the hydrophobicity of SS after modifying the surface using the same TEOS:MTES ratio used in the current work (57° to 87°), which are higher than the angles found in the present work, especially for the modified SS surface [12]. Other authors, including Santos et al. [45], Hosseinalipour et al. [44] and Wang et al. [46], have reported an increase in hydrophobicity. Wang et al. [46] demonstrated that an increase in the TEOS:MTES ratio increases hydrophobicity. They reported a contact angle value of 85° for a TEOS:MTES ratio similar to the current work. According to Wang et al. [46], such increase in surface hydrophobicity may be explained by the amount of methyl groups that might reduce the ability of the surface to absorb water. An additional explanation for this increase in surface hydrophobicity is that when roughness is created on a surface at the micro and nano scales, the water-air interface on a drop of liquid laying on such surface is enlarged and the capillary forces between the drop and the solid surface are reduced, which makes the drop take a spherical shape [23,47]. Nonetheless, the increase in hydrophobicity in this work appears to be related to the presence of silica on the modified surfaces rather than the presence of the features transferred from the *C. esculenta* leaf since its contact angle measurement was higher than that of the modified surfaces (131.2±7.1°), which is in agreement with the results of Burton and Bhushan [48]. This phenomenon may be explained by two facts: the leaf of *C. esculenta* has a hierarchical structure that provides pockets of air that lead to a high hydrophobicity [49] and the presence of a wax coating that covers the lamina of the leaf [48]. The transferring process used in the current investigation could not transfer the wax coating and only the external surface of the leaf could be copied and transferred, therefore, the complete hierarchical structure could not be transferred to the experimental surfaces.

Roughness was increased after modifying both substrates. SS 316L showed a higher increase in roughness than NiTi after modification. This increase in roughness from as-received to modified surfaces is in agreement with our previous work [12]. These results were expected since the patterns from *C. esculenta* showed a more voluminous topography that increased the roughness of the otherwise smooth untreated SS 316L and NiTi surfaces. Kim et al. [50] assessed the roughness of several orthodontic archwires and found a similar Ra for SS, even though they analyzed a different type of SS, and a higher Ra for the NiTi archwires. The leaf of *C. esculenta* showed higher roughness than the modified surfaces, which may be explained by the fact that a natural surface such as *C. esculenta*´s is not homogeneous since the hierarchical structures might have different heights.

Bacterial adhesion to the surface showed a reduction that had been observed in our previous work [12]. The work of Satou et al. [51] showed that the bacterial surface of *Streptococcus mutans* is rather hydrophilic and they also observed that hydrophilic bacteria show higher adhesion to hydrophilic surfaces. In the current work, the observed reduction in *S. mutans* adhesion might be explained by the combination of a higher hydrophobicity from the silica and the presence of surface features from the *C. esculenta*, which might disrupt the way this bacterial species arrange on the surface, since the modified surfaces showed lower bacterial adhesion.

The relation between bacterial adhesion and surface roughness is not as clear. Kim et al. [50] suggested that surface tension plays a more significant role in bacterial adhesion than surface roughness. Hydrophobicity is influenced by roughness, but changes in the Ra should be larger than 0.1 μm for the effect to be exhibited, as suggested by Busscher et al. [52]. The difference in the current work for SS 316L and NiTi was under 0.1 μm, so roughness did not seem to play an important role for the increase in hydrophobicity. Therefore, the most likely explanation for the increase in hydrophobicity is the presence of the methyl groups as explained. According to this suggestion [50], roughness was not the most important factor for bacterial adhesion, since the smoothest surfaces showed higher bacterial adhesion. Therefore, the presence of the features copied from *C. esculenta* and the increase in hydrophobicity played the predominant roles for reduction of *S. mutans* adhesion to these surfaces.

The results of the current work showed a reduction of around 87% in *S. mutans* adhesion to SS 316L at 2 h and 81% at 6 h and around 50% at 2 h and 66% at 6 h on NiTi. These results are in agreement with the results of Chung et al. [25], who showed an 87% reduction in the adhesion of *S. aureus* at 14 days and May et al. [53], who found a 95% to 99% reduction in adhesion by different bacterial species. Our previous work found a 95% reduction in *S. mutans* adhesion to modified stainless steel plates [12]. These results are relevant for a short-time study and demonstrate that reduction of bacterial adhesion might be obtained using different substrates.

5. Conclusions and Considerations

Within the limitations of this work, the results showed a reduction in adhesion of *S. mutans* to modified stainless steel and nickel-titanium surfaces in a short-time study. These results are promising to consider biomimetics as an alternative approach to fabricate surfaces with antibacterial properties in the biomedical field. Limitations included the use of a single bacterial species, chemical characterization of the *C. esculenta* leaf, mechanical characterization of the coatings and longer periods of evaluation. It is also important to consider that orthodontic archwires are commercially available in different sections and sizes, which might have an influence on the results. All of these limitations should be addressed in future works, including the evaluation of other natural surfaces.

Author Contributions: Conceptualization, S.A.-S.; Data curation, J.S.-G. and J.F.; Formal analysis, J.S.-G. and J.F.; Investigation, S.A.-S., C.G., A.A., A.C. and S.C.; Methodology, S.A.-S., C.G., A.A., A.C. and S.C.; Project administration, S.A.-S.; Supervision, S.A.-S. and J.S.-G.; Writing – original draft, S.A.-S., C.G., A.A., A.C. and S.C.; Writing – review & editing, S.A.-S., J.S.-G. and J.F. All authors have read and agreed to the published version of the manuscript.

Funding: This research was funded by CONADI (2439).

Acknowledgments: The authors would like to thank Ph.D. Claudia García from Universidad Nacional de Colombia for her assistance with the sol-gel method and M.Sc. Johanna Gutiérrez from Tecnoacademia for her assistance with the AFM.

Conflicts of Interest: The authors declare no conflict of interest.

References

1. Arango-Santander, S.; Ramírez-Vega, C. Titanio: Aspectos del material para uso en ortodoncia. *Rev. Nac. Odontol.* **2016**, *12*, 63–71. [CrossRef]
2. Arango Santander, S.; Luna Ossa, C.M. Stainless Steel: Material Facts for the Orthodontic Practitioner. *Rev. Nac. Odontol.* **2015**, *11*, 71–82. [CrossRef]

3. Bahije, L.; Benyahia, H.; El Hamzaoui, S.; Ebn Touhami, M.; Bengueddour, R.; Rerhrhaye, W.; Abdallaoui, F.; Zaoui, F. Behavior of NiTi in the presence of oral bacteria: Corrosion by Streptococcus mutans. *Int. Orthod.* **2011**, *9*, 110–119. [CrossRef] [PubMed]
4. Renner, L.D.; Weibel, D.B. Physicochemical regulation of biofilm formation. *MRS Bull.* **2011**, *36*, 347–355. [CrossRef] [PubMed]
5. Campoccia, D.; Montanaro, L.; Arciola, C.R. A review of the biomaterials technologies for infection-resistant surfaces. *Biomaterials* **2013**, *34*, 8533–8554. [CrossRef] [PubMed]
6. Patil, P.; Kharbanda, O.P.; Duggal, R.; Das, T.K.; Kalyanasundaram, D. Surface deterioration and elemental composition of retrieved orthodontic miniscrews. *Am. J. Orthod. Dentofac. Orthop.* **2015**, *147*, S88–S100. [CrossRef]
7. Mystkowska, J.; Niemirowicz-Laskowska, K.; Łysik, D.; Tokajuk, G.; Dąbrowski, J.R.; Bucki, R. The role of oral cavity biofilm on metallic biomaterial surface destruction–corrosion and friction aspects. *Int. J. Mol. Sci.* **2018**, *19*, 743. [CrossRef]
8. Øilo, M.; Bakken, V. Biofilm and dental biomaterials. *Materials (Basel)* **2015**, *8*, 2887–2900. [CrossRef]
9. Al Qahtani, W.M.S.; Schille, C.; Spintzyk, S.; Al Qahtani, M.S.A.; Engel, E.; Geis-Gerstorfer, J.; Rupp, F.; Scheideler, L. Effect of surface modification of zirconia on cell adhesion, metabolic activity and proliferation of human osteoblasts. *Biomed. Tech.* **2017**, *62*, 75–87. [CrossRef]
10. Carvalho, A.; Pelaez-Vargas, A.; Gallego-Perez, D.; Grenho, L.; Fernandes, M.H.; De Aza, A.H.; Ferraz, M.P.; Hansford, D.J.; Monteiro, F.J. Micropatterned silica thin films with nanohydroxyapatite micro-aggregates for guided tissue regeneration. *Dent. Mater.* **2012**, *28*, 1250–1260. [CrossRef]
11. Laranjeira, M.S.; Carvalho, Â.; Pelaez-Vargas, A.; Hansford, D.; Ferraz, M.P.; Coimbra, S.; Costa, E.; Santos-Silva, A.; Fernandes, M.H.; Monteiro, F.J. Modulation of human dermal microvascular endothelial cell and human gingival fibroblast behavior by micropatterned silica coating surfaces for zirconia dental implant applications. *Sci. Technol. Adv. Mater.* **2014**, *15*, 025001. [CrossRef] [PubMed]
12. Arango-Santander, S.; Pelaez-Vargas, A.; Freitas, S.C.; García, C. Surface Modification by Combination of Dip-Pen Nanolithography and Soft Lithography for Reduction of Bacterial Adhesion. *J. Nanotechnol.* **2018**, *2018*, 10. [CrossRef]
13. Biswas, A.; Bayer, I.S.; Biris, A.S.; Wang, T.; Dervishi, E.; Faupel, F. Advances in top-down and bottom-up surface nanofabrication: Techniques, applications & future prospects. *Adv. Colloid Interface Sci.* **2012**, *170*, 2–27. [PubMed]
14. Arango, S.; Peláez-Vargas, A.; García, C. Coating and surface treatments on orthodontic metallic materials. *Coatings* **2013**, *3*, 1–15. [CrossRef]
15. Weibel, D.B.; DiLuzio, W.R.; Whitesides, G.M. Microfabrication meets microbiology. *Nat. Rev. Microbiol.* **2007**, *5*, 209–218. [CrossRef]
16. Xia, Y.; Whitesides, G.M. Soft Lithography. *Angew. Chemie Int. Ed.* **1998**, *37*, 550–575. [CrossRef]
17. Butler, R.T.; Ferrell, N.J.; Hansford, D.J. Spatial and geometrical control of silicification using a patterned poly-l-lysine template. *Appl. Surf. Sci.* **2006**, *252*, 7337–7342. [CrossRef]
18. Pelaez-Vargas, A.; Gallego-Perez, D.; Fernandes, M.H.; Hansford, D.; Monteiro, F.J. Microstructured coatings to study the behavior of osteoblast-like cells on hard materials. *Bone* **2011**, *48* (supp.2), s106. [CrossRef]
19. Kitzmiller, J.; Beversdorf, D.; Hansford, D. Fabrication and testing of microelectrodes for small-field cortical surface recordings. *Biomed. Microdevices* **2006**, *8*, 81–85. [CrossRef]
20. Pelaez-Vargas, A.; Ferrel, N.; Fernandes, M.H.; Hansford, D.J.; Monteiro, F.J. Cellular Alignment Induction during Early In Vitro Culture Stages Using Micropatterned Glass Coatings Produced by Sol-Gel Process. *Key Eng Mater* **2009**, *396–398*, 303–306. [CrossRef]
21. Ferrell, N.; Woodard, J.; Hansford, D. Fabrication of polymer microstructures for MEMS: Sacrificial layer micromolding and patterned substrate micromolding. *Biomed. Microdevices* **2007**, *9*, 815–821. [CrossRef] [PubMed]
22. Tran, K.T.M.; Nguyen, T.D. Lithography-based methods to manufacture biomaterials at small scales. *J. Sci. Adv. Mater. Devices* **2017**, *2*, 1–14. [CrossRef]
23. Solga, A.; Cerman, Z.; Striffler, B.F.; Spaeth, M.; Barthlott, W. The dream of staying clean: Lotus and biomimetic surfaces. *Bioinspir Biomim* **2007**, *2*, S126–S134. [CrossRef] [PubMed]
24. Koch, K.; Barthlott, W. Superhydrophobic and superhydrophilic plant surfaces: An inspiration for biomimetic materials. *Philos. Trans. R. Soc. A Math. Phys. Eng. Sci.* **2009**, *367*, 1487–1509. [CrossRef]

25. Chung, K.K.; Schumacher, J.F.; Sampson, E.M.; Burne, R.A.; Antonelli, P.J.; Brennan, A.B. Impact of engineered surface microtopography on biofilm formation of *Staphylococcus aureus*. *Biointerphases* **2007**, *2*, 89–94. [CrossRef]
26. Bixler, G.D.; Theiss, A.; Bhushan, B.; Lee, S.C. Anti-fouling properties of microstructured surfaces bio-inspired by rice leaves and butterfly wings. *J. Colloid Interface Sci.* **2014**, *419*, 114–133. [CrossRef]
27. Bhadra, C.M.; Khanh Truong, V.; Pham, V.T.H.; Al Kobaisi, M.; Seniutinas, G.; Wang, J.Y.; Juodkazis, S.; Crawford, R.J.; Ivanova, E.P. Antibacterial titanium nano-patterned arrays inspired by dragonfly wings. *Sci. Rep.* **2015**, *18*, 16817. [CrossRef]
28. Lim, T.K. *Edible medicinal and non-medicinal plants. Modifed stems, roots, bulbs*; Springer: London, UK, 2015; Volume 12.
29. Neinhuis, C.; Barthlott, W. Characterization and distribution of water-repellent, self-cleaning plant surfaces. *Ann. Bot.* **1997**, *79*, 667–677. [CrossRef]
30. Hüger, E.; Rothe, H.; Frant, M.; Grohmann, S.; Hildebrand, G.; Liefeith, K. Atomic force microscopy and thermodynamics on taro, a self-cleaning plant leaf. *Appl. Phys. Lett.* **2009**, *95*, 033702. [CrossRef]
31. Bhushan, B.; Jung, Y.C. Wetting, adhesion and friction of superhydrophobic and hydrophilic leaves and fabricated micro/nanopatterned surfaces. *J. Phys. Condens. Matter* **2008**, *20*, 225010. [CrossRef]
32. Vasudevan, R.; Kennedy, A.J.; Merritt, M.; Crocker, F.H.; Baney, R.H. Microscale patterned surfaces reduce bacterial fouling-microscopic and theoretical analysis. *Colloids Surfaces B Biointerfaces* **2014**, *117*, 225–232. [CrossRef] [PubMed]
33. Hochbaum, A.I.; Aizenberg, J. Bacteria pattern spontaneously on periodic nanostructure arrays. *Nano Lett.* **2010**, *10*, 3717–3721. [CrossRef] [PubMed]
34. Xu, L.C.; Siedlecki, C.A. Submicron-textured biomaterial surface reduces staphylococcal bacterial adhesion and biofilm formation. *Acta Biomater.* **2012**, *8*, 72–81. [CrossRef]
35. Arango-Santander, S.; Pelaez-Vargas, A.; Freitas, S.C.; García, C. A novel approach to create an antibacterial surface using titanium dioxide and a combination of dip-pen nanolithography and soft lithography. *Sci. Rep.* **2018**, *8*, 15818. [CrossRef] [PubMed]
36. Durán, A.; Conde, A.; Gómez Coedo, A.; Dorado, T.; García, C.; Ceré, S. Sol-gel coatings for protection and bioactivation of metals used in orthopaedic devices. *J. Mater. Chem.* **2004**, *14*, 2282–2290. [CrossRef]
37. Schneider, C.A.; Rasband, W.S.; Eliceiri, K.W. NIH Image to ImageJ: 25 years of image analysis. *Nat. Methods* **2012**, *9*, 671–675. [CrossRef]
38. Horcas, I.; Fernández, R.; Gómez-Rodríguez, J.M.; Colchero, J.; Gómez-Herrero, J.; Baro, A.M. WSXM: A software for scanning probe microscopy and a tool for nanotechnology. *Rev. Sci. Instrum.* **2007**, *78*, 013705. [CrossRef]
39. Naghili, H.; Tajik, H.; Mardani, K.; Razavi Rouhani, S.M.; Ehsani, A.; Zare, P. Validation of drop plate technique for bacterial enumeration by parametric and nonparametric tests. *Vet. Res. forum an Int. Q. J.* **2013**, *4*, 179–183.
40. Sfondrini, M.F.; Debiaggi, M.; Zara, F.; Brerra, R.; Comelli, M.; Bianchi, M.; Pollone, S.R.; Scribante, A. Influence of lingual bracket position on microbial and periodontal parameters in vivo. *J Appl Oral Sci.* **2012**, *20*, 357–361. [CrossRef]
41. Türkkahraman, H.; Sayin, M.O.; Bozkurt, F.Y.; Yetkin, Z.; Kaya, S.; Onal, S. Archwire ligation techniques, microbial colonization, and periodontal status in orthodontically treated patients. *Angle Orthod.* **2005**, *75*, 231–236.
42. Hepyukselen, B.G.; Cesur, M.G. Comparison of the microbial flora from different orthodontic archwires using a cultivation method and PCR: A prospective study. *Orthod Craniofac Res.* **2019**, *22*, 354–360. [CrossRef] [PubMed]
43. Yang, H.; Pi, P.; Cai, Z.Q.; Wen, X.; Wang, X.; Cheng, J.; Yang, Z. Facile preparation of super-hydrophobic and super-oleophilic silica film on stainless steel mesh via sol-gel process. *Appl. Surf. Sci.* **2010**, *256*, 4095–4102. [CrossRef]
44. Hosseinalipour, S.M.; Ershad-langroudi, A.; Hayati, A.N.; Nabizade-Haghighi, A.M. Characterization of sol-gel coated 316L stainless steel for biomedical applications. *Prog. Org. Coatings* **2010**, *67*, 371–374. [CrossRef]

45. Santos, O.; Nylander, T.; Rosmaninho, R.; Rizzo, G.; Yiantsios, S.; Andritsos, N.; Karabelas, A.; Müller-Steinhagen, H.; Melo, L.; Boulangé-Petermann, L.; et al. Modified stainless steel surfaces targeted to reduce fouling - Surface characterization. *J. Food Eng.* **2004**, *64*, 63–79. [CrossRef]
46. Wang, M.; Wang, Y.; Chen, Y.; Gu, H. Improving endothelialization on 316L stainless steel through wettability controllable coating by sol-gel technology. *Appl. Surf. Sci.* **2013**, *268*, 73–78. [CrossRef]
47. Herminghaus, S. Roughness-induced non-wetting. *Europhys. Lett.* **2000**, *52*, 165. [CrossRef]
48. Burton, Z.; Bhushan, B. Surface characterization and adhesion and friction properties of hydrophobic leaf surfaces. *Ultramicroscopy* **2006**, *106*, 709–719. [CrossRef]
49. Grewal, H.S.; Cho, I.J.; Yoon, E.S. The role of bio-inspired hierarchical structures in wetting. *Bioinspiration Biomim.* **2015**, *10*, 026009. [CrossRef]
50. Kim, I.H.; Park, H.S.; Kim, Y.K.; Kim, K.H.; Kwon, T.Y. Comparative short-term in vitro analysis of mutans streptococci adhesion on esthetic, nickel-titanium, and stainless-steel arch wires. *Angle Orthod.* **2014**, *84*, 680–686. [CrossRef]
51. Satou, J.; Fukunaga, A.; Satou, N.; Shintani, H.; Okuda, K. Streptococcal Adherence on Various Restorative Materials. *J. Dent. Res.* **1988**, *67*, 588–591. [CrossRef]
52. Busscher, H.J.; van Pelt, A.W.J.; de Boer, P.; de Jong, H.P.; Arends, J. The effect of surface roughening of polymers on measured contact angles of liquids. *Colloids and Surfaces* **1984**, *9*, 319–331. [CrossRef]
53. May, R.M.; Hoffman, M.G.; Sogo, M.J.; Parker, A.E.; O'Toole, G.A.; Brennan, A.B.; Reddy, S.T. Micro-patterned surfaces reduce bacterial colonization and biofilm formation in vitro: Potential for enhancing endotracheal tube designs. *Clin. Transl. Med.* **2014**, *3*, 8. [CrossRef] [PubMed]

© 2020 by the authors. Licensee MDPI, Basel, Switzerland. This article is an open access article distributed under the terms and conditions of the Creative Commons Attribution (CC BY) license (http://creativecommons.org/licenses/by/4.0/).

Article

Active Packaging—Poly(Vinyl Alcohol) Films Enriched with Tomato By-Products Extract

Katalin Szabo [1], Bernadette-Emoke Teleky [1], Laura Mitrea [1,2], Lavinia-Florina Călinoiu [1,2], Gheorghe-Adrian Martău [1,2], Elemer Simon [2], Rodica-Anita Varvara [2] and Dan Cristian Vodnar [1,2,*]

[1] Institute of Life Sciences, University of Agricultural Sciences and Veterinary Medicine, Calea Mănăștur 3–5, 400372 Cluj–Napoca, Romania; katalin.szabo@usamvcluj.ro (K.S.); bernadette.teleky@usamvcluj.ro (B.-E.T.); laura.mitrea@usamvcluj.ro (L.M.); lavinia.calinoiu@usamvcluj.ro (L.-F.C.); adrian.martau@usamvcluj.ro (G.-A.M.)

[2] Faculty of Food Science and Technology, University of Agricultural Sciences and Veterinary Medicine, Calea Mănăștur 3–5, 400372 Cluj–Napoca, Romania; simon.elemer@usamvcluj.ro (E.S.); varvara.anita@yahoo.com (R.-A.V.)

* Correspondence: dan.vodnar@usamvcluj.ro; Tel.: +40-747-341-881

Received: 31 December 2019; Accepted: 3 February 2020; Published: 4 February 2020

Abstract: Active films were prepared from poly(vinyl alcohol) (PVA) blended with itaconic acid (Ia), and with chitosan (Ch), enriched with tomato processing by-products extract (TBE) in order to develop new bioactive formulations for food packaging. The effects of two biopolymers (Ch, Ia) and of the incorporated TBE—containing phenolic compounds and carotenoids—were studied regarding the physical and antimicrobial properties of films; in addition, their influence on the total phenolic content, viscosity, and flow behavior on the film-forming solutions was investigated. The results showed increased physical properties (diameter, thickness, density, weight) of the films containing the TBE versus their control. TBE and Ch conferred significant antimicrobial effects to PVA films toward all the tested microorganisms, whereas the best inhibition was registered against *S. aureus* and *P. aeruginosa*, with a minimum inhibitory concentration of <0.078 mg DW/mL. The Ia-PVA films also exhibited some antibacterial activity against *P. aeruginosa* (2.5 mg DW/mL). The total phenolic content of the film-forming solutions presented the highest values for the TBE and Ch-added PVA samples (0.208 mg gallic acid/100 mL film-forming solution). These results suggest that the PVA + Ch film containing TBE can be used for the development of intelligent and active food packaging materials.

Keywords: food packaging; poly(vinyl alcohol); by-product; carotenoids; itaconic acid; chitosan

1. Introduction

The mission of food packaging is to maintain the quality and safety of food products during storage and transportation by avoiding circumstances such as dangerous microorganisms, external physical forces, chemical compounds, sunlight, volatile permeable compounds, or oxygen; plastic packaging has these properties [1]. The worldwide use of plastics and plastic packaging (~26% of the total plastics) represents an indispensable element of the global economy, with an overall financial worth of 260 billion USD in 2013 and with an increase in the estimated industrial production from 78 million tons in 2013 to 350 million tons in 2017 [2,3]. Although plastics present valuable functional advantages, like low cost, versatile design, and light weight, it also has a number of negative features, such as freshwater pollution [4] and ocean pollution [5]. Over 90% of plastics are made from fossil resources, which have a high impact on greenhouse gases (GHGs) and other serious adverse effects on the environment [6,7].

The European Commission disclosed the "European Strategy for Plastics in a Circular Economy" at the beginning of 2018, with a specific highlight on plastic production, reuse, and recycling. The main priority of the strategy is to abandon the use of fossil resources and to reduce GHG emissions through plastic manufacturing [8]. The goals of "Plastics 2030 Voluntary Commitment" are to achieve 60% recycling, reduction, and/or reuse of plastic until 2030, and 100% by 2040 [9].

Bio-based materials help to reduce the use of non-renewable feedstock, which has decreased negative environmental effects (GHG emission) [10]. An efficient carbon-neutral alternative is the use of renewable biomass-wastes for biochemical production with efficient recyclability (mechanically, chemically, or through microbial degradation) [11]. The principal aspect is the replacement of fossil resources, but taking into consideration another significant aspect like biodegradability [12]. The characterization of biodegradability refers to thorough degradation to water and CO_2 with the help of different types of microorganisms like fungi, bacteria, and algae [13,14].

Poly(vinyl alcohol) (PVA), presented in Figure 1A, is a polymer that has a backbone comprising only carbon atoms. PVA is water-soluble and biodegradable under both aerobic and anaerobic conditions. Indeed, the physical characteristics of PVA are deeply related to its method of preparation by complete or partial hydrolysis of polyvinyl acetate. Therefore, PVA can be classified according to the degree of hydrolysis: Fully hydrolyzed and partially hydrolyzed, and the partially hydrolyzed PVA is known to be used in foods [15]. For film development, PVA can be used due to its characteristics of having good film-forming capacity, complete biodegradability, crystal modulus, and wide-ranging crystallinity [16]. In addition, PVA is used in the food industry, as a binder, thickener, and/or stabilizer [1,17].

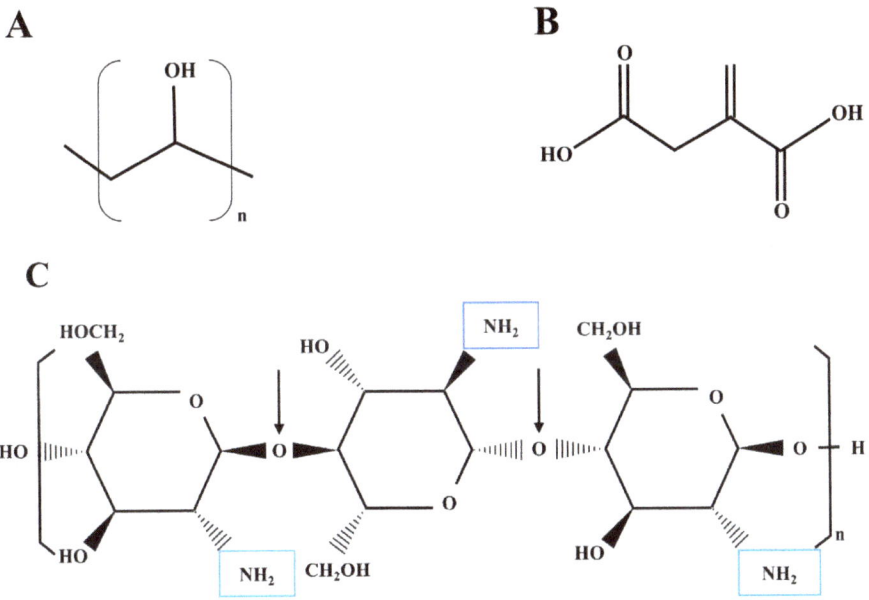

Figure 1. Chemical structures of poly(vinyl alcohol) (PVA) (**A**), itaconic acid (**B**), and chitosan (**C**).

Renewable biomaterials (like tomato processing by-products) are green options to reduce environmental pollution and waste formation [18,19]. In recent years, remarkable and creative ways of utilizing tomato by-products have delivered a continuous development for sustainable bioeconomy and biotechnology [20–22]. Tomatoes (*Solanum lycopersicum*), with an annual production of approximately 180 million tons (FAOSTAT, 2017), are one of the most popular vegetables worldwide. The processing of tomatoes for various foods such as sauces, ketchup, or juice generates significant quantities of by-products. About a quarter of total tomato production is subject to processing, which means that tomato peels, seeds, and small quantities of pulp are removed; these by-products can add around 5–30% of the main product [23–25]. Phytochemicals found in industrial tomatoes and their by-products have been shown to include valuable compounds such as carotenoids, polyphenols, tocopherols, terpenes, and sterols. These bioactive compounds seem to resist industrial treatments [24,26–28].

As tomato peels have lycopene, β-carotene, lutein, and various phenolic compounds in their composition, they could be a source of natural bioactive molecules, applicable in the food industry, especially for bioactive packaging. In addition, the methanolic extracts of tomato peels showed antimicrobial activity against *Staphylococcus aureus* and *Bacillus subtilis* [18].

Itaconic acid (Ia), presented in Figure 1B, is one of the most favorable platform chemicals with relevant usage as a monomer or in monomer synthesis [29]. Due to versatile applications, the US Department of Energy included this bio-based unsaturated dicarboxylic acid in the first twelve building block chemicals obtainable from lignocellulose biomass [30]. Biotechnologically, this 5-C dicarboxylic acid is produced through fermentation of biomass with fungi like *Aspergillus terreus* and *Ustilago maydis* or with different metabolically engineered bacteria [31]. The production of Ia (approx. 41,000 tons/year), as a consequence of increased requests for bio-based materials, is in constant growth and is anticipated to surpass 216 million USD in 2020. Ia and its derivatives are efficiently used in the production of various innovative polymers and present an essential organic acid for manufacturing active packages in food applications [32]. Although Ia-based polymers present various applicabilities, the study of Ia incorporation in active packaging for the food industry is not well studied yet. However, the production of active packaging through the use of Ia presents an efficient solution for the incorporation and release of bioactive molecules with antimicrobial properties that can be used as food preservatives [33].

The use of existing molecules in nature (biopolymers or by-product extracts) to create active packaging is a future method for protecting food, stabilizing and protecting compounds against degradation or against different environmental factors. Biopolymers from different sources have been studied in recent years for food, biomedical, and pharmaceutical applications [34–36].

The global biopolymer market is expected to reach around 10 billion USD by 2021, increasing by almost 17% over the forecast period 2017–2021 [37]. Western Europe has the largest market segment comprising 41.5% of the global market [35]. This development is due to the increasing use of biopolymers in various fields. Chitosan, presented in Figure 1C, is a non-toxic, biocompatible, and biodegradable biopolymer [38]. It can be used in the food industry, in applications for food packaging, for coating fresh and cut fruits or vegetables, and for the pharmaceutical industry as a microencapsulating agent or drug coating [39–44]. Numerous food components have been successfully protected, such as vitamins, antioxidants, enzymes, and minerals, due to chitosan's characteristics [45–50]. The observed antimicrobial activity of chitosan in numerous studies has led to the creation of biodegradable packaging or labels [51–54].

With increasing demand by consumers for safe and minimally processed food and a modern distribution system that requires a proper shelf life, active packaging is a viable option. The choice of active component must correspond to several factors, such as economy, safety, availability, and naturality. Herbal extracts have been widely used in the food industry as a substitute for synthetic additives. They are preferred for film development, mainly for their natural origin and phytochemical properties. The interaction between plant extracts and biopolymers could also influence the physico-chemical and techno-functional properties of the polymer [17].

The present paper aimed to investigate the antimicrobial potential of active films prepared from poly(vinyl alcohol) (PVA) with itaconic acid (Ia), and with chitosan (Ch), enriched with tomato processing by-products extract (TBE) in order to develop new formulas for active food packaging. The effects of two biopolymers (Ch, Ia) and of TBE, containing phenolic compounds and carotenoids, were studied on the physical and antimicrobial properties and the phenolic composition of films, as well as their influence on the viscosity and flow behavior on the film-forming solutions.

2. Materials and Methods

2.1. Materials

Acetic acid, acetonitrile, itaconic acid, chitosan, ethanol, Folin Ciocalteu reagent, and other reagents implied in the experiments were of analytical grade, purchased from Sigma-Aldrich (Steinheim, Germany). PVA with high molecular weight, 98%–99% hydrolyzed, was purchased from ThermoFisher (Kandel, Germany). Carotenoid standards (β-carotene), as well as chemicals used for antimicrobial assays, Mueller–Hinton agar, Mueller–Hinton broth, peptone special, tryptic soy broth, and resazurin, were purchased from BioMerieux (Marcy l'Etoile, France).

2.2. Preparation and Characterization of Tomato Waste Extracts

2.2.1. Ultrasound-Assisted Extraction (UAE)

By-products derived from tomato processing (tomato peels, seeds, and small amounts of pulp), were dehydrated in the dark, to avoid carotenoid loss, for 48 h. Following dehydration, the by-products were finely ground and extracted with ethanol 98% (1:5, w/v) in an ultrasonic unit for 30 min. The extracts were separated from the solid phase in a centrifuge at 10,000 rpm, and the supernatant was filtered through a Millipore 0.45 μm. The obtained tomato by-product extract, further named TBE, was subjected to high-performance liquid chromatography (HPLC) coupled to mass spectrometry (MS) in order to determine the total and individual carotenoids and phenolic contents.

2.2.2. Quantitative and Qualitative Analysis of Carotenoids (Lycopene, B-Carotene, and Lutein) by HPLC/DAD

The TBE was injected into an Agilent 1200 HPLC system with a Diode Array Detector (Agilent Technologies, Santa Clara, CA, USA), and individual carotenoids were determined using a reversed-phase EC 250/4.6 Nucleodur 300-5 C-18 ec. Column (250 mm × 4.6 mm) of 5 μm (Macherey-Nagel, Düren, Germany). Mobile phase A consisted of a mixture of acetonitrile:water (9:1, v/v) with 0.25% triethylamine, and mobile phase B was formed by ethyl acetate with 0.25% triethylamine. The gradient started with 90% mobile phase A at 0 min decreased to 50% A at 10 min and finalized with 10% mobile phase A at 20 min. The flow rate was 1 mL/min, and the chromatograms were monitored at 450 nm, common to carotenoids. The peaks were identified using carotenoid standards (lycopene, β-carotene, and lutein), and quantified on the basis of the calibration curve of a β-carotene standard.

2.2.3. Qualitative and Quantitative Analysis of Phenolic Compounds by HPLC-DAD-ESI-MS

Total and individual phenolic compounds of the TBE were determined on an HPLC-DAD-ESI-MS system consisting of an Agilent 1200 HPLC with a DAD detector, coupled to an MS-detector single-quadrupole Agilent 6110. The separation of phenolic compounds was performed at 25 °C on an Eclipse column, XDB C18 (4.6 mm×150 mm, 5 mm), with a binary gradient. Mobile phase A consisted of a mixture of 0.1% acetic acid:acetonitrile (99:1) in distilled water (v/v) and mobile phase B was formed by 0.1% acetic acid in acetonitrile (v/v), following the elution program used by Dulf et al. (2015) [55], at a flow rate of 0.5 mL/min. For MS fragmentation, the ESI (+) module was used with a capillary voltage of 3000 V, a nitrogen flow of 8 L/min, and a scanning range situated between 100 and 1000 m/z at 350 °C.

The phenolic compounds were monitored by DAD, and the absorption spectra (200–600 nm) were collected continuously during each run. Data analysis was performed using Agilent ChemStation Software (Rev B.04.02 SP1, Palo Alto, CA, USA).

2.3. Film Preparation

First, three different PVA concentrations were tested for polymerization (1, 2, and 3% wt.), and the optimum ratio for further experiments was selected to be PVA 3% wt. The PVA solution (3% wt./v) was prepared in hot distilled water (90 °C) under continuous stirring for 2 h. Further, the polymer films were prepared by the solvent casting technique.

PVA + Chitosan (PVA + Ch) solution (1% v/wt.) was prepared by dissolving chitosan in acidified PVA solution at 80 °C with continuous stirring (500 rpm) for two hours.

PVA + Itaconic acid (PVA + Ia) solution (1% v/wt.) was prepared by adding itaconic acid to the PVA solution at 65 °C with continuous stirring for 30 min.

To the prepared PVA + Ch and PVA + Ia film-forming solutions, we added 9% v/v TBE under continuous stirring (500 rpm) at room temperature, and the resulting samples were subjected to antimicrobial capacity tests, rheological measurements, and total phenolic content determination.

Further, the TBE-added film-forming solutions together with control PVA solutions (15 mL) were cast in plastic Petri dishes (8.5 cm) and evaporated at room temperature (21 °C) for 48 h. The obtained films were peeled off and stored for further investigations.

2.4. Shear Viscosity Measurement of Film Solutions

To examine the influence of tomato extract, Ch, and Ia on PVA viscosity, the viscosity flow behavior of the PVA, PVA + TBE, PVA + Ia + TBE, and PVA + Ch + TBE solutions was measured through an Anton Paar MCR 72 rheometer (Anton Paar, Graz, Austria), equipped with a concentric cylinder system and a double-gap 42 measuring system (temperature range 5 °C to 150 °C). For each sample solution, approx. 15 mL was poured in the double-gap system of the rheometer. Each measurement was operated at 21 °C, 30 °C, and 37 °C with a shear rate of 5–300 (logarithmic ramp) and determination value in 0–100 s^{-1} (logarithmic ramp).

2.5. Total Phenolics

Determination of the total phenolic content was made by the Folin–Ciocalteu assay, which estimates the total content of all phenolic compounds present in the analyzed samples and measures the total reducing capacity of a sample. In order to test their reducing power, aliquots of 25 µL PVA + TBE, PVA + Ia + TBE, and PVA + Ch + TBE samples were mixed with 1.8 mL distilled water in a 24well microplate. An aliquot of 120 µL of Folin–Ciocalteu reagent was added to the wells and mixed, followed by the addition of 340 µL Na_2CO_3 (7.5% wt./v) solution, to create basic conditions for the redox reaction between phenolic compounds and Folin–Ciocalteu reagent. After incubation in the dark for 90 min at room temperature, the absorbance was read at 750 nm using a microplate reader (BioTek Instruments, Winooski, VT, USA). The analysis was repeated three times and the results were expressed as gallic acid equivalent/100 mL sample.

2.6. Antimicrobial Activity of the Film-Forming Solutions

For this bioassay, five bacterial strains were used: *Staphylococcus aureus* (ATCC 49444), *Pseudomonas aeruginosa* (ATCC 27853), *Salmonella enterica* serovar Typhimurium (ATCC 14028) and serovar Enteritidis (ATCC 31194), and *Escherichia coli* (ATCC 25922). The microorganisms were obtained from the Food Biotechnology Laboratory of the University of Agricultural Sciences and Veterinary Medicine Cluj-Napoca, Romania. The strains were cultured on Mueller–Hinton agar, stored at 4 °C, and sub-cultured once a month.

To evaluate the antimicrobial potential of the film-forming solutions, the modified microdilution technique was used, described previously by Vodnar et al. (2017) [56]. In a 96-well plate, fresh overnight cell suspensions of the tested microorganisms were adjusted with sterile saline solution to a concentration of approximately 2×10^5 CFU/mL in a final volume of 100 µL per well. The inoculum was stored at 4 °C until further use. Further, serial dilutions of the film-forming solutions were attained in the wells containing 100 µL of nutrient broth. Afterward, 10 µL of inoculum was added to all the wells. The microplates were incubated for 24–48 h at 37 °C.

The minimum inhibitory concentration (MIC) was determined as the lowest concentration of the film-forming solutions (mg DW/mL), inhibiting the visual growth of the tested microorganisms on the microplate. The MIC of the samples was determined after the addition of 20 µL (0.2 mg/mL) of resazurin solution to each well, as a color reactive (intense blue), and the plates were incubated for 2 h at 37 °C. A color change from blue to pink indicated the reduction of resazurin and, therefore, bacterial growth. The final concentrations of the tested film-forming solutions were approximately 27.3 mg DW/mL for PVA, 9.1 mg DW/mL for Ch/Ia, and 0.5 mg DW/mL for TBE.

In order to interpret the results, we applied the guideline described by O'Donnell et al. (2010) [57], where the antimicrobial activity of the tested compounds is defined as follows: No bioactivity (MIC > 1000 µg/mL), mild (MIC = 501–1000 µg/mL), moderate (MIC = 126–500 µg/mL), good (MIC = 26–125 µg/mL), strong (MIC = 10–25 µg/mL), and very strong bioactivity (MIC < 10 µg/mL).

2.7. Solid Film Characterization

2.7.1. Solid Film Measurements

A ruler was used to measure the diameter of the obtained films, while the thickness was measured with a digital caliper. The weight of the film samples was measured with an analytical balance with 0.1 mg precision. The density of the films was established by using the following equation:

$$\text{Density (g/cm}^3) = m/s \cdot t \tag{1}$$

where m is the mass of sample in grams; s is the surface of the sample in cm^2; and t is the thickness of sample in cm.

All measurements were performed at room temperature (21 °C).

2.7.2. Fourier-Transform Infrared analysis

The solid samples were analyzed using Fourier-transform infrared (FTIR, Shimadzu IR Prestige-21, Kyoto, Japan) equipped with an attenuated total reflectance (ATR) module with a single reflection from PIKE provided with the press. Films were applied directly to the horizontal accessory after which infrared absorption spectra were recorded in the wavelength range of 600–4000 cm^{-1}, at a resolution of 4 cm^{-1}, and with 64 scans for one spectrum. The spectral data obtained were processed using the IR Solution Software program Overview (Shimadzu) and OriginR 7SR1 Software (OriginLab Corporation, Northampton, UK).

2.8. Statistical Analysis

The measurements regarding physical properties and total phenolic contents were made in triplicate and the values are expressed as mean values ±SD, $n = 3$. Only the variation in total phenolic content of the studied film-forming solutions was analyzed by one-way ANOVA, and the differences were examined by Tuckey's multiple comparison test ($p < 0.05$). Different letters on this particular chart indicate significant differences ($p < 0.05$) between the tested films (GraphPad Prism Version 8.0.1, Graph Pad Software, Inc., San Diego, CA, USA).

3. Results and Discussion

3.1. Characterization of Tomato By-Products Extracts

First, 11 bioactive compounds were identified in TBE belonging to the carotenoids and phenolic groups. From the carotenoids class, three main compounds were found: Lutein, lycopene, and β-carotene, while from the phenolics class, eight compounds were registered: Caffeic acid-glucoside isomer (CG), 5-caffeoylquinic acid (5-CQA), quercetin-diglucoside (QdiG), quercetin-glucuronide (QGl), quercetin-3-rutinoside (Q3R), di-caffeoylquinic acid (di-CQA), tri-caffeoylquinic acid (tri-CQA), and naringenin chalcone (NGC). The individual carotenoids and phenolic compounds identified in TBE's composition are illustrated in Figure 2 via HPLC chromatograms.

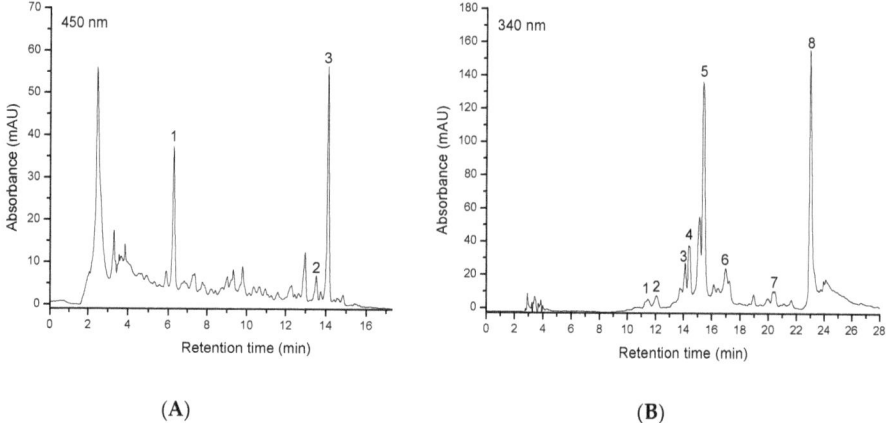

Figure 2. HPLC chromatograms of ethanolic tomato by-products extract for carotenoids (A) and for phenolic compounds (B).

Data regarding retention time, the wavelength along the absorption spectrum, or the mass to charge ratio (m/z) of the identified components are presented in Table 1.

Table 1. Identification of carotenoids and phenolic compounds in tomato processing by-products extract (TBE).

Class of Compounds	Peak No.	R_t (min)	λ_{max} (nm)	$[M + H]^+$ (m/z)	Compound
Carotenoids	1	6.41	448, 474		Lutein
	2	13.42	446, 473		Lycopene
	3	14.51	455, 480		β-Carotene
Phenolics	1	11.12	292, 245	343	Caffeic acid-glucoside isomer
	2	11.93	326, 248	355	5-Caffeoylquinic acid
	3	14.09	355, 259	627	Quercetin-diglucoside
	4	14.87	355, 259	478	Quercetin-glucuronide
	5	15.34	354, 256	611	Quercetin-3-rutinoside
	6	16.98	328, 250	517	Di-Caffeoylquinic acid
	7	20.33	328, 250	679	Tri-Caffeoylquinic acid
	8	23.10	366, 250	273	Naringenin chalcone

The content of each carotenoid compound identified in TBE is presented in Table 2, where β-carotene had the highest content, followed by lutein. Considering that the aim of the study is to find new formulas for food packaging and/or biomedical applications, a food-grade solvent was used to extract the bioactive components from the tomato processing by-products, in one-step, which explains the lower concentration of carotenoids compared to earlier findings. Previous studies reported a lycopene content in the range from 14.9 to 28.3 mg/100 g DW; however, the used protocol was tailored for carotenoid extraction by organic solvents [58].

Table 2. Individual carotenoids content of TBE expressed in mg/100 DW (β-carotene equivalent).

Peak	Carotenoids	mg/100 DW
1	Lutein	1.549 ± 0.04
2	Lycopene	0.127 ± 0.01
3	β-Carotene	1.597 ± 0.01
	Sum of identified carotenoids	3.273 ± 0.05

The carotenoids content, 3.273 mg/100g DW, can increase the biological activity potential of TBE-based films considering their previously reported antioxidant and antimicrobial properties. In the study of Strati, Gogou, and Oreopoulou, the total carotenoids reported in tomato by-products extracts were in the range from 0.36 to 16.52 mg/100 g DW, considering the existing differences in the extraction method and tomato varieties [59].

The individual phenolic contents of ethanolic TBE are reported in Table 3. Naringenin chalcone, belonging to the flavonoid class, was the major phenolic compound identified, followed by quercetin-glucuronide (flavonol sub-class).

Table 3. Individual phenolic compounds content of TBE, expressed in mg/100 g DW (equivalents of chlorogenic acid and rutin).

Peak	Phenolic Compounds	mg/100 DW
1	Caffeic acid-glucoside isomer	2.284 ± 0.02
2	5-Caffeoylquinic acid	2.623 ± 0.02
3	Quercetin-diglucoside	7.150 ± 0.05
4	Quercetin-glucuronide	24.427 ± 0.04
5	Quercetin-3-rutinoside	6.061 ± 0.03
6	Di-Caffeoylquinic acid	2.092 ± 0.02
7	Tri-Caffeoylquinic acid	1.782 ± 0.01
8	Naringenin chalcone	34.178 ± 0.02
	Sum of identified phenolics	80.596 ± 0.20

3.2. Shear Viscosity Measurement of Film-Forming Solutions

The semi-crystalline polymer PVA consists of a hydroxyl group (–OH) that generates intra- and intermolecular hydrogen bonding that has a strong influence on the mechanical and rheological characteristics of the polymer solutions [60]. Figure 3A–C present the relationship between viscosity and the shear rate of the PVA-based biopolymers at various temperatures.

The films PVA + Ch and PVA + Ch + TBE for the investigated shear rate range presented pseudo-plastic (shear-thinning) behavior, and as can be seen in the figures (Figure 3A–C), for all three temperatures, the viscosity decreased slowly with intensifying shear rate, which is a typical feature of polymer solutions. These pseudo-plastic materials at increasing shear rate have reduced shear viscosity, due to the alignment of the large molecular chains in the course of higher shear rate, which generates lower resistance [61,62]. However, with the increase in temperature, the viscosity presented higher values, and at higher shear rates, the viscosity also decreased exponentially.

The other films (PVA, PVA + TBE, PVA + Ia, PVA + Ia + TBE) presented a dilatant (shear-thickening) behavior. These dilatant materials with increasing shear rate have increased viscosity, but at shear rates higher than 30 s^{-1}, the viscosity remained generally constant. The addition of Ia to the biofilms at 37 °C increased the viscosity, and at a temperature of 21 °C, they presented a lower viscosity. The simple biofilms, which contained only PVA and PVA + TBE (small molecular-weight compounds) presented inverse viscosities. The addition of TBE did not influence the viscosity of the biofilms.

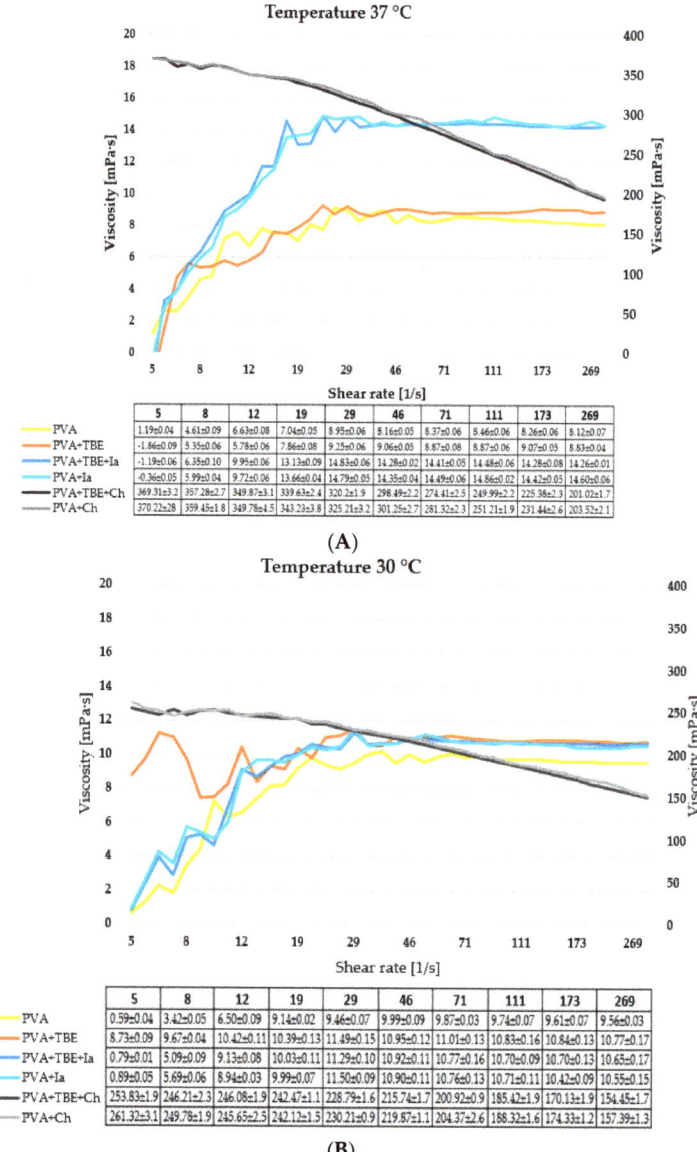

Figure 3. Cont.

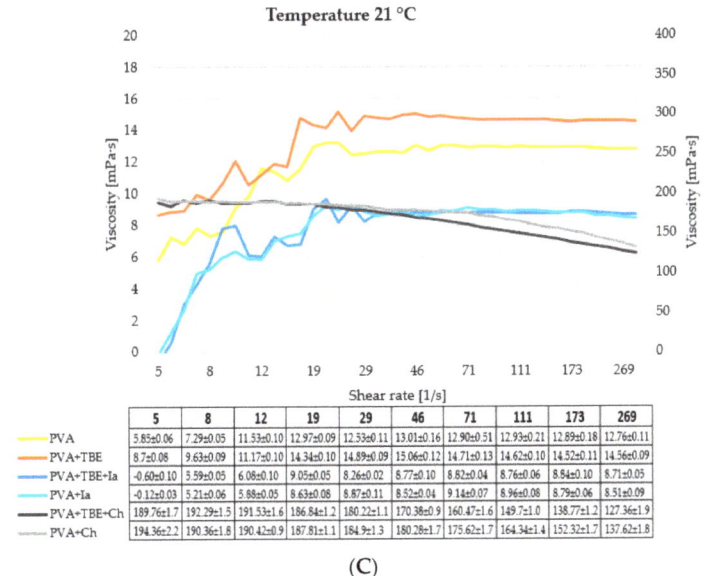

	5	8	12	19	29	46	71	111	173	269
PVA	5.85±0.06	7.29±0.05	11.53±0.10	12.97±0.09	12.53±0.11	13.01±0.16	12.90±0.51	12.93±0.21	12.89±0.18	12.76±0.11
PVA+TBE	8.7±0.08	9.63±0.09	11.17±0.10	14.34±0.10	14.89±0.09	15.06±0.12	14.71±0.13	14.62±0.10	14.52±0.11	14.56±0.09
PVA+TBE+Ia	−0.60±0.10	3.59±0.05	6.08±0.10	9.05±0.05	8.26±0.02	8.77±0.10	8.82±0.04	8.76±0.06	8.84±0.10	8.71±0.05
PVA+Ia	−0.12±0.03	5.21±0.06	5.88±0.05	8.63±0.08	8.87±0.11	8.52±0.04	9.14±0.07	8.96±0.08	8.79±0.06	8.51±0.09
PVA+TBE+Ch	189.76±1.7	192.29±1.5	191.53±1.6	186.84±1.2	180.22±1.1	170.38±0.9	160.47±1.6	149.7±1.0	138.77±1.2	127.36±1.9
PVA+Ch	194.36±2.2	190.36±1.8	190.42±0.9	187.81±1.1	184.9±1.3	180.28±1.7	175.62±1.7	164.34±1.4	152.32±1.7	137.62±1.8

(**C**)

Figure 3. Relationship between viscosity and shear rate at 37 °C (**A**), 30 °C (**B**), 21 °C (**C**) for PVA, PVA + TBE, PVA + TBE + itaconic acid (Ia), and PVA + TBE + Ch.

3.3. Total Phenolic Content (TPC)

Phenolic compounds are plant-derived metabolites. According to the literature, they have significant –OH structures for scavenging free radicals and, therefore, an important antioxidant capacity [21]. The PVA, PVA + Ia, and PVA + Ch films (without any added TBE) contained no phenolic compounds; therefore, data are not shown on the chart for these particular films. The total phenolic content of films containing TBE was in the range of 0.167–0.208 mg gallic acid/100 mL film-forming solution, the results of which are illustrated in Figure 4. The testing temperature was constant (room temperature). The content of phenolics determined in the film-forming solutions differed significantly according to the type of biopolymer used. The PVA + Ch films containing TBE registered the highest TPC, namely, 0.208 mg gallic acid/100 mL film-forming solution, being significantly different from the other two types of films.

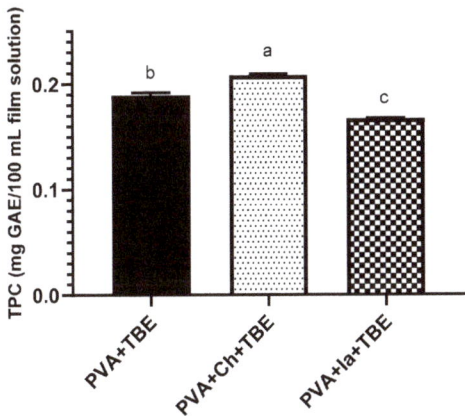

Figure 4. Total phenolic content (Folin–Ciocalteu method).

The total phenolic content of film-forming solutions may deliver important antioxidant activity, but was not tested in the present paper considering the focus of choosing the best candidate with the highest antimicrobial property for active packaging. Arcan and Yemenicioglu (2011) reported a significant antioxidant activity of zein films enriched with phenolic compounds such as catechin, gallic acid, p–hydroxybenzoic acid, and ferulic acid, justified by the existance of soluble phenolics in the film [63].

The total phenolic content of the film-forming solutions is expressed as gallic acid equivalents (GAE) in mg/100 mL sample. Values are expressed as mean values ±SD, $n = 3$, and are followed by different letters (a–c), indicating significant differences ($p < 0.05$) between the type of films (one-way ANOVA - Multiple comparison test - Tukey multiple range test ($p = 0.05$); GraphPad Prism Version 8.0.1, Graph Pad Software, Inc., San Diego, CA, USA).

3.4. Antimicrobial Activity

The antimicrobial properties of active packaging can be achieved by direct incorporation of natural antimicrobial compounds, with a high interest being the agro-industrial waste-derived bioactive compounds. The antimicrobial activity of the PVA, PVA + Ia, and PVA + Ch films, all containing TBE, and against the selected microorganisms, is presented in Table 4 below.

Table 4. Minimum inhibitory concentration (mg dry weight/mL) against *S. aureus*, *E. coli*, *P. aeruginosa*, *S. enterica* Enteritidis, and *S. enterica* Typhimurium.

Samples	G (+) Bacteria	G (−) Bacteria			
	S. aureus	*E. coli*	*P. aeruginosa*	*S. enterica* Enteritidis	*S. enterica* Typhimurium
PVA	n.b.	n.b.	n.b.	n.b.	n.b.
PVA + TBE	n.b.	n.b.	n.b.	n.b.	n.b.
PVA + Ch + TBE	<0.078	5.00	<0.078	0.312	5.00
PVA + Ch	0.156.	10	0.156	0.624	10
PVA + Ia + TBE	n.b.	n.b.	2.5	n.b.	n.b.
PVA + Ia	n.b.	n.b.	5.00	n.b.	n.b.

n.b.—no bioactivity.

In accordance with the guideline described by O'Donnell et al. (2010) [57], the PVA + Ch films containing TBE showed good antibacterial activity toward *S. aureus* and *P. aeruginosa*, with a MIC of <0.078 mg DW/mL, while toward *S. enterica* Enteritidis, they showed a moderate antimicrobial capacity, with a MIC of 0.312 mg DW/mL, and no bioactivity against the other tested strains were exhibited. The TBE considering its carotenoids and phenolics content, previously reported with good antimicrobial capacity [18], may significantly contribute to this antibacterial capacity. The PVA + Ch film registered a moderate antibacterial effect toward *S. aureus* and *P. aeruginosa* strains, with a MIC of 0.156 mg DW/mL, and a mild inhibition capacity toward *S. enterica* Enteritidis. The precise mechanism of Ch antibacterial capacity is not fully elucidated. The main explanation relies on the interaction between its positively charged amino group and the negatively charged microbial cell membranes, the interaction of which contributes to the leakage of proteinaceous and other intracellular constituents of the bacteria [64]. The PVA had no antimicrobial capacity, in line with a previously reported PVA-based film study [65]. The PVA + Ia film containing TBE registered a MIC of 2.5 mg DW/mL, and the PVA + Ia without TBE registered a MIC of 5 mg DW/mL against *P. aeruginosa*; therefore, both had no antibacterial effect. This fact could be explained by the low concentration of Ia used, precisely 1%, insufficient to exert its previously reported antimicrobial characteristics [66,67]. Both the inherent antibacterial characteristics and film forming capacity of Ch make it the best candidate for use as an active packaging material [68,69].

This enhanced antibacterial activity of TBE-containing films could be explained by the important antibacterial capacity of TBE previously reported in our study [58]. Other researchers have also found an enhanced antibacterial capacity of Ch-based films by incorporation of natural extracts, whereas the present study supports the idea of using the agro-industrial by-products extracts, precisely, the industrially derived tomato seeds and peels, for increasing their value and range of applications. Mathew and Abraham (2008) [70] showed that the shelf-life of food has been extended by ferulic acid-incorporated starch-chitosan blend films. In addition, the incorporation of garlic oil and nisin enhanced the antimicrobial activity of chitosan film [71].

3.5. Physical Characterization of Solid Films

The physical appearance of the films is illustrated in Figure 5. Six types of films were obtained and presented as PVA (control), PVA + Ia, PVA + Ch, PVA + TBE, PVA + Ia + TBE, and PVA + Ch + TBE. In addition, 15 mL of the PVA-based film solutions entirely covered the surface of the Petri dishes, and when solidified, the films easily peeled off the plates. Each film type was visually homogeneous with no bubbles inside. From all the obtained films, the PVA film was perfectly transparent compared to the other sample films, and it was the most fragile one. The addition of Ia, Ch, and TBE contributed substantially to the resistance of the PVA films, and at the same time, to the opacity and color of the films. Itaconic acid and chitosan reduced the transparency of the PVA films and conferred their rigidity by making them brittle. The addition of TBE reduced the films' rigidity induced by Ia and Ch by imprinting them with a slight elasticity. Considering the color of the films, TBE with a concentration of 9% imprinted yellow colors of different tones to all tested films. Similar studies point out that vegetal extracts such as mint extract or pomegranate peel extract might imprint color to the PVA-based films even at a low concentration of 0.1% [65]. The unevenness of the PVA films color might be associated with the different distribution of pigments existing in TBE (lutein, β-carotene) due to their different molecular mass, because of their Brownian motion, and because of colloidal interactions, until the films are completely solidified [72].

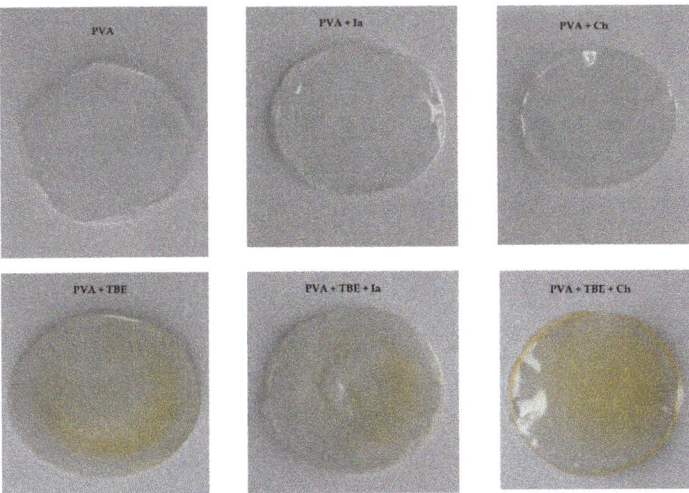

Figure 5. The physical appearance of the obtained films.

Considering the physical measurements (Table 5), all the obtained films had a diameter of less than 8.5 cm, a weight above 0.4 g, a thickness over 0.3 mm, and a density higher than 2.4 g/cm^3. The largest diameter was registered for PVA film (8.33 mm ± 0.15 mm), while the addition of bioactive compounds (Ia, Ch, and TBE) narrowed down the films' diameter, making them dense and compact.

As we expected, the addition of bioactive constituents increased the weight, thickness, and density, with the PVA + Ch + TBE films recording the highest values for these parameters.

Table 5. The physical measurements of the solid PVA-based films; the shown data represent the mean value of three replicates with the standard deviation (±).

Sample	Diameter (cm)	Weight (g)	Thickness (mm)	Density (g/cm^3)
PVA	8.33 ± 0.15	0.44 ± 0.00	0.03 ± 0.01	2.40 ± 0.79
PVA + TBE	7.95 ± 0.07	0.46 ± 0.00	0.04 ± 0.01	3.70 ± 1.35
PVA + Ia	8.27 ± 0.06	0.61 ± 0.04	0.04 ± 0.01	4.58 ± 1.33
PVA + Ia + TBE	7.87 ± 0.06	0.61 ± 0.02	0.04 ± 0.01	4.63 ± 1.44
PVA + Ch	8.00 ± 0.00	0.57 ± 0.01	0.06 ± 0.01	6.79 ± 1.08
PVA + Ch + TBE	8.07 ± 0.12	0.64 ± 0.03	0.06 ± 0.01	7.55 ± 1.40

FTIR is a useful method to identify specific molecular structures or the vibration of functional groups from certain matrices [73,74]. For the present study, FTIR analysis was performed for PVA-based films supplemented with TBE (Figure 6) in order to observe the influence of tomato waste-based extract on the PVA formulations. For all investigated samples, a large band with a moderate height around 3300 cm^{-1} was observed, vibrations that were attributed to the presence of inter- and intramolecular hydrogen-bonded –OH groups from the PVA matrix [75]. Peaks of different heights situated between 2800 and 3000 cm^{-1} were observed for all tested films (2906 and 2939 cm^{-1} for PVA, 2908 and 2939 cm^{-1} for PVA + TBE, 2910 and 3022 for PVA + Ia, 2854 and 2922 for PVA + Ia + TBE, and 2854 and 2924 cm^{-1} for PVA+Ch and PVA + Ch + TBE). Their presence was connected with vibrations of C–H bonding specific to polyalcohols, as the scientific literature suggests [74–76]. Moderate peaks associated with –C–C–C– vibrations were identified around 1400 cm^{-1} (at 1417 cm^{-1} for PVA, PVA + TBE, and PVA + Ia + TBE; and at 1411 cm^{-1} for PVA + Ch + TBE) [75]. The large peak situated around 1000 cm^{-1} identified in all samples (1087 cm^{-1} for PVA and PVA + Ia; 1085 cm^{-1} for PVA + TBE and PVA + Ia + TBE; 1083 cm^{-1} for PVA + Ch and PVA + Ch + TBE) emphasized the presence of C–O–C bonds from PVA [75].

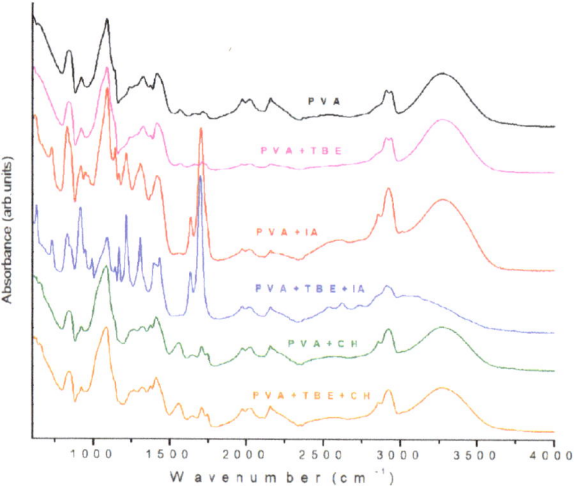

Figure 6. FTIR spectra of the solid films.

The spectra obtained for PVA + Ia + TBE and PVA + Ch + TBE films showed few differences toward the simpler matrices like PVA and PVA + TBE. The film containing itaconic acid showed sharp peaks at 1633 and 1699 cm^{-1}, peaks that can be linked to the presence of C=O bonds, specific to carboxylic groups [74,77]. The spectra obtained for the film containing chitosan showed absorbance at a wavelength of 1411 cm^{-1}, which is specific to C=N bonds [78]. In the case of PVA spectra, it showed a peak at 2160 cm^{-1}, which indicates the presence of C=O bonds [75], this particular peak being missing from the rest of the formulations that contained TBE. The incorporation of tomato waste-based extract into the PVA film solutions seemed to attenuate the presence of these aliphatic carbonyl groups.

4. Conclusions

Due to consumers' preference for minimally processed and natural products, active packaging is gaining importance; however, bio-packaging films generally exhibit weak mechanical attributes and high water solubility and permeability compared to commercially available polymers. In the present study, films containing poly(vinyl alcohol) (3% wt./v) and chitosan (1% wt./v) were enriched with TBE and showed good antibacterial activity toward *S. aureus* and *P. aeruginosa*, with a MIC of <0.078 mg DW/mL, also having important antimicrobial effects toward all the tested strains. The addition of Ia, Ch, and TBE significantly contributed to the resistance of the PVA films and modified the opacity and color of the films at the same time. Itaconic acid and chitosan reduced the transparency of the PVA films and conferred rigidity by making them brittle, and the addition of TBE imprinted a slight elasticity to the films. The rheological characteristics of the studied films presented shear thinning (chitosan) and shear thickening (itaconic acid and simple films) behaviors. In the first case, the viscosity decreased with the increase in shear rate, and in the second case, it increased and then remained constant with the increase in shear rate. In addition, the total phenolic content of TBE-enriched PVA + Ch film-forming solutions may deliver important antioxidant activity. Therefore, the produced films are suitable for the packaging of a large variety of foods. Considering the high importance of low permeability in developing ideal food packaging materials, our future research will focus on investigating the water permeability of active packaging films enriched with by-products extracts. In subsequent studies, we intend to also examine the anti-adhesive and anti-biofilm activity, in addition to biodegradability assessment, of the newly developed films.

Author Contributions: K.S. was the main author of this work. K.S., L.M., E.S., R.-A.V. and B.-E.T. worked on the experimental part of this research. K.S., L.M., B.-E.T., L.-F.C. and G.-A.M. worked on the documentation and writing part of the article. D.C.V. was the lead supervisor of the group on this project. All authors have read and agreed to the published version of the manuscript.

Funding: This work was supported by a grant of the Romanian National Authority for Scientific Research and Innovation, CCDI-UEFISCDI, project number 27/2018 CO FUND–MANUNET III-NON-ACT-2, within PNCDI III.

Conflicts of Interest: The authors declare no conflict of interest. The funders had no role in the design of the study; in the collection, analyses, or interpretation of data; in the writing of the manuscript, or in the decision to publish the results.

References

1. Garavand, F.; Rouhi, M.; Razavi, S.H.; Cacciotti, I.; Mohammadi, R. Improving the integrity of natural biopolymer films used in food packaging by crosslinking approach: A review. *Int. J. Biol. Macromol.* **2017**, *104*, 687–707. [CrossRef] [PubMed]
2. Nielsen, T.D.; Hasselbalch, J.; Holmberg, K.; Stripple, J. Politics and the plastic crisis: A review throughout the plastic life cycle. *Wires Energy Environ.* **2019**, *9*, e360. [CrossRef]
3. Raddadi, N.; Fava, F. Biodegradation of oil-based plastics in the environment: Existing knowledge and needs of research and innovation. *Sci. Total Environ.* **2019**, *679*, 148–158. [CrossRef] [PubMed]
4. Blettler, M.C.M.; Wantzen, K.M. Threats Underestimated in Freshwater Plastic Pollution: Mini-Review. *Water Air Soil Pollut.* **2019**, *230*, 174. [CrossRef]
5. Wang, M.H.; He, Y.; Sen, B. Research and management of plastic pollution in coastal environments of China. *Environ. Pollut.* **2019**, *248*, 898–905. [CrossRef] [PubMed]
6. Available online: http://www3.weforum.org/docs/WEF_The_New_Plastics_Economy.pdf (accessed on 4 February 2020).
7. Jagiello, Z.; Dylewski, L.; Tobolka, M.; Aguirre, J.I. Life in a polluted world: A global review of anthropogenic materials in bird nests. *Environ. Pollut.* **2019**, *251*, 717–722. [CrossRef]

8. Cherubini, F.; Ulgiati, S. Crop residues as raw materials for biorefinery systems—A LCA case study. *Appl. Energy* **2010**, *87*, 47–57. [CrossRef]
9. Available online: https://www.earthday.org/fact-sheet-single-use-plastics/ (accessed on 4 February 2020).
10. Weiss, M.; Haufe, J.; Carus, M.; Brandão, M.; Bringezu, S.; Hermann, B.; Patel, M.K. A Review of the Environmental Impacts of Biobased Materials. *J. Ind. Ecol.* **2012**, *16*, S169–S181. [CrossRef]
11. Hatti-Kaul, R.; Nilsson, L.J.; Zhang, B.; Rehnberg, N.; Lundmark, S. Designing Biobased Recyclable Polymers for Plastics. *Trends Biotechnol.* **2020**, *38*, 50–67. [CrossRef]
12. Zhu, Y.; Romain, C.; Williams, C.K. Sustainable polymers from renewable resources. *Nature* **2016**, *540*, 354–362. [CrossRef]
13. Iwata, T. Biodegradable and bio-based polymers: Future prospects of eco-friendly plastics. *Angew. Chem. Int. Ed. Engl.* **2015**, *54*, 3210–3215. [CrossRef] [PubMed]
14. Mitrea, L.; Vodnar, D.C. Klebsiella pneumoniae—A Useful Pathogenic Strain for Biotechnological Purposes: Diols Biosynthesis under Controlled and Uncontrolled pH Levels. *Pathogens* **2019**, *8*, 293. [CrossRef] [PubMed]
15. Available online: http://www.fao.org/fileadmin/templates/agns/pdf/jecfa/cta/61/PVA.pdf (accessed on 4 February 2020).
16. Wong, C.Y.; Wong, W.Y.; Loh, K.S.; Daud, W.R.W.; Lim, K.L.; Khalid, M.; Walvekar, R. Development of Poly(Vinyl Alcohol)-Based Polymers as Proton Exchange Membranes and Challenges in Fuel Cell Application: A Review. *Polym. Rev.* **2019**, 171–202. [CrossRef]
17. Kanatt, S.R.; Makwana, S.H. Development of active, water-resistant carboxymethyl cellulose-poly vinyl alcohol-Aloe vera packaging film. *Carbohydr. Polym.* **2020**, *227*, 115303. [CrossRef]
18. Szabo, K.; Diaconeasa, Z.; Catoi, A.F.; Vodnar, D.C. Screening of Ten Tomato Varieties Processing Waste for Bioactive Components and Their Related Antioxidant and Antimicrobial Activities. *Antioxidants* **2019**, *8*, 292. [CrossRef]
19. Călinoiu, L.F.; Mitrea, L.; Precup, G.; Bindea, M.; Rusu, B.; Szabo, K.; Dulf, F.V.; Ștefănescu, B.E.; Vodnar, D.C. Sustainable use of agro-industrial wastes for feeding 10 billion people by 2050. In *Professionals in Food Chains: Ethics, Roles and Responsibilities*; Wageningen Academic Publishers: Wageningen, The Netherlands, 2018.
20. Bilal, M.; Iqbal, H.M.N. Naturally-derived biopolymers: Potential platforms for enzyme immobilization. *Int. J. Biol. Macromol.* **2019**, *130*, 462–482. [CrossRef]
21. Calinoiu, L.F.; Catoi, A.F.; Vodnar, D.C. Solid-State Yeast Fermented Wheat and Oat Bran as A Route for Delivery of Antioxidants. *Antioxidants* **2019**, *8*, 372. [CrossRef]
22. Mitrea, L.; Calinoiu, L.F.; Precup, G.; Bindea, M.; Rusu, B.; Trif, M.; Stefanescu, B.E.; Pop, I.D.; Vodnar, D.C. Isolated Microorganisms for Bioconversion of Biodiesel-Derived Glycerol Into 1,3-Propanediol. *Bull. Univ. Agric. Sci. Vet. Med. Cluj-Napoca-Food Sci. Technol.* **2017**, *74*, 43–49. [CrossRef]
23. Saini, R.K.; Moon, S.H.; Keum, Y.S. An updated review on use of tomato pomace and crustacean processing waste to recover commercially vital carotenoids. *Food Res. Int.* **2018**, *108*, 516–529. [CrossRef]
24. Trif, M.; Vodnar, D.C.; Mitrea, L.; Rusu, A.V.; Socol, C.T. Design and Development of Oleoresins Rich in Carotenoids Coated Microbeads. *Coatings* **2019**, *9*, 235. [CrossRef]
25. Szabo, K.; Catoi, A.F.; Vodnar, D.C. Bioactive Compounds Extracted from Tomato Processing by-Products as a Source of Valuable Nutrients. *Plant Foods Hum. Nutr.* **2018**, *73*, 268–277. [CrossRef]
26. Calinoiu, L.F.; Vodnar, D.C. Whole Grains and Phenolic Acids: A Review on Bioactivity, Functionality, Health Benefits and Bioavailability. *Nutrients* **2018**, *10*, 1615. [CrossRef]
27. Strati, I.F.; Oreopoulou, V. Recovery of carotenoids from tomato processing by-products—A review. *Food Res. Int.* **2014**, *65*, 311–321. [CrossRef]
28. Calinoiu, L.F.; Vodnar, D.C. Thermal Processing for the Release of Phenolic Compounds from Wheat and Oat Bran. *Biomolecules* **2019**, *10*, 21. [CrossRef]
29. Kuenz, A.; Krull, S. Biotechnological production of itaconic acid-things you have to know. *Appl. Microbiol. Biotechnol.* **2018**, *102*, 3901–3914. [CrossRef]
30. Weastra SRO. Available online: https://www.igb.fraunhofer.de/content/dam/igb/en/documents/publications/BioConSepT_Market-potential-for-selected-platform-chemicals_ppt1.pdf (accessed on 4 February 2020).
31. Regestein, L.; Klement, T.; Grande, P.; Kreyenschulte, D.; Heyman, B.; Massmann, T.; Eggert, A.; Sengpiel, R.; Wang, Y.; Wierckx, N.; et al. From beech wood to itaconic acid: Case study on biorefinery process integration. *Biotechnol. Biofuels* **2018**, *11*, 279. [CrossRef]

32. Teleky, B.E.; Vodnar, D.C. Biomass-Derived Production of Itaconic Acid as a Building Block in Specialty Polymers. *Polymers* **2019**, *11*, 1035. [CrossRef]
33. Fuciños, C.F.P.; Amado, I.R.; Míguez, M.; Fajardo, P.; Pastrana, L.M.; Rúa, M.L. *Smart Nanohydrogels for Controlled Release of Food Preservatives*; Elseviser: Amsterdam, The Netherlands, 2016.
34. González-Henríquez, C.M.; Sarabia-Vallejos, M.A.; Rodriguez-Hernandez, J. Polymers for additive manufacturing and 4D-printing: Materials, methodologies, and biomedical applications. *Prog. Polym. Sci.* **2019**, *94*, 57–116. [CrossRef]
35. Top 5 Vendors in the Global Biopolymers Market From 2017–2021: Technavio. Available online: https://www.businesswire.com/news/home/20170112005066/en/Top-5-Vendors-Global-Biopolymers-Market-2017-2021 (accessed on 2 September 2019).
36. Calinoiu, L.F.; Vodnar, D.; Precup, G. A Review: The Probiotic Bacteria Viability under Different Conditions. Bulletin of University of Agricultural Sciences and Veterinary Medicine Cluj-Napoca. *Food Sci. Technol.* **2016**, *73*. [CrossRef]
37. Martau, G.A.; Mihai, M.; Vodnar, D.C. The Use of Chitosan, Alginate, and Pectin in the Biomedical and Food Sector-Biocompatibility, Bioadhesiveness, and Biodegradability. *Polymers* **2019**, *11*, 1837. [CrossRef]
38. Sinha, V.R.; Singla, A.K.; Wadhawan, S.; Kaushik, R.; Kumria, R.; Bansal, K.; Dhawan, S. Chitosan microspheres as a potential carrier for drugs. *Int. J. Pharm.* **2004**, *274*, 1–33. [CrossRef]
39. Cavallaro, G.; Lazzara, G.; Milioto, S. Sustainable nanocomposites based on halloysite nanotubes and pectin/polyethylene glycol blend. *Polym. Degrad. Stab.* **2013**, *98*, 2529–2536. [CrossRef]
40. Sanuja, S.; Agalya, A.; Umapathy, M.J. Studies on Magnesium Oxide Reinforced Chitosan Bionanocomposite Incorporated with Clove Oil for Active Food Packaging Application. *Int. J. Polym. Mater. Polym. Biomater.* **2014**, *63*, 733–740. [CrossRef]
41. Guerreiro, A.C.; Gago, C.M.L.; Miguel, M.G.C.; Faleiro, M.L.; Antunes, M.D.C. The influence of edible coatings enriched with citral and eugenol on the raspberry storage ability, nutritional and sensory quality. *Food Packag. Shelf Life* **2016**, *9*, 20–28. [CrossRef]
42. Kanetis, L.; Exarchou, V.; Charalambous, Z.; Goulas, V. Edible coating composed of chitosan and Salvia fruticosa Mill. extract for the control of grey mould of table grapes. *JSCI Food Agric.* **2017**, *97*, 452–460. [CrossRef]
43. Călinoiu, L.-F.; Ştefănescu, B.; Pop, I.; Muntean, L.; Vodnar, D. Chitosan Coating Applications in Probiotic Microencapsulation. *Coatings* **2019**, *9*, 194. [CrossRef]
44. Krisanti, E.A.; Naziha, G.M.; Amany, N.S.; Mulia, K.; Handayani, N.A. Effect of biopolymers composition on release profile of iron(II) fumarate from chitosan-alginate microparticles. In *IOP Conference Series: Materials Science and Engineering*; IOP Publishing: Bristol, UK, 2019.
45. Estevinho, B.N.; Rocha, F.; Santos, L.; Alves, A. Microencapsulation with chitosan by spray drying for industry applications—A review. *Trends Food Sci. Technol.* **2013**, *31*, 138–155. [CrossRef]
46. Gonçalves, A.; Estevinho, B.N.; Rocha, F. Microencapsulation of vitamin A: A review. *Trends Food Sci. Technol.* **2016**, *51*, 76–87. [CrossRef]
47. Lee, J.B.; Ahn, J.; Lee, J.; Kwak, H.S. The microencapsulated ascorbic acid release in vitro and its effect on iron bioavailability. *Arch. Pharm. Res.* **2003**, *26*, 874–879. [CrossRef]
48. Zhang, Z.; Zhang, R.; Chen, L.; McClements, D.J. Encapsulation of lactase (beta-galactosidase) into kappa-carrageenan-based hydrogel beads: Impact of environmental conditions on enzyme activity. *Food Chem.* **2016**, *200*, 69–75. [CrossRef]
49. Gupta, C.; Chawla, P.; Arora, S.; Tomar, S.K.; Singh, A.K. Iron microencapsulation with blend of gum arabic, maltodextrin and modified starch using modified solvent evaporation method—Milk fortification. *Food Hydrocoll.* **2015**, *43*, 622–628. [CrossRef]
50. Valenzuela, C.; Hernández, V.; Morales, M.S.; Neira-Carrillo, A.; Pizarro, F. Preparation and characterization of heme iron-alginate beads. *LWT Food Sci. Technol.* **2014**, *59*, 1283–1289. [CrossRef]
51. Nasui, L.; Vodnar, D.; Socaciu, C. Bioactive Labels for Fresh Fruits and Vegetables. Bulletin of University of Agricultural Sciences and Veterinary Medicine Cluj-Napoca. *Food Sci. Technol.* **2013**, *70*, 74–82. [CrossRef]
52. Dong, H.; Cheng, L.; Tan, J.; Zheng, K.; Jiang, Y. Effects of chitosan coating on quality and shelf life of peeled litchi fruit. *J. Food Eng.* **2004**, *64*, 355–358. [CrossRef]
53. Guo, Z.; Xing, R.; Liu, S.; Zhong, Z.; Ji, X.; Wang, L.; Li, P. The influence of molecular weight of quaternized chitosan on antifungal activity. *Carbohydr. Polym.* **2008**, *71*, 694–697. [CrossRef]

54. Athayde, A.J.A.A.; de Oliveira, P.D.L.; Guerra, I.C.D.; da Conceição, M.L.; de Lima, M.A.B.; Arcanjo, N.M.O.; Madruga, M.S.; Berger, L.R.R.; de Souza, E.L. A coating composed of chitosan and Cymbopogon citratus(Dc. Ex Nees) essential oil to control Rhizopus soft rot and quality in tomato fruit stored at room temperature. *J. Hortic. Sci. Biotechnol.* **2016**, *91*, 582–591. [CrossRef]
55. Dulf, F.V.; Vodnar, D.C.; Dulf, E.H.; Tosa, M.I. Total phenolic contents, antioxidant activities, and lipid fractions from berry pomaces obtained by solid-state fermentation of two Sambucus species with Aspergillus niger. *J. Agric. Food Chem.* **2015**, *63*, 3489–3500. [CrossRef]
56. Vodnar, D.C.; Calinoiu, L.F.; Dulf, F.V.; Stefanescu, B.E.; Crisan, G.; Socaciu, C. Identification of the bioactive compounds and antioxidant, antimutagenic and antimicrobial activities of thermally processed agro-industrial waste. *Food Chem.* **2017**, *231*, 131–140. [CrossRef]
57. O'Donnell, F.; Smyth, T.J.; Ramachandran, V.N.; Smyth, W.F. A study of the antimicrobial activity of selected synthetic and naturally occurring quinolines. *Int. J. Antimicrob. Agents* **2010**, *35*, 30–38. [CrossRef]
58. Szabo, K.; Dulf, F.V.; Diaconeasa, Z.; Vodnar, D.C. Antimicrobial and antioxidant properties of tomato processing byproducts and their correlation with the biochemical composition. *LWT* **2019**, *116*, 108558. [CrossRef]
59. Strati, I.F.; Gogou, E.; Oreopoulou, V. Enzyme and high pressure assisted extraction of carotenoids from tomato waste. *Food Bioprod. Process.* **2015**, *94*, 668–674. [CrossRef]
60. Ding, J.; Chen, S.-C.; Wang, X.-L.; Wang, Y.-Z. Preparation and Rheological Behaviors of Thermoplastic Poly(vinyl alcohol) Modified by Lactic Acid. *Ind. Eng. Chem. Res.* **2011**, *50*, 9123–9130. [CrossRef]
61. Muresan, V.; Danthine, S.; Racolta, E.; Muste, S.; Blecker, C. The Influence of Particle Size Distribution on Sunflower Tahini Rheology and Structure. *J. Food Process Eng.* **2014**, *37*, 411–426. [CrossRef]
62. Merlusca, I.P.; Ibanescu, C.; Tuchilus, C.; Danu, M.; Atanase, L.I.; Popa, I.M. Characterization of Neomycin-Loaded Xanthan-Chitosan Hydrogels for Topical Applications. *Cellul. Chem. Technol.* **2019**, *53*, 709–719. [CrossRef]
63. Arcan, I.; Yemenicioğlu, A. Incorporating phenolic compounds opens a new perspective to use zein films as flexible bioactive packaging materials. *Food Res. Int.* **2011**, *44*, 550–556. [CrossRef]
64. Shahidi, F.; Arachchi, J.K.V.; Jeon, Y.-J. Food applications of chitin and chitosans. *Trends Food Sci. Technol.* **1999**, *10*, 37–51. [CrossRef]
65. Kanatt, S.R.; Rao, M.S.; Chawla, S.P.; Sharma, A. Active chitosan-polyvinyl alcohol films with natural extracts. *Food Hydrocoll.* **2012**, *29*, 290–297. [CrossRef]
66. Birajdar, M.S.; Cho, H.; Seo, Y.; Choi, J.; Park, H. Surface conjugation of poly(dimethyl siloxane) with itaconic acid-based materials for antibacterial effects. *Appl. Surf. Sci.* **2018**, *437*, 245–256. [CrossRef]
67. Sakthivel, M.; Franklin, D.S.; Sudarsan, S.; Chitra, G.; Sridharan, T.B.; Guhanathan, S. Investigation on pH/salt-responsive multifunctional itaconic acid based polymeric biocompatible, antimicrobial and biodegradable hydrogels. *React. Funct. Polym.* **2018**, *122*, 9–21. [CrossRef]
68. Ouattara, B.; Simard, R.E.; Piette, G.; Begin, A.; Holley, R.A. Diffusion of Acetic and Propionic Acids from Chitosan-based Antimicrobial Packaging Films. *J. Food Sci.* **2000**, *65*, 768–773. [CrossRef]
69. Rao, M.S.; Kanatt, S.R.; Chawla, S.P.; Sharma, A. Chitosan and guar gum composite films: Preparation, physical, mechanical and antimicrobial properties. *Carbohydr. Polym.* **2010**, *82*, 1243–1247. [CrossRef]
70. Mathew, S.; Abraham, T.E. Characterisation of ferulic acid incorporated starch–chitosan blend films. *Food Hydrocoll.* **2008**, *22*, 826–835. [CrossRef]
71. Pranoto, Y.; Rakshit, S.K.; Salokhe, V.M. Enhancing antimicrobial activity of chitosan films by incorporating garlic oil, potassium sorbate and nisin. *LWT Food Sci. Technol.* **2005**, *38*, 859–865. [CrossRef]
72. Sand, A.; Kniivilä, J.; Toivakka, M.; Hjelt, T. Structure formation mechanisms in consolidating pigment coatings—Simulation and visualisation. *Chem. Eng. Process. Process Intensif.* **2011**, *50*, 574–582. [CrossRef]
73. Luzi, F.; Pannucci, E.; Santi, L.; Kenny, J.M.; Torre, L.; Bernini, R.; Puglia, D. Gallic Acid and Quercetin as Intelligent and Active Ingredients in Poly(vinyl alcohol) Films for Food Packaging. *Polymers* **2019**, *11*, 1999. [CrossRef]
74. Mitrea, L.; Ranga, F.; Fetea, F.; Dulf, F.V.; Rusu, A.; Trif, M.; Vodnar, D.C. Biodiesel-Derived Glycerol Obtained from Renewable Biomass—A Suitable Substrate for the Growth of Candida zeylanoides Yeast Strain ATCC 20367. *Microorganisms* **2019**, *7*, 265. [CrossRef]

75. Harun-or-Rashid, M.D.; Saifur Rahaman, M.D.; Enamul Kabir, S.; Khan, M.A. Effect of hydrochloric acid on the properties of biodegradable packaging materials of carboxymethylcellulose/poly(vinyl alcohol) blends. *J. Appl. Polym. Sci.* **2016**, *133*. [CrossRef]
76. Li, R.; Wang, Y.; Xu, J.; Ahmed, S.; Liu, Y. Preparation and Characterization of Ultrasound Treated Polyvinyl Alcohol/Chitosan/DMC Antimicrobial Films. *Coatings* **2019**, *9*, 582. [CrossRef]
77. Milosavljevicĺ, N.B.; Kljajevicĺ, L.M.; Popovicĺ, I.G.; Filipovicĺ, J.M.; Kalagasidis KrusĺŒicĺ, M.T. Chitosan, itaconic acid and poly(vinyl alcohol) hybrid polymer networks of high degree of swelling and good mechanical strength. *Polym. Int.* **2009**, *59*, 686–694. [CrossRef]
78. Liu, Y.; Wang, S.; Lan, W.; Qin, W. Fabrication and Testing of PVA/Chitosan Bilayer Films for Strawberry Packaging. *Coatings* **2017**, *7*, 109. [CrossRef]

 © 2020 by the authors. Licensee MDPI, Basel, Switzerland. This article is an open access article distributed under the terms and conditions of the Creative Commons Attribution (CC BY) license (http://creativecommons.org/licenses/by/4.0/).

Article

Preparation and Formula Analysis of Anti-Biofouling Titania–Polyurea Spray Coating with Nano/Micro-Structure

Yuanzhe Li [1], Boyang Luo [1], Claude Guet [1], Srikanth Narasimalu [2] and Zhili Dong [1,*]

1 School of Materials Science & Engineering, Nanyang Technological University, Singapore 639798, Singapore
2 Energy Research Institute @ NTU (ERI@N), CleanTech One, Singapore 637141, Singapore
* Correspondence: zldong@ntu.edu.sg; Tel.: +65-6790-6727

Received: 19 July 2019; Accepted: 28 August 2019; Published: 2 September 2019

Abstract: This paper proposes the preparation and formula analysis of anti-biofouling Titania–polyurea (TiO_2–SPUA) spray coating, which uses nano-scale antibacterial and photocatalytic agents, titanium dioxide, to construct regularly hydrophobic surface texture on the polyurea coating system. Through formulating analysis of anti-biofouling performance, it is found the causal factors include antibacterial TiO_2, surface wettability and morphology in order of their importance. The most optimized formula group is able to obtain uniform surface textures, high contact angle (91.5°), low surface energy (32.5 mJ/m^2), and strong hardness (74 A). Moreover, this newly fabricated coating can effectively prevent Pseudomonas aeruginosa and biofilm from enriching on the surface, and there is no toxins release from the coating itself, which makes it eco-friendly, even after long-time exposure. These studies provide insights to the relative importance of physiochemical properties of Titania–polyurea spray coatings for further use in marine, as well as bio medical engineering.

Keywords: Titania–polyurea (TiO_2–SPUA) spray coating; morphology study; surface wettability; anti-biofouling study; surface hardness; formula analysis

1. Introduction

Photocatalytic titanium dioxide (TiO_2) is a semi-conductive material, which could produce free hydroxyl and oxygen species (reactive oxygen species) with strong oxidation-reduction ability by photosynthesis reactions [1]. Organic compounds and inorganic oxides could be decomposed under illumination [2]. The photocatalytic character of TiO_2 has successfully made it widely applied in various types of bacteria and biofilms contamination abatement [3].

The conditioning of photoinduced bactericidal activity on TiO_2-films surface experiments by Pleskova et al., also states that reactive oxygen species (ROS) is the key to causing destruction of bacteria [4], and its effectiveness of bactericidal activity of TiO_2-films depends on several factors, such as time of UV irradiation and thermal and chemical treatment of films on the bactericidal activity [5]. Recently, Kim et al., reported the reaction of titanium dioxide on Streptococcus mutans biofilm, confirming that the antibacterial effect could be indicated by causing the photocatalytic reaction of TiO_2 in S. mutans biofilm, even at the wavelength of visible light (405 nm) [6]. Besides, titanium dioxide has extremely strong characters of sterilization, deodorant, mould proof, and antifouling function of self-cleaning [7].

Polyurea coatings with better performance than traditional polyurethane and extreme properties, such as rapid cure, insensitivity to humidity, flexibility, high hardness, tear and tensile strength [8], and chemical and water resistance provide a great platform for the fabrication of new type of composite coatings [9]. With proper primer and surface treatment, excellent adhesion to steel and other substrate materials could be achieved [10]. In addition to advantages of polyurea over other coatings, the titanium dioxide could form nano-micro scale regular textures on the coating surface, which might contribute to the hydrophobic and drag reduction performance and also the control of bio-adhesion [11]. Moreover, the combination of these two materials, which is designed for the prevention of biological attachment, could maintain the immersion structure, such as ships as well as the cleanness of sonar in marine, is seldom documented.

Typically, the hull antifouling paint is made of paint, toxic pigments, solvents, additives, and other components [12]. One state-of-the-art antifouling paint that is used in ships in seawater as well as fresh water reviewed by Lin et al., is indicated in Figure 1a. When the antifouling paint confronts with water, the toxic pigment cuprous oxide releases, leaving only the adhesive shell afterwards [13]. However, after a long enough period of exposure, the adhesive shell becomes so thick that the rate of toxins that release into the meager water layer is below the critical value for bactericidal effect and forms a thick, sealed, "sandwich" structure with thickness of 1000 to 1200 μm. Such a "sandwich" will produce a lot of interfacial stress, and peel from the substrate, resulting in very rough surface morphology [12,14]. Additionally, the toxin, cuprous oxide, also imposes the risk to the environment [15].

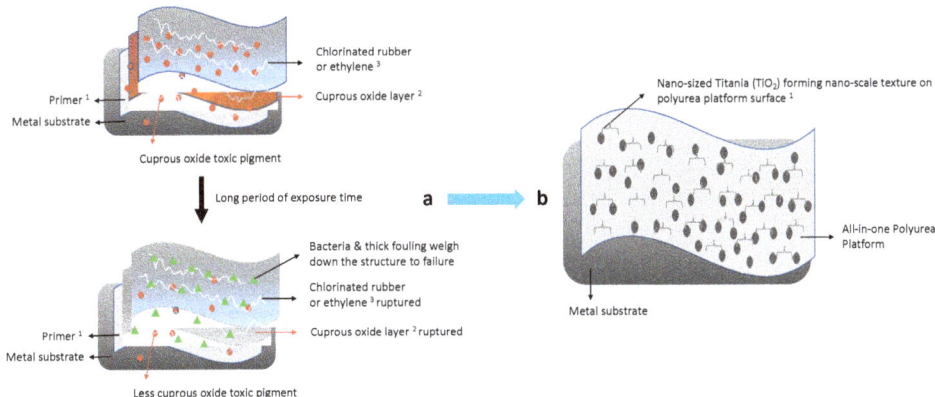

Figure 1. Schematic diagram of typically antifouling paint and all-in-one Titania–polyurea spray coating. (**a**) typically antifouling paint and ruptured structure after long-time exposure, and (**b**) all-in-one anti-biofouling Titania–Polyurea spray coating (TiO$_2$–SPUA).

In this paper, the preparation and formula analysis of such coatings according to their chemical composition, surface morphology, and wettability, which contributes to different bioactive properties, are discussed. Different from high consumption of nano-titanium dioxide and complicated forming process for better antibiofouling performance in most researches [16,17], this Titania–polyurea (TiO$_2$–SPUA) spray coating greatly reduces the use of nano-TiO$_2$ wt.% by forming all-in-one design coating system easily, as shown in Figure 1b. It is believed that the coating that we aimed at and developed can greatly slow down the biofouling process for marine structures, increase the efficiency of the ecosystem with less energy consumption, and extend the maintenance cycle.

Furthermore, this newly fabricated TiO$_2$–SPUA coating system, especially in its microstructure properties and formulating analysis of anti-biofouling performances, have not been recorded yet.

2. Materials and Methods

2.1. Materials

50–80 wt.% isocyanates (TDI)–component "A" and 50–90 wt.% polyether amines (Polyoxypropylenediamine)–component "B" were provided by Dragonsheild-BC^{TM}, Specialty Products Inc. (Washington, DC, USA). Versalink® P-1000, and it was alternative choice of Component "B" purchased from Versalink (Essen, Germany), Air Products and Chemicals Inc. (Allentown, PA, USA).

Low-surface-energy deformer, polydimethylsiloxane (PDMS), and isopropyl alcohol (IPA), from Sigma Aldrich (Bangkok, Thailand) were used as provided. The anatases nano-titanium dioxide with mean size of 39.6 nm and 0.293 polydispersity was fabricated by the sol-gel methods [2].

2.2. Experimental Procedure

2.2.1. Formulations and Fabrication of Titania–Polyurea (TiO_2–SPUA) Spray Coating

Table 1 indicated the three typical group formulations of TiO_2–SPUA coatings. The first two groups (G1 & G2) of TiO_2–SPUA coating formulations took consideration of the excess isocyanate compensates for the 'loss' of isocyanate-groups during storage due to humidity and/or application, as well as the 'losses' from additives in Part B [8], and the increased weight percentage of Part A slightly higher than Part B with index range 1.66 in Group 1 (G1) and relatively low index range 1.50 in Group 2 (G2) [10]. The Group 3 (G3) of TiO_2–SPUA coating prescription remained the index range 1.00, the only difference between T-PG3E1 (air-drying) and T-PG3E2 (oven-drying) is drying methods, which might contribute to the different configuration of the surface [17], as shown in the pre-experiment.

Table 1. Table of TiO_2–SPUA coating formulations (Group 1–3).

Formulation Code Name	Part A		Part B			
	Component "A"	Component "B"	P1000	TiO_2	PDMS	IPA
PG1E1	62.5	15.0	17.5	–	–	–
T-PG1E2	62.5	14.8	22.3	0.4	–	–
T-PG1E3	62.5	14.7	22.0	0.4	0.4	–
T-PG1E4	62.5	14.7	22.0	0.4	–	0.4
PG2E1	60.0	–	40.0	–	–	–
PG2E2	60.0	8.0	32.0	–	–	–
T-PG3E1	50.0	10.5	36.5	1.5	1.5	–
T-PG3E2	50.0	10.5	36.5	1.5	1.5	–

Note: T is for formulation obtained TiO_2, P is short for "Polyurea"; G is short for "Group" with same wt.% of Part A/Component "A"; E is short for "Experiment" with different formulation of Part B; e.g., T-PG1E2 means TiO_2–polyurea Group 1 (62.5 wt.% Component "A") Experiment 2.

For a better mixture of components, both of the chemicals were raised to 70 °C to decrease their viscosity [18]. Subsequently, Part A and Part B were put into Kakuhunter SK-300TVSII mixer (Shashin Kagaku Pte Ltd., Kyoto, Japan) for 180 s with vacuum level of 0.5 kPa to remove bubbles. The revolution speed and rotation speed were set at 580 rpm and 1700 rpm, respectively [10]. After mixing, the TiO_2–SPUA coating was sprayed with 2.0 mm thickness into the Teflon mould with dimension of $100 \times 100 \times 10$ mm^3. Finally, all of the samples were put into oven under 70 °C (except for T-PG3E1) for curing of 48 h.

2.2.2. Surface Characterizations

The surface morphology and chemical composition of TiO_2–SPUA coating were characterized by scanning electron microscopy (Hitachi S-4700, California, CA, USA) [17] that was attached with a Bruker AXS Quantax 4010 energy dispersive X-ray spectrometer (EDX, Karlsruhe, Germany) [19]. OCA15 plus (Dataphysics, Filderstadt, Germany) was employed to measure the contact angles while using distilled water. Surface energy was measured using three different liquids, diiodomethane, ethylene glycol, and distilled water, and SCA 20 software (Dataphysics) was used to analyze data from the regression line in a suitable plot [20,21].

2.2.3. Preparation of Bio-Medium and Bioassays

All of the tests were conducted while using a nutrient-rich artificial fresh water medium referred to Yuan et al. [22]. A well-studied aerobic Pseudomonas sp. NCIMB 2021 bacterium was obtained from the National Collection of Marine Bacteria (Sussex, UK). CDC biofilm reactor (BioSurface Technologies Corp., Bozeman, MT, USA) was used to grow Pseudomonas enriched cultures from Sentosa (SG) seawater under different shear force and continuous flow conditions at 30.0 °C [7,22]. The source of UV-irradiation was a Philip TUV15W lamp (Eindhoven, Netherlands) with power densities at 32.12 mW/cm^2, i.e., the maximum intensity was at 353 nm, which is under the UV-A region (315–400 nm). The distance between the light and specimens was set to 7.0 cm, and the interval for each UV treatment was 8 h/day to simulate the practical sunlight radiation. Specimens that were cut from coating samples were served as removable coupons and ASTM method and E2562-12 Manual was followed. Live/Dead BacLight^{TM} Bacterial Viability Kit for microscopy (Thermo Fisher Scientific, Waltham, MA, USA) was employed to undertake the detection of biofilms [23]. Pyruvate and organic carbon were inserted into the carboy lid to speed up the enriched culture in CDC biofilm reactor [24–26]. CVD titanium dioxide coated on AlZr (5.25% Zr) 2 µm 316 L polished substrate [27] and fluoro-modified elastomeric polyurethane, followed by previous drag-reduction work [11], were also involved as two control groups for the analysis of biofilm-attachment mechanism and formula optimization.

2.2.4. Surface Hardness Test

ASTM-D-2240-00 Type A Teclock durometer was used to test the hardness of TiO$_2$–SPUA coating and its impact resistance [17,21].

3. Result and Discussion

3.1. Formulating of Surface Features

The prescription added more Part A than Part B, which was considered the possibility of self-healing and might have un-reacted isocyanates inside the fracture surface, to have continuous reaction with water as per the following chemical reaction [10,28]:

$$R\text{-}NCO + H_2O \rightarrow R\text{-}NH_2 + CO_2 \tag{1}$$

This off-white colored Titania–polyurea (TiO$_2$–SPUA) spray coating, as shown in Figure 2, successfully inherited the fast reactivity of the polyurea. The curing time of all polyurea spray coating was within one hour. In the meantime, vacuumed mixer obtained by the spraying equipment could contribute to realize suitable mixing and remove the waste product and carbon dioxide shortly [29]. Besides, different drying methods, air-drying and oven-drying, were also tried under Group 3, would have an impact on the curing time of TiO$_2$–SPUA coating surface in the mould [30]. As for foam inside such a coating, both technology-based methods (e.g., vacuum degree inside the spraying equipment) and chemical-based prescription (e.g., defoamer, such as PDMS and IPA) have been conducted to address this issue [10]. To conclude, the ideal solution for the foam issue should be preventing the CO$_2$ generation and diffusion from the reaction between moisture in the air [31]. The shorter drying time TiO$_2$–SPUA coating needs, the smaller amount of foam caused by carbon dioxide that it would generate.

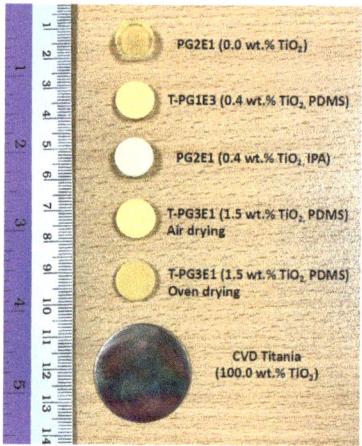

Figure 2. Photographs of Titania–polyurea spray coating samples.

3.1.1. Morphological and Chemical Composition Analysis

For a polyurea coating group without titanium dioxide, the morphological image was relatively flat when compared with TiO$_2$–SPUA coatings, even other liquid additives, such as PDMS and IPA solution, would not cause any modification of the coating surface structure [17,32]. While, for the SEM images shown in Figure 3a–c, only the increment in nano-titanium dioxide could directly lead to the surface textures change, and these surface structures might have a good orientation of the water flow and detachment of biofilm [33], which was caused by the spraying and curing during preparation. Additionally, it was gratifying to observe that the nano-TiO$_2$ particles did not generate agglomeration and had a favorable dispersion in polyurea coatings. As the nano-titanium dioxide weight percentage went higher, from 0.4 wt.% TiO$_2$ shown in Figure 3b to 1.5 wt.% TiO$_2$ shown in Figure 3c, textures that formed by nano-titanium dioxide would become more distinct.

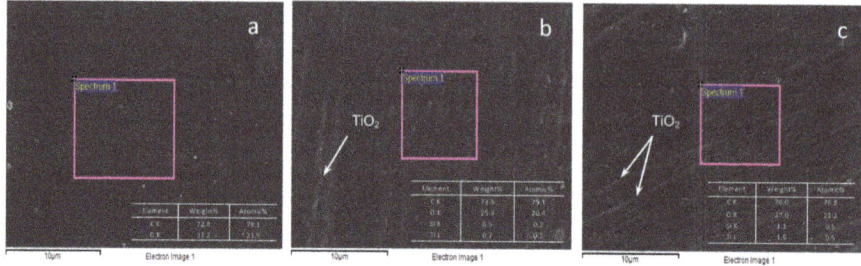

Figure 3. Scanning Electron Micrographs showing surface morphologies of sample (**a**) PG2E2, (**b**) T-PG1E2, and (**c**) T-PG3E2.

It could be summarized that, as the nano-TiO$_2$ weight percentage continuously going high, the surface morphology would cohere hydrophobic and homogeneous surfaces along with drag reduction character at first. It is important to add that, as the weight percentage continuously increased, the hydrophilic character from nano-TiO$_2$ would gradually take the lead, and its internal agglomerating force would end up with rough coating surface morphology as compared to the original smooth finish. Moreover, the profile degree, force orientation, and wettability of TiO$_2$–SPUA coatings would also be affected [34]. From the investigations of contact angle (CA) and surface energy (SFE) analysis subsection, these inferences were also confirmed. From the imaging of EDX spectra, the Ti and Si signal appeared, and the signal intensity would continuously increase along with the increment in nano-TiO$_2$ and defoamer PDMS.

3.1.2. Surface Wettability

Pure titanium, as well as titanium dioxide surfaces, performed as hydrophilic. However, from the overall surface energy (SFE) and contact angle (CA) result in Table 2, the hydrophobic character behavior of TiO$_2$–SPUA coating was exactly different. The contact angle changed by the increase of TiO$_2$ wt.% was also indicated, respectively, in Figure 4a–c. It could be not only due to its small weight percentage and its minor texture structure on the coating surface, but its dispersion inside the polyurea base coatings and use of low surface energy defoaming agent could also influence the polar component, CA, SFE, and even biofilm test result [34].

Table 2. Values (mean ± standard deviation) of contact angle (CA) and surface energy (SFE).

Formulations & Effect	PG1E1	T-PG1E2	T-PG1E3	T-PG1E4	PG2E1	PG2E2	T-PG3E1	T-PG3E2
TiO$_2$ (wt.%)	–	0.4	–	0.4	–	–	1.5	1.5
PDMS (wt.%)	–	–	0.4	–	–	–	1.5	1.5
CA (°)	61.5 ± 2.7	66.9 ± 2.9	73.0 ± 4.9	64.4 ± 1.4	68.4 ± 6.9	63.8 ± 1.9	88.5 ± 6.4	91.5 ± 2.6
SFE (mJ/m^2)	57.3	49.4	45.8	51.5	47.1	52.3	37.2	32.5

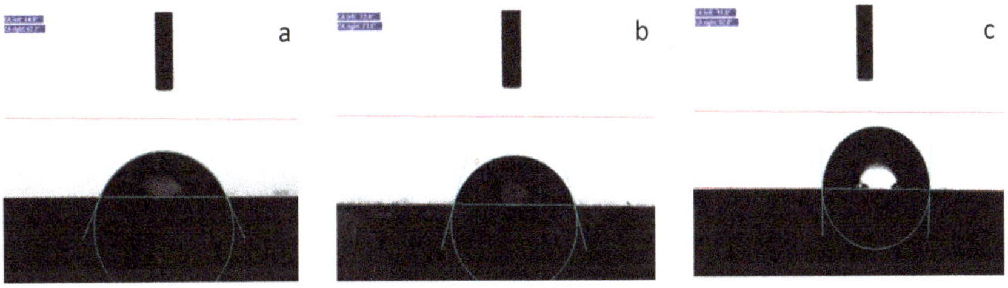

Figure 4. Surface contact angle (CA) of coating surface (**a**) contact angle of PG2E2 (0.0 wt.%TiO$_2$) 63.8° ± 1.9°, (**b**) contact angle of T-PG1E2 (0.4 wt.%TiO$_2$) 66.9° ± 2.9°, and (**c**) contact angle of T-PG3E2 (1.5 wt.%TiO$_2$) 91.5° ± 2.6°.

These CA and SFE results should also be related to morphology. It is worth noting that increment scenarios of CA would only happen at small weight percentage scale of nano titanium dioxide. As the increase of nano-titanium dioxide particles wt.%, the nano-texture and low surface energy structure would gradually disappear. High weight percentage of hydrophilic nano-titanium dioxide at 5% and 10% would lead to the decrement of CA to 68.0° and SFE would raise as high as 50.3 mJ/m^2.

Otherwise, based on the fact that low surface energy chemistry and nano-textured morphology of the hydrophobic coating could result in reduced protein adsorption and bacterial attachment [35], it was not found that there was low attachment of bacteria and biofilms in the following biofouling assays. Additionally, the hydrophobicity of such coating surface might reduce the flow resistance and also offer a new research direction for drag reduction.

3.2. Formulating of Anti-Biofouling Performance

Cells with compromised membranes that are considered to be dead or dying will stain brightly red, whereas the cells with an intact membrane will stain brightly green. The dark spot and slightly blur background were due to the artifacts of substrate structures, and even surface reflection/refraction under the Bacterial Viability Kit, respectively, rather than biofilms. In addition, more quantified information of Pseudomonas aeruginosa in this pilot bio-assays is provided in Table S1.

3.2.1. Nano-Titanium Dioxide vs. Photocatalytic and/or EPS Degradation & Morphology

During the early stage of the Pseudomonas aeruginosa attached, reversible attachment is secreted by bacteria with tightly-bound EPS (TB-EPS), which keeps cells together in clusters and loosely-bound EPS (LB-EPS) bonding different bacteria clusters together to form stable micro-colonies [6]. The continuous production of EPS by the bacteria community (made up of protein, polysaccharides, eDNA, bacterial lytic products, and compounds from the host) provides the biofilm structural integrity [36,37].

- High weight percentage of nano-titanium dioxide (TiO_2 wt.%) in the coating system may cause the photocatalytic degradation and EPS degradation to inhibit the reversible attachment of biofilm.

The strategy to tackle Pseudomonas for high weight percentage of nano-titanium dioxide (TiO_2 wt.%) is the photodegradation of the EPS matrix [38]. Through the observation of the live cells while using the Bacterial Viability Kit, there was no sign of any live/dead Pseudomonas or any ruptured structure of microbial cells under higher weight percentage of the nano-TiO_2, i.e., there was no indication of attached biofilm or any other microbial cells on the coating surface, except for the artificial defects (dark areas) that are caused by long term exposure and etching from the surface as shown in Table 3.

Table 3. Live/dead microscopy of control group with pure nano-scale TiO_2.

Formulation Code Name	Surface Features	Nano-TiO_2 wt.%	60 ± 5 rpm/10 days (Low Shear Force)	240 ± 5 rpm/10 days (High Shear Force)
CVD TiO_2 Surface (Control Group 1)-Live	Super hydrophilic (CA < 5°)	100.0		
CVD TiO_2 Surface (Control Group 1)-Dead	Super hydrophilic (CA < 5°)	100.0		

Under high shear force, the etching of CVD titanium dioxide surface was even more serious than the ones under low shear force, as indicated in Table 3. Through characterized by 90 plus nanoparticle size analyzer (Brookhaven, New York, NY, USA) for a leaching test, nano-titanium dioxide leaking from CVD titanium dioxide surface was observed at the region between 35.0 and 50.0 nm in the seawater medium after 10 days. Hence, this nano-titanium dioxide from etching in the CDC system was supposed to work on the photodegradation of Pseudomonas [39]. The schematic diagram in Figure 5a demonstrated the detailed processing of how nano-TiO_2 particles produce reactive oxygen species (ROS) [40], inhibit DNA form replication, photodegrade EPS and many other proteins, damage cell membrane, and even interrupt the electron transportation of cells [41].

- Low weight percentage of nano-titanium dioxide (TiO_2 wt.%) in the coating system may use photocatalytic degradation to inhibit the attachment of biofilm by damaging microbial membrane and quorum sensing.

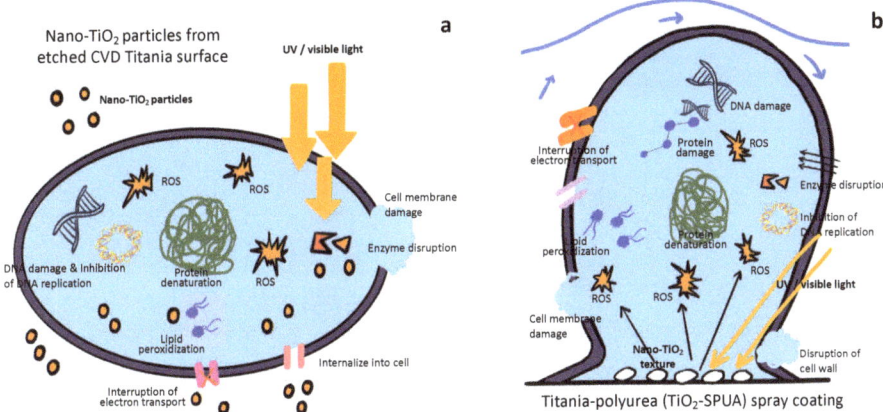

Figure 5. Schematic diagram of different anti-biofouling processing caused by nano-TiO_2 photodegradation (**a**) nano-TiO_2 particles from etched CVD TiO_2 surface, and (**b**) Titania–polyurea (TiO_2–SPUA) spray coating.

At these lower formulation groups of 0.4 wt.%TiO_2 (T-PG1E2), 1.5 wt.%TiO_2 (T-PG3E2), and even 5.0% wt.%TiO_2, with the decrement of weight percentage of nano-titanium dioxide (TiO_2 wt.%), the Pseudomonas inhibition and photodegradation rate would become lower than the one with higher TiO_2 wt.% (CVD TiO_2 Surface). The first-stage EPS would be already stabilized, and biofilm might also reach the irreversible attachment [37,42]. However, the good news was that, through leaching test by nanoparticle size analyzer, only fewer nano-particles could be detected than CVD ones. The result also indicated its potential application for long term exposure in the biofouling environment.

A broken structure of microbial cells and biofilm (DNA of Pseudomonas), which were stained red, could be clearly observed. The internal substances and degraded components [43] from microbial cells are indicated in the Table 4. Though Pseudomonas structure was still destabilized from the photocatalytic degradation and reactive oxygen species (ROS) from the nano-TiO_2, as indicated in Figure 5b, the biofilm was already mature when comparing with the high TiO_2 wt.% groups. Additionally, the free adhesion of biofilm would possibly result from the communication of the Pseudomonas clusters, which is so called quorum sensing/quenching [44].

Table 4. Live/dead microscopy of control group indicating the effect from nano-TiO_2 wt.%.

Formulation Code Name	Surface Features	Nano-TiO_2 wt.%	60 ± 5 rpm/10 days (Low Shear Force)	240 ± 5 rpm/10 days (High Shear Force)
T-PG1E2	Hydrophilic (5° < CA < 90°)	0.4		
T-PG1E2	Hydrophilic (5° < CA < 90°)	0.4		

Table 4. *Cont.*

Formulation Code Name	Surface Features	Nano-TiO$_2$ wt.%	60 ± 5 rpm/10 days (Low Shear Force)	240 ± 5 rpm/10 days (High Shear Force)
T-PG3E2	Hydrophobic (90° < CA < 150°)	1.5		
T-PG3E2	Hydrophobic (90° < CA < 150°)	1.5		

Moreover, the coating samples with regular roughness and morphology would also be conducive to the detachment of biofilm [45] by changing the contact area and air-water interface [46] between the cells and coating surface, as demonstrated in Figure S1a. For nano/micro-scale surface textures (<10 µm), which is smaller than the cell size of Pseudomonas or its cluster (>20 µm), the liquid mobile phase biofilm could shear freely at the air-water interface with small resistance, which resulted in wall slippage [47] and lower friction coefficient and flow resistance. Such a slip effect indicated in Figure S1b would also contribute to the drag reduction effect of microfluid, as proven in previous research [11,48].

At present, there are few studies on substrate interface with certain morphological characteristics, especially those on nanometer or micron level morphological characteristics and substrate interface film covering. Machado et al. [49] used chemical etching techniques to study the surface energy of nano-scale PVC materials and their effects on initial cell adhesion. It was found that the surface energy of the material and the initial adhesion of cells to the interface would be changed due to the change of roughness and morphology and the ability to attach was reduced. Furthermore, since the cell sizes are different, there is no uniform standard for different bacteria [50], and the relationship between roughness and morphology and cell adhesion remains to be further studied.

3.2.2. Hydrophobicity/Hydrophilicity vs. Biofilm Adhesion

- The hydrophobic surface may reduce the adhesion of the biofilm more significantly than hydrophilic ones under high shear force.

For most of the hydrophobic surfaces, the electrostatic interaction and cohesive strength between the biofilm and the surface are weaker than the hydrophilic [51], so the biofilm on the hydrophobic/superhydrophobic surface can be even easier to fail and lead to the detachment event than the hydrophilic one [52]. PDMS, as a polymeric matrix of silicone coatings, are also well known for their smoother and non-stick surfaces relative to other polymeric matrices. Likewise, through vertical comparison between super hydrophobic/hydrophilic surface without effect of any nano-titanium dioxide in the Table 5, it could be found the attachment of biofilm (Pseudomonas) at the superhydrophobic surface (fluoro-modified elastomeric polyurethane) was relatively lower than the super hydrophilic (concrete coupon) ones.

Table 5. Live/dead microscopy of control group excluding the effect from nano-TiO$_2$.

Formulation Code Name	Surface Features	Nano-TiO$_2$ wt.%	60 ± 5 rpm/10 days (Low Shear Force)	240 ± 5 rpm/10 days (High Shear Force)
Concrete coupon of CDC biofilm reactor-Live	Super hydrophilic (CA < 5°)	0.0		
Concrete coupon of CDC biofilm reactor-Dead	Super hydrophilic (CA < 5°)	0.0		
Fluoro-modified elastomeric polyurethane (Control Group 2)-Live	Superhydrophobic (CA > 150°)	0.0		
Fluoro-modified elastomeric polyurethane (Control Group 2)-Dead	Superhydrophobic (CA > 150°)	0.0		

Additionally, it should be noted that biofilms on both of these two groups of polyurea coating samples were also shown to become micro-colony formation and biofilm maturation, which were definitely irreversible [53].

- Hydrophobic and hydrophilic coating surface may both potentially reduce the adhesion of the biofilm under high shear force by convection factors.

According to the horizontal comparison of each sample, the rotating speed (rpm) and shear force could result in difference adhesive image of live/dead Pseudomonas cells, which is mainly because of the convection around the biofilm. Under slow rotating speed (60 ± 5 rpm), the shear force was relatively low, which enable the rapid growth of the Pseudomonas, as fluid could contribute to the diffusive transport of the metabolic substrate and the surrounding environment. As the rotating speed went higher (240 ± 5 rpm), fluid would surround Pseudomonas cell clusters, but was still too weak to flush through them. Complex secondary flows were able to occur under this situation. Continuously increased rotating speed would gradually weaken the cohesive strength inside the biofilm cells and oscillating streamers would form on the downstream edge of a cell cluster [54]. Additionally, it is not hard to understand why under higher rotating speed (rpm), the attached biofilm for both coating samples were decreased more or less.

To conclude, setting all of the photodegradation factors aside, there is a preference between hydrophobic and hydrophilic coating surface for the attachment of the Pseudomonas. Although the critical influence would be taken by the coating materials themselves, hydrophobic coatings will still show more significant reduction for the biofilm adhesion, especially at high shear force, than 5h3 hydrophilic ones [55]. Moreover, these reduction or inhibition of biofilm adhesion should be non-selective.

3.3. Formulating of Surface Hardness

As most of the coating surface might suffer different types of impacts from marine organisms as well as an undercurrent, an investigation of durability and impact resistance of TiO_2–SPUA coating was also very important [17,56]. Through supplementary optimizing design and numerical simulation of these newly fabricated coatings, rather than PDMS, deformer IPA did not exhibit significant increment of other mechanical performance, including surface hardness and even compression and tensile strength, which had no means to exhausted all.

The surface hardness results are shown in Table 6. Besides, the compression behaviour for these coatings, as shown in Figure S2, was also consistent with the hardness test results. It is not difficult to find that three main factors: 1) proportion of Component "B" (long chain) to P1000 (short chain) wt.%, 2) Deformer PDMS wt.%, and 3) TiO_2 wt.% would have the main influence for the performance of coating surface hardness. The increment of PDMS wt.% and P1000wt.% would decrease the surface hardness, while the surface hardness would significantly increase as TiO_2 wt.% and Component "B" wt.% grows. More information of the influence factors for these three main components to the surface hardness, including many other formulations, is also indicated in Figure S3 by regression lines.

Table 6. Values of surface hardness.

Formulations & Effect	PG1E1	T-PG1E2	T-PG1E3	T-PG1E4	PG2E1	PG2E2	T-PG3E1	T-PG3E2
Component "B": P1000 (wt.%)	1.5	1.5	1.5	1.5	0.0	4.0	3.5	3.5
PDMS (wt.%)	0.0	0.0	0.4	0.0	0.0	0.0	1.5	1.5
TiO_2 (wt.%)	0.0	0.4	0.0	0.4	0.0	0.0	1.5	1.5
SH (A)	77	98	48	75	50	92	72	74

To sum-up of the results that are indicated above, nano-scale TiO_2 and low-surface-energy PDMS were the most two critical factors for the optimization of the coating formulation. By using the matrix shown in Figure 6, performance that is derived from different weight percentage (wt.%) is indicated.

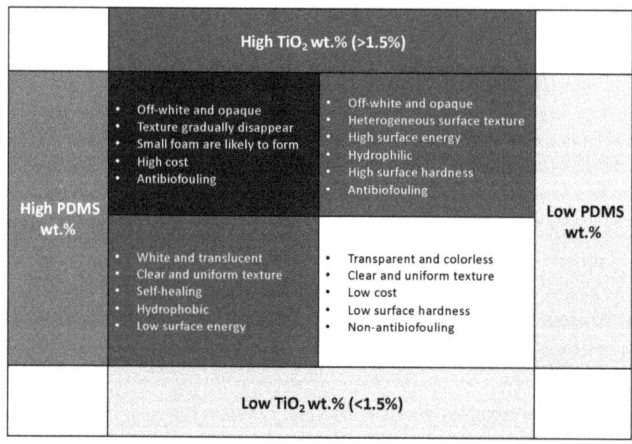

Figure 6. Different basic performance derived from coating formulation depending on weight percentage of TiO_2 and polydimethylsiloxane (PDMS).

Based on comprehensive consideration, the formulation group with high weight percentage (wt.%) of nano-titanium dioxide is not a good alternative to get high mechanical performance, as it might end up with relatively heterogenous texture and a poor dispersion of the nano-titanium dioxide [17]. On the contrary, the formulation group with lower nano TiO_2 wt.% and higher PDMS wt.% could not only prevent the biofouling attachment as previous group, but also obtain homogeneous texture, low surface energy, and many other good physicochemical performances. Additionally, these results will also become the good references for the practical applications and the optimization of the formulations.

For the formulation of such anti-biofilm and regular morphology TiO_2–SPUA coating fabrication, a Gaussian distribution should also be agreed with and the peak of its performance was supposed to be located at 1.0 to 5.0 wt.% of nano-titanium dioxide. One, in particular, is these biofilm attachment assays should be treated as pilot experiments at present. Further experiments are in progress to justify the relation among antibacterial character applied in practically marina environment.

This newly fabricated Titania–polyurea (TiO_2–SPUA) spray coating has the potential to replace the traditional antifouling paint, and its concept of all-in-one and rapid-cure polyurea platform would greatly shorten the painting and curing time and extend the maintenance intervals. There would be no more internal stress and peeling, resulting in very rough surface underwater, caused by the ruptured structure of the "sandwich" coating system after long time exposure. Moreover, there is no more toxins leaking from this Titania–polyurea (TiO_2–SPUA) spray coating, which is even more eco-friendly, and there is no worry about reaching the critical point for the antibacterial agents.

4. Conclusions

In this study, antibacterial Titania–polyurea (TiO_2–SPUA) spray coating with low consumption of TiO_2 (1.5 wt.% only) are fabricated. Regular surface textures and morphology, hydrophobic wettability, and low surface energy could be obtained during the process.

Through formulating analysis of free adhesion of Pseudomonas aeruginosa, it is found that the directly causal factors include: (1) Antibacterial TiO_2, (2) Surface wettability, and (3) Textures and morphology in order of their importance. Nano-titanium dioxide (TiO_2 wt.%) in the coating system may use photocatalytic degradation to inhibit the attachment of biofilm by damaging microbial membrane; hydrophobic wettability might reduce the adhesion of the biofilm more significantly than hydrophilic ones, especially under high shear force; and, nano-texture may also potentially reduce the biofouling adhesion. Moreover, the surface hardness is affected by: (1) the proportion of Component "B" (long chain) to P1000 (short chain) wt.%, (2) deformer PDMS wt.%, and (3) TiO_2 wt.%, which also has internal relationship.

The root cause of such features would be the two main additives, nano-titanium dioxide (TiO_2) and low surface energy defoamer (PDMS) inside the coating system. All of these studies would also provide good reference of formulation design and promising application for further application in marine, as well as bio-medical engineering.

Supplementary Materials: The following are available online at http://www.mdpi.com/2079-6412/9/9/560/s1, Figure S1: Schematic diagram of Pseudomonas detachment influenced by surface roughness and morphology (a) surface roughness/morphology and non-shear air-water interface, and (b) non-shear air-water interface and slip effect of TiO2-SPUA spray coating, Figure S2: Graph of Compression Stress vs Strain (a) Compression Stress vs Strain by PDMS wt.%, and (b) Compression Stress vs Strain by TiO2 wt.%, Figure S3: Influence factors for three main components to the surface hardness, Table S1: Bactericidal effects of different surfaces on Pseudomonas biofilms.

Author Contributions: Y.L. designed the experiments; Z.D. and S.N. contributed reagents and materials; Y.L. and B.L. conducted the formulation design, analysis, and fabrication; Y.L. wrote the paper; Z.D., S.N., and C.G. reviewed and provided corrections on the original draft.

Funding: This research was funded by MOE Academic Research Fund (AcRF) Tier 1 Grant Call (Call 1/2018)_MSE (EP Code EP5P, Project ID 122018-T1-001-077), Ministry of Education (MOE), Singapore.

Conflicts of Interest: The authors declare no conflict of interest.

References

1. Fu, L.; Hamzeh, M.; Dodard, S.; Zhao, Y.H.; Sunahara, G.I. Effects of TiO_2 nanoparticles on ROS production and growth inhibition using freshwater green algae pre-exposed to UV irradiation. *Environ. Toxicol. Pharmacol.* **2015**, *39*, 1074–1080. [CrossRef] [PubMed]
2. Li, Y. Fabrication and indoor air purification of composite photocatalyst spray with nano-titanium dioxide and chlorophytum leach solution. *China Sci. Technol.* **2014**, *9*, 7–8.
3. Jiang, X.; Yang, L.; Liu, P.; Li, X.; Shen, J. The photocatalytic and antibacterial activities of neodymium and iodine doped TiO_2 nanoparticles. *Colloid Surf. B Biointerfaces* **2010**, *79*, 69–74. [CrossRef] [PubMed]
4. Pleskova, S.N.; Golubeva, I.S.; Verevkin, Y.K.; Pershin, E.A.; Burenina, V.N.; Korolichin, V.V. Photoinduced bactericidal activity of TiO_2 films. *Appl. Biochem. Microbiol.* **2011**, *47*, 23–26. [CrossRef]
5. Tang, H.; Chun, Y. TiO_2 photocatalytic reaction mechanism and dynamics research. *Prog. Chem.* **2002**, *14*, 662–672.
6. Kim, C.-H.; Lee, E.-S.; Kang, S.-M.; de Josselin de Jong, E.; Kim, B.-I. Bactericidal effect of the photocatalystic reaction of titanium dioxide using visible wavelengths on streptococcus mutans biofilm. *Photodiagnosis Photodyn. Ther.* **2017**, *18*, 279–283. [CrossRef] [PubMed]
7. Garvey, M.; Rabbitt, D.; Stocca, A.; Rowan, N. Pulsed ultraviolet light inactivation of pseudomonas aeruginosa and staphylococcus aureus biofilms: Pulsed UV inactivation of microbial biofilms. *Water Environ. J.* **2014**, *29*, 36–42. [CrossRef]
8. Johnston, J.A. Spray Polyurea Coating Systems. U.S. Patent US6797798, 28 September 2004.
9. Broekaert, M. Polyurea spray coatings: The technology and latest developments. (International). *Paint. Coat. Ind.* **2002**, *18*, 80–93.
10. Shen, C. *Spray Polyurea Waterproofing Coating*; Chemical Industry Press CN: Beijing, China, 2010; pp. 18–20.
11. Li, Y.; Wang, Q.; Li, S.; Yang, M. Preparation of super-hydrophobic surfaces with modified polyurethane and drag reduction in microchannel. *Acad. J. Nanjing Tech. Univ.* **2013**, *11*, 19–24.
12. El Saeed, A.M.; El-Fattah, M.A.; Azzam, A.M.; Dardir, M.M.; Bader, M.M. Synthesis of cuprous oxide epoxy nanocomposite as an environmentally antimicrobial coating. *Int. J. Biol. Macromol.* **2016**, *89*, 190–197. [CrossRef]
13. Lin, X.; Huang, X.; Zeng, C.; Wang, W.; Ding, C.; Xu, J.; He, Q.; Guo, B. Poly(vinyl alcohol) hydrogels integrated with cuprous oxide–tannic acid submicroparticles for enhanced mechanical properties and synergetic antibiofouling. *J. Colloid Interface Sci.* **2019**, *535*, 491–498. [CrossRef] [PubMed]
14. Palraj, S.; Venkatachari, G.; Subramanian, G. Bio-fouling and corrosion characteristics of 60/40 brass in Mandapam waters. *Anti-Corros. Methods Mater.* **2002**, *49*, 194. [CrossRef]
15. Gurianov, Y.; Nakonechny, F.; Albo, Y.; Nisnevitch, M. Antibacterial composites of cuprous oxide nanoparticles and polyethylene. *Int. J. Mol. Sci.* **2019**, *20*, 439. [CrossRef] [PubMed]
16. Vatanpour, V.; Madaeni, S.S.; Moradian, R.; Zinadini, S.; Astinchap, B. Novel antibiofouling nanofiltration polyethersulfone membrane fabricated from embedding TiO_2 coated multiwalled carbon nanotubes. *Sep. Purif. Technol.* **2012**, *90*, 69–82. [CrossRef]
17. Hosokawa, M.; Naito, M.; Nogi, K.; Yokoyama, T. Structural control of nanoparticles. In *Nanoparticle Technology Handbook*; Elsevier: Amsterdam, The Netherlands, 2012; Volume 49, pp. 51–112. [CrossRef]
18. Broekaert, M. Polyurea spray-applied systems: For concrete protection. (International). *Paint Coat. Ind.* **2003**, *19*, 70–80.
19. Gill, P.K.S. Assessment of biodegradable magnesium alloys for enhanced mechanical and biocompatible properties. *Proquest* **2014**, *17*, 201–208.
20. Zhao, A.; An, J.; Yang, J.; Yang, E.-H. Microencapsulated phase change materials with composite titania-polyurea (TiO_2–PUA) shell. *Appl. Energy* **2018**, *215*, 468–478. [CrossRef]
21. Chen, P.; Xie, H.; Huang, F.; Huang, T.; Ding, Y. Deformation and failure of polymer bonded explosives under diametric compression test. *Polym. Test.* **2006**, *25*, 333–341. [CrossRef]
22. Yuan, S.; Hu, Z.; Wang, X. Evaluation of formability and material characteristics of aluminum alloy friction stir welded tube produced by a novel process. *Mater. Sci. Eng. A* **2012**, *543*, 210. [CrossRef]
23. Godoy-Gallardo, M.; Wang, Z.; Shen, Y.; Manero, J.M.; Gil, F.J.; Rodriguez, D.; Haapasalo, M. Antibacterial coatings on titanium surfaces: A comparison study between single-species and multispecies biofilm. *ACS Appl. Mater. Interfaces* **2015**, *7*, 5992–6001. [CrossRef]

24. Garvey, M.; Andrade Fernandes, J.P.; Rowan, N. Pulsed light for the inactivation of fungal biofilms of clinically important pathogenic Candida species: Inactivation of Candida biofilms using pulsed light. *Yeast* **2015**, *32*, 422. [CrossRef] [PubMed]
25. Park, S.; Hu, J.Y. Assessment of the extent of bacterial growth in reverse osmosis system for improving drinking water quality. *J. Environ. Sci. Health Part A* **2010**, *296*, 968–977. [CrossRef] [PubMed]
26. Fujoka, K. Toxicity test: Fluorescent silicon nanoparticles. *J. Phys. Conf. Ser.* **2011**, *304*, 1–5. [CrossRef]
27. Lee, H.; Park, S.H.; Kim, S.J.; Kim, B.H.; Yoon, H.S.; Kim, J.S.; Jung, S.C. The effect of combined processes for advanced oxidation of organic dye using CVD TiO_2 film photo-catalysts. *Prog. Org. Coat.* **2012**, *74*, 758–763. [CrossRef]
28. Saunders, K.J. *Organic Polymer Chemistry*; Springer: Berlin, Germany, 1988; pp. 358–387.
29. Youssef, G.; Brinson, J.; Whitten, I. The effect of ultraviolet radiation on the hyperelastic behavior of polyurea. *J. Polym. Environ.* **2018**, *26*, 183–190. [CrossRef]
30. Li, B.; Zhang, Z.; Wang, X.; Liu, X. Investigation on the debonding failure model of anchored polyurea coating under a high-velocity water flow and its application. *Sustainability* **2019**, *11*, 1261. [CrossRef]
31. Sasan, P.; Houssam, T. Monotonic and fatigue performance of RC beams strengthened with a polyurea coating system. *Constr. Build. Mater.* **2015**, *101*, 22–29.
32. Zhang, P.S.; Wang, R.-M.; He, Y.; Song, P.; Wu, Z. Waterborne polyurethane-acrylic copolymers crosslinked core-shell nanoparticles for humidity-sensitive coatings. *Prog. Org. Coat.* **2013**, *76*, 729–735. [CrossRef]
33. Pons, L.; Délia, M.-L.; Bergel, A. Effect of surface roughness, biofilm coverage and biofilm structure on the electrochemical efficiency of microbial cathodes. *Bioresour. Technol.* **2011**, *102*, 2678–2683. [CrossRef]
34. Wang, B.; Wang, F.; Kong, Y.; Wu, Z.; Wang, R.-M.; Song, P.; He, Y. Polyurea-crosslinked cationic acrylate copolymer for antibacterial coating. *Colloids Surf. A* **2018**, *549*, 122–129. [CrossRef]
35. Vatanpour, V.; Madaeni, S.S.; Moradian, R.; Zinadini, S.; Astinchap, B. Fabrication and characterization of novel antifouling nanofiltration membrane prepared from oxidized multiwalled carbon nanotube/polyethersulfone nanocomposite. *J. Membr. Sci.* **2011**, *375*, 284–294. [CrossRef]
36. Kong, H.; Song, J.; Jang, J. Photocatalytic antibacterial capabilities of TiO_2-biocidal polymer nanocomposites synthesized by a surface-initiated photopolymerization. *Environ. Sci. Technol.* **2010**, *44*, 5672–5676. [CrossRef] [PubMed]
37. Namkung, E.; Rittmann, B.E. Soluble microbial products (SMP) formation kinetics by biofilms. *Water Resour.* **1986**, *20*, 795–806. [CrossRef]
38. Zhang, X.; Guo, Q.; Cui, D. Recent advances in nanotechnology applied to biosensors. *Sensors* **2009**, *9*, 1033–1053. [CrossRef] [PubMed]
39. Choi, H.; Stathatos, E.; Dionysiou, D.D. Photocatalytic TiO_2 films and membranes for the development of efficient wastewater treatment and reuse systems. *Desalination* **2007**, *202*, 199. [CrossRef]
40. Wandiyanto, J.V.; Linklater, D.; Tharushi Perera, P.G.; Orlowska, A.; Truong, V.K.; Thissen, H.; Ghanaati, S.; Baulin, V.; Crawford, R.J.; Juodkazis, S.; et al. Pheochromocytoma (PC12) Cell response on mechanobactericidal titanium surfaces. *Materials* **2018**, *11*, 605. [CrossRef] [PubMed]
41. Shahadat, M.; Teng, T.T.; Rafatullah, M.; Arshad, M. Titanium-based nanocomposite materials: A review of recent advances and perspectives. *Colloids Surf. B Biointerfaces* **2015**, *126*, 121–137. [CrossRef]
42. Nielsen, P.H.; Jahn, A.; Palmgren, R. Conceptual model for production and composition of exopolymers in biofilms. *Water Sci. Technol.* **1997**, *36*, 9–11. [CrossRef]
43. Khorshidi, B.; Biswas, I.; Ghosh, T.; Thundat, T.; Sadrzadeh, M. Robust fabrication of thin film polyamide-TiO_2 nanocomposite membranes with enhanced thermal stability and anti-biofouling propensity. *Sci. Rep. (Nature Publisher Group)* **2018**, *8*, 1–10. [CrossRef]
44. Horswill, A.R.; Stoodley, P.; Stewart, P.S.; Parsek, M.R. The effect of the chemical, biological, and physical environment on quorum sensing in structured microbial communities. *Anal. Bioanal. Chem.* **2007**, *387*, 371–380. [CrossRef]
45. Alpkvist, E.; Klapper, I. Description of mechanical response including detachment using a novel particle model of biofilm/flow interaction. *Water Sci. Technol.* **2007**, *55*, 265–273. [CrossRef] [PubMed]
46. Brindle, E.R.; Miller, D.A.; Stewart, P.S. Hydrodynamic deformation and removal of Staphylococcus epidermidis biofilms treated with urea, chlorhexidine, iron chloride, or DispersinB. *Biotechnol. Bioeng.* **2011**, *108*, 2968–2977. [CrossRef] [PubMed]

47. Davison, W.M.; Pitts, B.; Stewart, P.S. Spatial and temporal patterns of biocide action against Staphylococcus epidermidis biofilms. *Antimicrob. Agents Chemother.* **2010**, *54*, 2920–2927. [CrossRef] [PubMed]
48. Cogan, N.G. Two-fluid model of biofilm disinfection. *Bull. Math. Biol.* **2008**, *70*, 800–819. [CrossRef] [PubMed]
49. Machado, M.C.; Webster, T.J. Decreased Pseudomonas aeruginosa biofilm formation on nanomodified endotracheal tubes: A dynamic lung model. *Int. J. Nanomed.* **2016**, *2016*, 3825–3831.
50. Adán, C.; Marugán, J.; Mesones, S.; Casado, C.; van Grieken, R. Bacterial inactivation and degradation of organic molecules by titanium dioxide supported on porous stainless steel photocatalytic membranes. *Chem. Eng. J.* **2017**, *318*, 29–38. [CrossRef]
51. Eberl, H.J.; Sudarsan, R. Exposure of biofilms to slow flow fields: The convective contribution to growth and disinfection. *Theor. Biol.* **2008**, *253*, 788–807. [CrossRef]
52. Lappin-Scott, H.M.; Stoodley, P. The influence of fluid shear on the structure and material properties of sulphate-reducing bacterial biofilm. *Ind. Microbiol. Biotechnol.* **2002**, *29*, 347–353.
53. Besemer, K.; Singer, G.; Hödl, I.; Battin, T.J. Bacterial community composition of stream biofilms in spatially variable-flow environments. *Appl. Environ. Microbiol.* **2009**, *75*, 7189–7195. [CrossRef]
54. Dockery, J.; Klapper, I. Finger formation in biofilm layers. *SIAM J. Appl. Math.* **2001**, *62*, 853–869.
55. Marrelli, M.; Pujia, A.; Palmieri, F.; Gatto, R.; Falisi, G.; Gargari, M.; Caruso, S.; Apicella, D.; Rastelli, C.; Nardi, G.M.; et al. Innovative approach for the in vitro research on biomedical scaffolds designed and customized with cad-cam technology. *Int. J. Immunopathol. Pharmacol.* **2016**, *29*, 778–783. [CrossRef] [PubMed]
56. Asaro, R.J.; Lattimer, B.; Mealy, C.; Steele, G. Thermo-physical performance of a fire protective coating for naval ship structures. *Compos. Part A* **2009**, *40*, 11–18. [CrossRef]

© 2019 by the authors. Licensee MDPI, Basel, Switzerland. This article is an open access article distributed under the terms and conditions of the Creative Commons Attribution (CC BY) license (http://creativecommons.org/licenses/by/4.0/).

Review

Recent Progress in Functional Edible Food Packaging Based on Gelatin and Chitosan

Bianca Eugenia Ștefănescu [1], Carmen Socaciu [2] and Dan Cristian Vodnar [1,2,*]

[1] Institute of Life Sciences, University of Agricultural Sciences and Veterinary Medicine Cluj-Napoca, Calea Mănăștur 3-5, 400372 Cluj-Napoca, Romania
[2] Faculty of Food Science and Technology, University of Agricultural Science and Veterinary Medicine Cluj-Napoca, Calea Mănăștur 3-5, 400372 Cluj-Napoca, Romania
* Correspondence: dan.vodnar@usamvcluj.ro; Tel.: +40-(747)-341-881

Abstract: Nowadays, edible and eco-friendly packaging applications have been studied as an alternative to conventional/synthetic packaging due to the great interest of consumers in healthy, safe, and natural food, and of researchers in meeting the needs of consumers and producers. Various biopolymers are being extensively explored as potential materials for food packaging. The edible biopolymers utilized so far for packaging applications include proteins, lipids, and polysaccharides. Occasionally, these biopolymers have incorporated different bioactive substances to enhance the composite films' characteristics. Gelatin and chitosan are two of the most important biopolymers for the production of films. Different biopolymers or bioactive substances have been incorporated into the matrix to enhance the gelatin-based and chitosan-based films. By incorporating other biopolymers and bioactive compounds, the composite films' overall physicochemical and mechanical characteristics are improved. Additionally, by incorporating bioactive compounds (polyphenolic compounds, natural extracts, and essential oils), the composite films present important biological properties, such as antioxidant and antimicrobial activities.

Keywords: gelatin; chitosan; films; bioactive compounds; biopolymers; packaging application

Citation: Ștefănescu, B.E.; Socaciu, C.; Vodnar, D.C. Recent Progress in Functional Edible Food Packaging Based on Gelatin and Chitosan. *Coatings* **2022**, *12*, 1815. https://doi.org/10.3390/coatings12121815

Academic Editor: Domingo Martínez-Romero

Received: 30 October 2022
Accepted: 22 November 2022
Published: 24 November 2022

Publisher's Note: MDPI stays neutral with regard to jurisdictional claims in published maps and institutional affiliations.

Copyright: © 2022 by the authors. Licensee MDPI, Basel, Switzerland. This article is an open access article distributed under the terms and conditions of the Creative Commons Attribution (CC BY) license (https://creativecommons.org/licenses/by/4.0/).

1. Introduction

The main direction of food packaging is to preserve the quality and aspect of products, and this can be obtained by reducing lipid oxidation, inhibiting microbial growth, and therefore extending the shelf life of food products. The conventional packaging is mostly made from petroleum-based plastics [1,2]. The manufacturing of plastic globally has grown in the last decades, and 40% of the plastic produced is utilized in packaging applications [3]. Even though plastic is advantageous as a packaging element because it is low-priced, it has a light weight, and its facility in form molding, excellent mechanical strength, and thermal sealing, the sizeable utilization of plastic packaging may also result in unfavorable outcomes for the environment [4–6]. These adverse environmental effects of plastic packaging are related to its low biodegradability and reduced reuse and recycling. Therefore, large quantities of plastic may cause world contamination and pollution [7,8].

Nowadays, edible and eco-friendly packaging applications have been studied as an alternative to conventional/synthetic packaging, due to the great interest of consumers in healthy, safe, and natural food, and of researchers to meet the needs of consumers and producers and to obtain biodegradable and nontoxic films/coatings for the food industry. Therefore, various biopolymers are being extensively explored as potential materials for food packaging. The edible biopolymers tested so far for packaging applications include proteins, lipids, polysaccharides, and all achievable mixtures among these. Occasionally, these biopolymers or their combinations have incorporated different additives, such as antioxidants, antimicrobials, flavors, or colors, to enhance the characteristics of the films [4,9].

Additionally, controlling the migration of components from packaging materials to foods is essential because it could result in the transfer of undesirable compounds that could reduce the safety of food for consumption or change its sensory and nutritional properties [10]. However, migration may also be desirable whenever the incorporated compounds are meant to be released gradually over time to preserve the food from any unfavorable chemical reactions, and thus extending the shelf life [11]. For this reason, it is essential to also research the components that can be passed from the packaging to the food.

Protein and polysaccharides are the most suitable and renewable biopolymers for food packaging applications [12,13]. Films/coatings based on gelatin and chitosan have been studied intensively in the last few years [14–17].

This review highlights the recent progress in food packaging based on gelatin and chitosan or on combinations of them or with other biopolymers and bioactive compounds. The overview also provides the most important physical, chemical, mechanical, and biological characteristics of the obtained composite films and their possible applications for the food industry.

2. Gelatin

Gelatin is one of the important components of protein-based packaging, and it can be obtained from collagen by its partial hydrolysis. It possesses the capacity to form adequate films for the food packaging industry [18,19]. The primary rheological properties of gelatin are bloom and viscosity, and these properties are typically the outcome of the production process utilized. The average molecular weight, amino acid content, and chain polymerization level are all connected to the viscoelastic characteristics [20]. Commonly, gelatin is obtained from certain mammals, like pork and cow, or poultry [19,21].

Nowadays, alternative gelatin sources, such as gelatin from various fish species, are being explored [19,22]. Due to its functional qualities, such as its capacity to bind water, produce gels, operate as a gas barrier, form films, create foam, and have an emulsification property, gelatin is widely utilized in the food, pharmaceutical, photographic, and cosmetics sectors [23].

Although gelatin exhibits excellent gas barrier and swelling properties, it has poor mechanical resistance and is permeable to water vapor. Gelatin's poor water vapor barrier characteristic limits its application as a packing material. The limitations of gelatin can be improved by combining it with other valuable components [4,9,14,24].

According to previous research, the performance of gelatin for food packaging applications has been improved by combining gelatin with other biopolymers, such as chitosan, starch, soy protein, pectin, and carboxymethylcellulose (CMC) [25–30], or with other natural compounds, such as polyphenols [31–35] and essential oils [36–45]. For instance, combining gelatin with chitosan, starch, and tapioca starch improved the mechanical characteristics [25,26,29]. Incorporating polyphenols in the gelatin matrix was reported to add antioxidant and antimicrobial activities for the composite films [19,31,32,35]. Also, a gelatin–essential oil composite presented low WVP in comparison with the native gelatin films [37–39,41,42]. Figure 1 contains a schematic representation of gelatin types and origination, and the compounds that can be incorporated in the gelatin matrix.

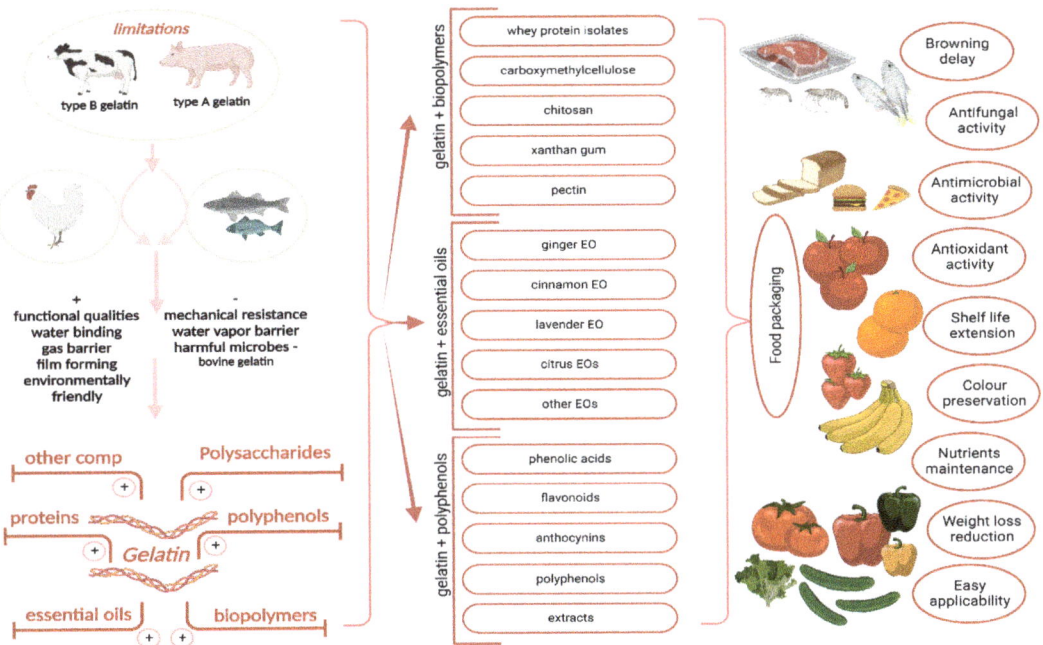

Figure 1. Schematic representation of gelatin matrix incorporated with functional materials for improving the characteristics of the composite films.

2.1. Origination of Gelatin

2.1.1. Gelatin Obtained from Mammals

Most gelatin sources originate from mammals, particularly cattle and pigs, with pig skin representing 46% of all sources, bovine hide approximately 29%, and pork and cattle bones 23% [46]. Due to their abundant availability, bovine and porcine skin gelatins are used extensively in the food sectors.

Gelatin is produced by partial hydrolysis of collagen, and is classified into two types.

Gelatin from bovine skin is usually defined as type B gelatin and is prepared through an alkaline method. In contrast, gelatin from porcine skin is defined as type A gelatin and is prepared through an acidic method. Type A gelatin has an isoelectric point at pH 7–9.4, and has higher amino acid content compared to type B, which has an isoelectric point at pH 4.8–5.5. Due to its greater gel characteristics (gel strength and viscosity), and powerful film-forming properties, mammalian gelatin is more often used than other sources [24].

Mammal gelatins, however, have considerable limitations and issues regarding religious concerns, as Muslims, Jews, or Hindus cannot use or ingest them for various reasons [47]. Furthermore, substitutes for porcine and bovine gelatin replacement have also been prioritized and taken into consideration due to the possible risk of transmitting harmful microbes from bovine spongiform encephalopathy, known as mad cow disease, and from foot and mouth diseases [48].

Therefore, due to these limitations of mammalian gelatin and the need to use gelatin from different sources, researchers' interest in gelatin obtained from other sources, like poultry and aquatic species, has grown considerably [47].

2.1.2. Gelatin Obtained from Poultry

Poultry skin, foot, and bone represent an alternative to mammalian gelatin. Duck, chicken, and turkey species are among the poultry sources used for this purpose. According to several studies, poultry gelatin is similar to mammalian gelatin in terms of its amino acids, secondary structure, and molecular weight (285 g/mol for poultry and 350 g/mol for mammal gelatin) [49,50].

Compared to bovine gelatin, the gel derived from chicken skin and chicken feet seems to have a much higher bloom value, according to Sarbon et al. [49] and Rahman and Jamalulail [51].

Meanwhile, according to Nik Muhammad et al. [52], commercial bovine gelatin had a bloom value of 217 g, but duck feet gelatin obtained by diverse acids treatment had a greater bloom strength (226–334 g). A higher percentage of cross-linked ß and α chain components results in high bloom strength and leads to higher melting temperatures and viscosity. Additionally, it was observed that the gelatin from chicken skin and duck feet contained amino acids like glycine, proline, hydroxyproline, and alanine, which helped increase the gel's strength and stability. Interestingly, it was found that the imino acid content (e.g., proline, hydroxyproline) of duck-feet- and chicken-skin-derived gelatins was higher than that of the bovine gelatin [37,49].

Gelatin derived from poultry products has good film-forming characteristics due to its high imino acid concentration and high bloom value [49,52].

2.1.3. Gelatin Obtained from Aquatic Species

An alternative to mammalian gelatin comes from marine sources, such as warm- and cold-water fish (skins, bones, and fins). Marine gelatin sources are not connected to the risk of bovine spongiform encephalopathy epidemics. In addition, it is suitable for usage by Muslims, Jews, or Hindus, where mammalian gelatin is prohibited [47].

Fish processing byproducts can be used as an alternative raw material for preparing high-protein ingredients, since protein makes up the majority of components of most fish. This is especially true for producing food-grade gelatin due to the significant amounts of collagen in fish [22].

Compared to mammalian gelatin's bloom values, fish gelatin often has a lower bloom value due to the variances in proline and hydroxyproline content, which depends on the fish species and environment temperature. Depending on the type of fish, the environment, and the extraction technique utilized, there may be variations in viscosity values. Proline and hydroxyproline concentrations in fish gelatins are typically lower than those in the mammalian gelatins [24].

However, several studies have reported that warm-water fish gelatins have higher imino acid levels when compared to cold-water ones [53,54].

Regarding film characteristics, fish gelatin shows potential linked to attributes such as remaining translucent, nearly colorless, water-soluble, and very extensible [55].

2.2. Gelatin-Based Composites
2.2.1. Combined Gelatin and Other Biopolymers

The variety of biopolymer combinations' physical, chemical, and textural features have been the subject of intense research in the last years to generate novel products. The formulation and characterization of combined gelatin–biopolymers films are summarized in Table 1.

In a study published by Howell, three ways were outlined that proteins might be described in terms of how they interact with other biopolymers: synergistic interactions, aggregation, and phase separation. These three characteristics may lead to fascinating and technically valuable applications. It has been noted that synergistic interactions can improve gelation qualities beyond those of the individual protein utilized alone [56], and they were observed in gelatin–whey protein isolate [57] and whey–egg albumen mixtures [58]. Proteins aggregation may occur due to electrostatic interactions, and it could be useful for improving gelation in case of β-lactoglobulin [59]. Combining two biopolymers may also occur with phase separation. In the composite obtained, the biopolymers produced separate phase networks. Phase separation has been reported in protein–protein and protein–polysaccharide mixtures [57].

Sarbon et al. [57] investigated the physical, thermal, and microstructural properties of the gelatin–whey protein isolate mixture by using a large deformation rheological test (heating gelation followed by cooling and compression). All combinations of gelatin and whey protein isolate produced gel strength values that were higher than anticipated, indicating a synergistic interaction and improvement of the gelling properties of both the gelatin and whey proteins.

Several studies also outlined the possible mixture between protein and polysaccharides. Gelatin–chitosan composite was prepared and characterization was performed of its physical and mechanical properties. The findings showed that adding chitosan significantly increased the elastic modulus (EM) and tensile strength (TS), making the films stronger than gelatin films. However, adding chitosan significantly lowered the elongation at break (EAB) characteristic. According to the structural characteristics examined, gelatin and chitosan interacted to create a novel material with improved mechanical performance [25].

Incorporating gelatin with CMC also highlighted some important modifications, such as increased TS and puncture test of the films, water vapor permeability (WVP), reduced EAB, opacity, and UV-light penetration of the films, and it increased the thermal stability. By using Fourier transforms infrared spectroscopy (FTIR) and X-ray diffraction (XRD) analyses, it has been confirmed that the functional groups of gelatin interact strongly with CMC. With the addition of CMC to gelatin-based film, crosslinking and intermolecular bonds were established and improved some of the mechanical and physical properties of the film [60].

The mechanical and physical characteristics of edible films were found to be affected by the addition of potato starch to gelatin-based film. With increasing potato starch concentrations, the TS, EM, transparency, thermal characteristics, WVP, ultraviolet, microscopy, and visible light barrier transmission improved, while the EAB lowered. This may suggest promising developments for the insertion of potato starch as a potential crosslinking agent to enhance the mechanical and physical characteristics of gelatin-based films, particularly in the context of the production of food packaging materials [61].

A formulation of three polymers was also studied. Gelatin was combined with CMC and chitosan [62,63]. Jahit et al., showed that chitosan and the CMC addition greatly impact the film's characteristics. The film's amorphous nature was minimized by making it more crystalline as the chitosan concentrations increased. Given that the formulation's gelatin/CMC/chitosan ratio of 60/30/10 exhibited the second-lowest WVP (2.250×10^{-7} g·mm·h^{-1}·cm^{-2}·Pa^{-1}) and the highest biodegradability rate, it seems ideal for prospective usage in the food packaging [62].

Table 1. The characteristics of the gelatin–biopolymers mixtures.

Formulation	Physical/Chemical/Mechanical/Biological Characteristics	References
Gelatin, whey protein isolate	synergistic interaction ↑ gelling properties, EM	[57]
Gelatin, soy protein isolate	↑ mechanical properties when the weight ratio of soy protein isolate: gelatin is 1:3	[64]
Gelatin, soy protein isolate	↑ TS, EAB, EM, flexibility	[27]
Gelatin, chitosan	↑ mechanical properties ↓ permeability good UV-light protection qualities	[25]
Gelatin, CMC	↑ TS, puncture test of film, thermal stability, WVP, ↓ EAB, opacity, and UV-light penetration of the films	[60]
Gelatin, CMC, chitosan	↓ WVP ↑ biodegradability	[62]
Gelatin, CMC, chitosan	↑ flexibility, EAB, WVP, thickness ↓ TS and puncture force	[63]
Gelatin, chitosan, xanthan gum	↑ thickness, WVP, UV-light protection, thermal stability, ↓ TS, EAB, VIS light transparency	[28]
Gelatin, starch	↑ mechanical strength, water solubility (WS), WVP, thickness ↓ opacityimproved appearance of refrigerated Red Crimson grapes	[26]
Gelatin, potato starch	↑ TS, EM, WVP, melting temperature, UV–VIS light protection ↓ WS, EAB	[61]
Gelatin, tapioca starch	↑ TS, EAB, thickness, WVP, UV-light protection, thermal stability visible light transmission, film transparency	[29]
Gelatin, pectin	↑ thickness, TS, antioxidant, and antibacterial activities ↓ WVP, EAB	[30]

↑-increased values of the tested characteristics, ↓-decreased values of the tested characteristics.

2.2.2. Combined Gelatin and Polyphenols/Extracts Rich in Polyphenols

An extensive and increasing list of bioactive substances have been or are now being integrated into films, with phenolic compounds (polyphenols, phenolic acids, flavonoids, anthocyanin) (Table 2) being the most prevalent [31,32,65–68].

Plant extracts represent an important source of polyphenols. These compounds have antioxidant and antimicrobial effects; therefore, the incorporation of polyphenols in the biopolymers matrix leads to composite films with antioxidant and antimicrobial activities. Based on that result, several studies have used these compounds in bioactive and biodegradable films [19,35,66,69–72].

The capacity of the active groups in gelatin-based films to quench radicals plays a significant role in the antioxidant activity of those films. The gelatin protein's amino acid groups provide to native gelatin films poor antioxidant activity [73,74]. However, a gelatin film's antioxidant activity is increased when it is conjugated with diverse phenolic compounds [19,31,32,67,71,74].

Both 2,2-azino-di-3-ethylbenzthiazoline-6-sulfonate (ABTS) and 2,2-diphenyl-1-picrylhydrazyl (DPPH) radical scavenging and reducing power analyses are used to measure the antioxidant activity of gelatin-based films. Using the ABTS and DPPH scavenging methods, Hanani et al., evaluated the antioxidant capacity of gelatin-based films. The radical scavenging ability for the control (only gelatin) film was 32% for ABTS and 53% for DPPH assay. However, the film's antioxidant activity was significantly enhanced by adding 1% pomegranate peel powder, and the ABTS and DPPH scavenging activities rose to 48% and 60%, respectively. The antiradical activity was significantly improved with the addition of pomegranate peel powder, as seen by the increase in ABTS and DPPH radical scavenging activity as pomegranate peel powder concentration increased in the gelatin-based films. The ABTS and DPPH radical scavenging activity of the gelatin film with 5% pomegranate peel powder was the highest, with 80% and 72%, respectively. Pomegranate is well known for its high content of bioactive substances, including phenolic compounds and anthocyanins, which are strong antioxidants and may scavenge DPPH and ABTS radicals [71].

Gelatin-based films are also tested for their antioxidant properties using the reducing power assay. Wu et al., observed that the reducing power activity of gelatin film was low and gelatin film with 0.7% green tea extract exhibits a reducing power of 65% of 1.0 mg/mL vitamin C. These findings showed that the gelatin–green tea extract film's antioxidant activity was enhanced in a concentration-dependent manner in comparison with the gelatin film without the extract [69].

Similar results were obtained when gelatin films incorporated rosmarinic acid [67], chlorogenic acid [65] and grape seed extract, and gingko leaf extract [70].

Gelatin-based films with antimicrobial properties are a crucial barrier in preventing the spread of foodborne infections. Incorporating rosmarinic acid into gelatin film provides antimicrobial action with a lengthy half-life [74]. Pathogenic bacteria such as *Escherichia coli* and *Staphylococcus aureus* aggregated following treatment with gelatin-rosmarinic acid films, exhibiting a morbid morphology and afterward being fully lysed. Even after three months of storage, it was discovered that gelatin–rosmarinic acid films possessed substantial antimicrobial activity, indicating that these films offer major benefits in food packaging [74].

The antimicrobial activity of gelatin-based films containing phenolic compounds or plant extracts rich in polyphenolic content was also reported by several studies. Gelatin–protocatechuic acid [31], gelatin–epigallocatechin gallate [32], gelatin–tannic acid [68], gelatin–mangrove extract [19], gelatin–pomegranate peel extract [71], and gelatin–date by-products [72] films displayed good antimicrobial activity against Gram-negative (*E. coli*) and Gram-positive (*S. aureus*) bacteria.

Fu et al., obtained a gelatin–chlorogenic acid film with antioxidant and antimicrobial activity against *E. coli*, *Pseudomonas aeruginosa*, *Listeria monocytogenes*, and *S. aureus*, with potential applications in fresh seafood preservation [65].

Moreover, phenolic compounds can form hydrogen and hydrophobic interactions. Hydrogen and hydrophobic interactions between the hydroxyl groups present in the aromatic rings of phenolic compounds and the carboxyl groups of gelatin side chains might improve gelatin film functional characteristics [73].

One study designed a gelatin film including haskap berries extract, where the phenolic components of the extract, mainly anthocyanins and phenolic acids, generated hydrogen crosslinking between the hydroxyl groups of the phenolic compounds and the amino/hydroxyl groups of the gelatin. This crosslinking improved mechanical strength, flexibility, air, WVP, film brightness, and WS. Consequently, haskap berries extract enhanced the capacity of gelatin composite films for application in active packaging [75].

The incorporation of phenolic compounds in gelatin-based films led to the improvement of the functional properties. The protection and tamper-resistance of food packaging are significantly influenced by the TS of the packaging materials. Higher tensile strengths are typically chosen for a range of packing items as they provide a stronger seal with safe load stability and help to produce higher-quality products for the customer [24].

Several studies showed that phenolic compounds increased gelatin-based film TS yield [19,66–69,71,75]. EAB is the ratio of the modified length to the starting length when the sample is damaged. It refers to the ability of a plastic sample to withstand shape changes without developing cracks [24].

Various researchers concluded that adding natural extracts, such as phenolic compounds, to gelatin-based films increased the films' extensibility and EAB values. These compounds may have a plasticizing effect on the resulting films [19,31,67,74,75].

Good oxygen and moisture protection properties are essential for food packaging films since too much oxygen or moisture can cause lipid oxidation and microbiological degradation of food during transport or storage. Packed products' quality and storage life can be significantly enhanced when the packaging films serve as practical barriers against oxygen or water. As a result, it is essential to keep WVP as low as achievable. Currently, gelatin film linked with phenolic compounds can produce composite films with WVP lower than simple gelatin films [19,31,35,67–70,74,75].

Table 2. The characteristics of gelatin–polyphenol mixtures.

Formulation	Physical/Chemical/Mechanical Characteristics	Biological Properties	Applications	References
Gelatin, protocatechuic acid	↑ thickness, EAB achieved fine look, ↓ light transmittance, TS, WVP	Antioxidant activity (DPPH), antimicrobial activity against *E. coli* and *S. aureus*, with high protocatechuic acid amounts.	Beef preservation	[31]
Gelatin, epigallocatechin gallate (EGCG)	↑ bloom strength	Antioxidant activity (DPPH (50%–99%), FRAP (200–662 μg Vc/g)), antimicrobial activity against *E. coli* and *S. aureus*	Active packaging	[32]
Gelatin, *Galla chinensis* extract	↑ gel strength and thermal stability, ↓ swelling of gelatin	Not determined	Packaging	[33]
Gelatin, eugenol/β-cyclodextrin emulsion	not determined	Reduced the H_2S-producing bacteria, total viable *Pseudomonas* spp. and Psychrophilic counts, total volatile basic nitrogen, K value, free fatty acids	Chinese Seabass during superchilling storage	[34]
Gelatin, mango peel	↓ WVP, solubility films more rigid and less flexible	Antioxidant activity (DPPH 70%–85%)	Active packaging	[35]
Gelatin, green tea extract grape seeds extract gingko leaf extract	↓ TS, EAB, lowest WVP lowest TS, EAB, ↓ WVP ↓ TS, EAB, WVP	All the films presented antioxidant activity (DPPH)	Active food packaging	[70]
Gelatin, *Fructus chebulae* extract	↑ gel strength, thermal stability	Not determined	Packaging	[76]
Gelatin, chlorogenic acid	not determined	Antioxidant activity (ABTS), antimicrobial activity against *E. coli*, *P. aeruginosa*, *L. monocytogenes*, and *S. aureus*	Fresh seafood preservation	[65]
Gelatin, epigallocatechin gallate	↑ TS, EM, ↓EAB	Antioxidant activity (DPPH 67%)	Reduce the oxidation of cod-liver oil	[66]
Gelatin, green tea powder	↓ TS, EM, EAB with high amounts of green tea powder	Antioxidant activity (DPPH 77%)	Reduce the oxidation of cod-liver oil	[66]
Gelatin, green tea extract	↑ TS↓ EAB, WS, WVP	Antioxidant activity (DPPH 15%–55%)	Active packaging	[69]
Gelatin, rosmarinic acid	↑ thickness, TS, EAB, light protection, ↓ WS, WVP	Antioxidant activity (DPPH 75%–90%)	Bacon preservation	[67]
Gelatin, rosmarinic acid	↑ EAB, ↓ TS, EM, WVP	Antioxidant activity (ABTS), antimicrobial activity against *E. coli* and *S. aureus*	Active packaging	[74]
Gelatin, tannic acid	↑ TS ↓ EAB, WVP, oxygen permeability	Antimicrobial activity against *E. coli* and *S. aureus*	Cherry tomatoes, grapes	[68]
Gelatin, mangrove extracts	↑ thickness, EAB, TS ↓ WVP	Antioxidant activity (DPPH 15%–60%), antimicrobial activity against *S. aureus*, *E. coli*, *Bacillus subtillis*, *Salmonella* sp.	Active packaging	[19]
Gelatin, pomegranate peel powder	↑ thickness, WVP, TS ↓ film solubility, EAB	Antioxidant activity (DPPH 59%–72%, ABTS 48%–80%), antimicrobial activity against *S. aureus*, *L. monocytogenes*, and *E. coli*	Active packaging	[71]
Gelatin, haskap berries extract	↑TS, EAB ↓WVP, WS	Antioxidant activity (DPPH)	Shrimp spoilage	[75]
Gelatin, date by-products	↓water holding capacity, WS color change	Antimicrobial activity against *E. coli* and *S. aureus*	Active packaging	[72]

↑-increased values of the tested characteristics, ↓-decreased values of the tested characteristics.

2.2.3. Combined Gelatin and Essential Oil

The antimicrobial and antioxidant capacities of the essential oils from plants and spices make them valuable food additives. Moreover, by reducing lipid oxidation, essential oils can increase the shelf life of food products. Terpenic and phenolic compounds, biologically active substances, are abundant in essential oils. Additionally, most of them are declared to be Generally Recognized as Safe. However, due to their robust flavor, their application as food preservatives is frequently restricted. Therefore, essential oils can be added to the edible film to avoid this issue [77,78]. The formulation and characterization of combined gelatin–essential oil films are summarized in Table 3.

Table 3. The characteristics of gelatin–essential oil mixtures.

Formulation	Physical/Chemical/Mechanical Characteristics	Biological Properties	Applications	References
Gelatin, ginger essential oil	↑ thickness, WVP, EAB, ↓ TS	Antimicrobial activity against *E. coli* and *S. aureus*	Antimicrobial active packaging	[36]
Gelatin, cinnamon leaf oil	↓ TS, slightly decreased WVP	Antimicrobial activity against *Salmonella typhimurium*, *E. coli*, *L. monocytogenes*, and *S. aureus*	Cherry tomatoes	[37]
Gelatin, oregano essential oil	Insignificant modification	Antioxidant activity (DPPH 12%–60%, FRAP), antimicrobial activity against *E. coli* and *S. aureus*	Food active packaging	[38]
Gelatin, lavender essential oil	↓ WVP, TS	Antioxidant activity (DPPH 1%–9%, FRAP), antimicrobial activity against *E. coli* and *S. aureus*	Food active packaging	[38]
Gelatin, thyme essential oil	↑ EAB, ↓ TS, WVP	Antimicrobial activity against *L. monocytogenes* and *E. coli*	Chicken tenderloin packaging	[39]
Gelatin, citrus essential oils (bergamot, kaffir lime, lemon, lime)	↑TS, ↓EAB, WVP (glycerol 20%)↑ EAB, ↓ TS (glycerol 30%)	Antioxidant activity (DPPH, ABTS, FRAP)	Active packaging	[40]
Gelatin, root essential oils (ginger, turmeric, plai)	↑ EAB, ↓ TS and WVP	Antioxidant activity (DPPH, ABTS)plai > turmeric > ginger essential oils	Active packaging	[41]
Gelatin, *Zataria multiflora* (thyme-like plant) essential oil	↑ WVP, EAB, light barrier properties, ↓ TS	Antioxidant activity (ABTS), antimicrobial activity against *P. aeruginosa*, *E. coli*, *S. aureus*, *B. subtilis*	Antioxidant, antimicrobial active packaging	[43]
Gelatin, essential oils (bergamot, lemongrass)	↓ TS, EAB, WVP (lemongrass), solubility, transparency ↑ heat stability	Antimicrobial activity Lemongrass: *E. coli*, *L. monocytogenes*, *S. aureus*, *S. typhimurium* Bergamot: *L. monocytogenes*, *S. aureus*	Active packaging	[42]
Gelatin, essential oils (clove, garlic, origanum)	↓ thickness, WS, EABslightly decreased WVP	Antioxidant activity (DPPH 38%–72%), antimicrobial activity against *Brochothrix thermosphacta*, *Listeria innocua*, *L. monocytogenes*, *Shewanella putrefaciens*	Biodegradable food packaging systems	[44]
Gelatin, sage essential oil	↓ WVP, ↑ thickness	Antimicrobial activity against *E. coli*, *S. aureus*, *L. innocua*, *Saccharomyces cerevisiae*, *Penicillium expansum*	Fruits, vegetables, and meat packaging	[45]

↑-increased values of the tested characteristics, ↓-decreased values of the tested characteristics.

Essential oils added to edible films, in this case, gelatin films, lead to an increase in the gelatin film's biological activity and water resistance [78].

In recent years, researchers have focused on the analysis of incorporating essential oils into gelatin-based films. In a study conducted by Tongnuanchan et al., citrus essential oils were added to a gelatin-based film, which decreased WVP, and the obtained films displayed antioxidant activity [40].

Similar results, such as decreased WVP, were obtained when the gelatin-based films incorporated cinnamon leaf essential oil (0.5 %) [37], lavender essential oil (2000–6000 ppm) [38], thyme essential oil (0.5, 1, 1.5%) [39], root essential oil (ginger, turmeric, plai, different levels, 25%, 50%, and 100%, based on protein content) [41], lemongrass essential oil (5%–25% (w/w protein)) [42], garlic and clove essential oil (1 µL/cm^2 of plates) [44], and sage essential oil (2 mL/100 mL distilled water) [45].

Li et al., reported that ginger essential oil was incorporated into gelatin-based films and led to increased WVP when the ratio of oil/gelatin rose from 0% to 12.5% [36].

Kavoosi et al., reported a similar result after the WVP of a gelatin–*Zataria multiflora* composite increased when the oil/protein ratio changed from 0 to 8% [43].

Given that the WVP relies on the hydrophilic/hydrophobic ratio of the film's compounds, the specific composition of essential oils may be responsible for the observed variations in reported different WVP. Even so, adding a hydrophobic material will not necessarily diminish the WVP of the films; it also depends on how the added lipids affect the microstructure of the composite film [44]. An essential quality of food packaging materials is WVP. Loss of textural characteristics and subsequent microbiological growth in foods could result from moisture from the atmosphere moving into food products. In light of this, a lower WVP may offer good water barrier properties in the gelatin–essential oil films [39].

The antimicrobial activity of the gelatin-based films incorporating essential oils was reported for several strains (Table 3). Li et al., prepared gelatin-based films by combining it with low-content (0%–1%) ginger essential oil. The antimicrobial activity of films that incorporated ginger essential oil was tested on *E. coli* as Gram-negative and *S. aureus* as Gram-positive bacteria. A higher log CFU/mL value indicates better antimicrobial efficiency. Low antimicrobial activity was detected on gelatin-based film for both strains. As the amount of ginger essential oil in the films grew, so did their antibacterial activity. The 1% gelatin–ginger essential oil film reported the greatest antimicrobial activity, with values of 2.65 log CFU/mL against *E. coli* and 5.63 log CFU/mL against *S. aureus* [36].

Yang et al., discovered similar results when they investigated gelatin-based films' antimicrobial activity with cinnamon leaf essential oil against *E. coli*, *S. typhimurium*, *L. monocytogenes*, and *S. aureus*. In contrast to the gelatin film, which did not prevent bacterial pathogens from growing, the inhibitory zone grew in proportion to the cinnamon leaf essential oil concentration. Additionally, the antimicrobial activities were more efficient against Gram-positive bacteria than against Gram-negative ones [37].

Similar results, where antimicrobial activities are more efficient against Gram-positive bacteria than against Gram-negative ones, were obtained when gelatin-based films incorporated other essential oils, such as thyme essential oil [39], *Zataria multiflora* essential oil [43], bergamot and lemongrass essential oils [42], clove and garlic essential oils [44], and sage essential oil [45].

The mechanism of action of oils against bacteria is attributed to cytoplasm loss due to phospholipid cellular wall degradation, or due to interactions among oils and cell enzymes. Because an external lipopolysaccharide wall or proteins protect the peptidoglycan cell wall in the outer membrane, Gram-negative bacteria are more resistant to oil attack [45,79].

3. Chitosan

Chitosan (Figure 2) is a polysaccharide-related chemical compound and a copolymer of N-acetylglucosamine and glucosamine residues linked by -1,4-glycosidic bonds. Because of its availability and low price, biocompatibility, non-toxicity, biodegradability, and film-forming properties, chitosan is regarded as the most promising replacement for conventional plastics in the production of films/coatings, with a wide range of uses in many fields, including food application, pharmaceuticals, agriculture, and beauty products [80–83].

Chitosan products are very viscous, closely resembling natural gums with antimicrobial activities due to active amino groups, and they can constitute clear films to improve the quality and shelf life of processed and fresh foods [6,84]. Due to its capacity to form a partially permeable, durable, and flexible film, chitosan can be used to create edible films that can change the internal atmosphere, reduce water loss, and postpone the spoilage of fruits and vegetables. These characteristics give chitosan advantages over other edible coatings [85,86].

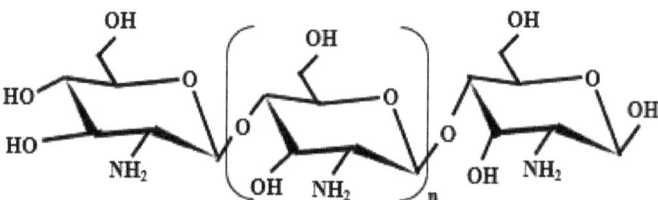

Figure 2. The chemical structure of chitosan.

In contrast to these associated advantages, chitosan-based films have drawbacks such as low UV-light barrier properties and reduced mechanical characteristics. In addition, the hydrophilicity of chitosan films makes them extremely susceptible to moisture, which is a significant disadvantage for packaged food products with high water content. Despite chitosan's implicit antioxidant and antimicrobial properties, these might not be sufficient to avoid severe growth of microorganisms and oxidation in the ambient environment. Hence, adding natural compounds, such as phenolic compounds, plant extracts, and essential oils or other biopolymers, may provide better antioxidant and antimicrobial activities and enable the development of packaging films with improved mechanical, physical, and biological characteristics [87].

3.1. Origination of Chitosan

Chitosan is obtained from chitin. Chitin (Figure 3), after cellulose, is the second-most prevalent structural polysaccharide in nature. Because of its acetyl groups, chitin has few applications; however, the deacetylation process transforms chitin into chitosan. The acetyl group in chitin is changed into hydroxyl and amino groups in the chitosan during the deacetylation process [88,89].

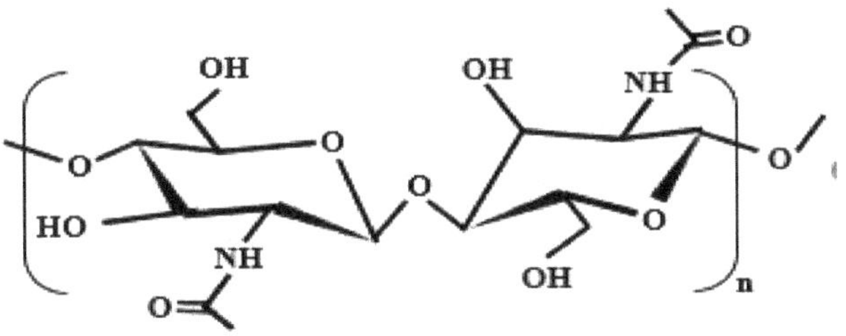

Figure 3. The chemical structure of chitin.

Chitin is naturally found as organized crystalline microfibrils that serve as structural elements for funguses or yeast cell walls and the external skeletons of arthropods. Crab and shrimp shells are currently the primary commercial sources of chitin, where it is present in the α-chitin form [90]. Another significant source of chitin is squid, found in the β-form, which has been reported to be more susceptible to deacetylation. Due to the substantially weaker intermolecular hydrogen link caused by the parallel configuration of the major chains, this chitin also exhibits increased properties such as higher solubility and reactivity, and a better affinity for solvents and swelling than the α-chitin [91]. Chitin can be found in the γ-form, mainly in fungi and yeast, as a mixture of the α- and β-forms instead of a distinct polymorph [92]. Algae, fungi, bacteria, and some species of insects can also serve as substitute sources of chitin and chitosan [88,89,92].

To extract chitin from the shell, protein and minerals must be removed by deproteinization and demineralization. Additionally, a discoloration step is added [88,92].

Typically, "chitosan" is a group of polymers created following the variable degrees of chitin deacetylation [90]. In reality, chitin and chitosan are distinguished by the degree of deacetylation, which represents the equilibrium of the two types of residues. Chitosan is a product that has a deacetylation degree greater than 50% [92]. Deacetylation also results in a depolymerization process, as shown by modifications in chitosan's molecular weight. Using an enzymatic or a chemical procedure, chitin can be transformed into chitosan [88]. Because of their low costs and efficiency for mass production, chemical techniques are frequently utilized to produce chitosan for commercial usage [90].

3.2. Chitosan-Based Composites

3.2.1. Combined Chitosan and Other Biopolymers

Films and coatings produced from chitosan have some disadvantages, including low water resistance, low UV-light barrier properties, and reduced mechanical characteristics, when compared to films formed by mixing two or more biopolymers, rendering them unsuitable for use in films/coatings production in the food industry [17,87].

Several naturally occurring biopolymers, such as polysaccharides (e.g., cellulose, pectin, starch, or alginate) and proteins (e.g., protein isolate, gelatin, or collagen), can be combined with chitosan to create films (Figure 4). As the produced films are affordable, stable, and display improved properties (water and thermal stability, mechanical or biological properties), polysaccharide blends generally provide several advantages over other biopolymer blends [17]. The physical, mechanical, and biological properties of chitosan films that have incorporated other biopolymers for packaging materials have been investigated (Table 4) [93–99].

Table 4. The characteristics of the chitosan–biopolymers mixtures.

Formulation	Physical/Chemical/Mechanical Characteristics	Biological Properties	Applications	References
Chitosan, corn starch	↑ WS, TS, EAB, ↓ WVP by comparison with corn starch film color change	Not determined	Active packaging	[93]
Chitosan, starch	↑ thickness and WS, ↓ WVP	Antimicrobial activity against *L. innocua*	Active packaging	[94]
Chitosan, sporopollenin	↓ thickness, light transmittance, ↑ TS, EAB, Young's modulus successfully incorporate sporopollenin into chitosan, enhanced hydrophobicity of films	Antifungal activity against *Aspergillus niger*, antioxidant activity	Active packaging	[95]
Chitosan, pectin	↑ thickness, WVP, WS, TS, EAB, Young's modulus ↓ density and opacity	Not determined	Packaging	[96]
Chitosan, nanocellulose	↑ thermal stability, oxygen barrier properties, thickness, WVP, TS, Young's modulus, ↓ film's transparency	Antimicrobial activity against *S. aureus*, *E. coli*, and *Candida albicans*	Chicken meat	[97]
Chitosan, Sardinella protein isolate	↑ thickness, moisture content, opacity, UV–VIS light barrier, WS, ↓ WVP, TS, and EAB, color change	Antioxidant activity (DPPH), antimicrobial activity against *S. aureus*, *Micrococcus luteus*, *L. monocytogenes*, *Bacillus cereus*, *Salmonella enterica*, *P. aeruginosa*, *E. coli*, *Klebsiella pneumoniae*	Shrimp packaging	[98]
Chitosan, CMC, sodium alginate	The optimal contents of the chitosan, CMC, and sodium alginate for the preparation of this composite film were 1.5%, 0.5%, and 1.5%. ↑ TS, EAB, WVP	Antimicrobial activity against *E. coli* and *S. aureus*	Packaging	[99]

↑-increased values of the tested characteristics, ↓-decreased values of the tested characteristics.

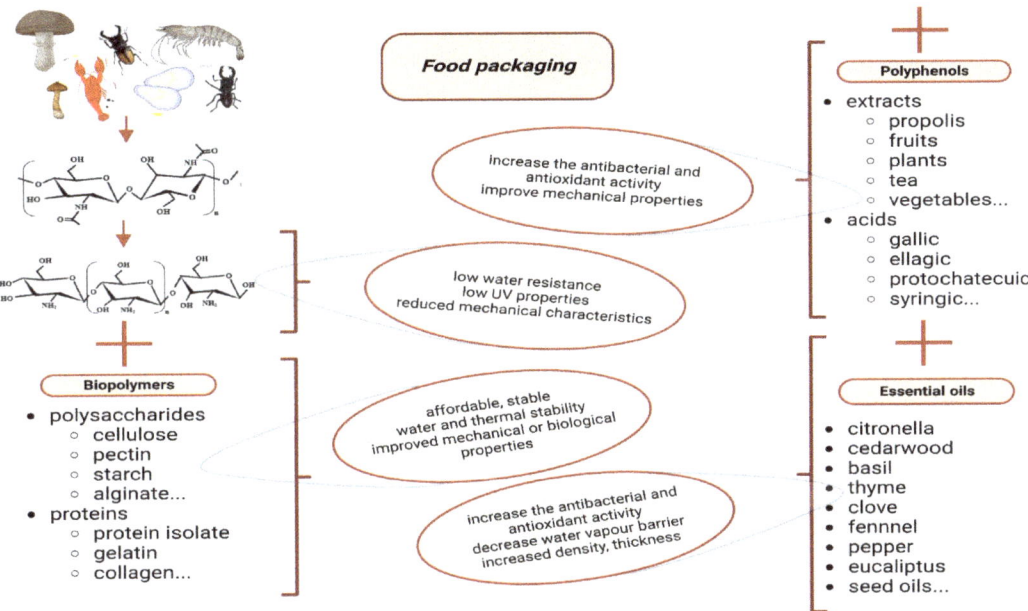

Figure 4. Schematic representation of chitosan matrix incorporating functional materials for improving the characteristics of the composite films.

For example, chitosan was combined with corn starch. The cross sections of films exhibited a continuous surface devoid of a blurry and porous structure, devoid of phase separation among the two polymeric materials, and a compact design in the polymer composite [93]. The study's findings demonstrated that chitosan could interact with corn starch to establish hydrogen bonds between the NH_3^+ of the chitosan and the OH^- of the corn starch, which enhanced the composite film's mechanical characteristics, TS and EAB, while lowering its WVP, properties needed for films used in the food industry packaging [93].

Similar results were obtained in the study published by Escamilla-García et al., where chitosan was combined with starch to enhance the composite film's physical, mechanical, and biological properties. Additionally, the chitosan–starch films presented antimicrobial activity against *L. innocua*, which indicates that these composites could be used to ensure the safety of food products in the packaging industry [94].

Films produced from the sporopollenin–chitosan blend were developed and characterized for the first time in a study published by Kaya et al. [95]. Sporopollenin is a biopolymer obtained from plant pollens; in that study, pollens of *Betula pendula* were used, that possess outstanding properties, including biocompatibility with other materials, nontoxicity, and biodegradability, as well as good thermal and strong acid and basic solutions resistance. To take advantage of these essential benefits, and the fact that this biopolymer is easy to collect and available in nature, sporopollenin samples were mixed into chitosan film to obtain a composite with improved characteristics. The incorporation of sporopollenin into the chitosan matrix has been confirmed by several analyses. The incorporation of growing quantities of sporopollenin into chitosan-based films was favorable, considering that physical, chemical, and mechanical characteristics were improved, and the films' hydrophobicity and biological (antioxidant and antifungal) properties were enhanced. These results indicate that sporopollenin could be recommended as a material for manufacturing chitosan-based composites [95].

Younis et al. [96] have developed a biodegradable packaging material by combining chitosan with pectin. While chitosan and pectin may be used separately due to their capacity to produce films, the current research found that mixing chitosan and pectin could develop a composite film with better characteristics than either of its parts. Combining chitosan with pectin may have synergistic effects that enhance several film properties, notably mechanical characteristics. The chitosan and pectin intermolecular interactions (hydrogen bonds, hydrophobic interactions, and ionic complexation) significantly enhanced the network of polymers in the film matrix. They allowed the TS, Young's modulus, and EAB to increase, which were likely responsible for these synergistic effects [96].

Recently, Costa et al. [97] evaluated the characteristics of chitosan–nanocellulosebased films. Combining nanocellulose into other biopolymeric matrices makes possible the development of greater composites while maintaining their biodegradability and improving their mechanical properties and barrier characteristics [100]. Therefore, in this study, the addition of nanocellulose enhanced the thermal stability and oxygen barrier properties and slightly increased WVP. The improvement of the mechanical characteristics of the chitosan/nanocellulose-based films was noticed by the rise in TS and Young's modulus of the composite. Additionally, these composite films exhibited antimicrobial activity against *E. coli*, *S. aureus*, and *Candida albicans*. The reduction of total volatile basic nitrogen on the surface of the chicken meat by chitosan/nanocellulose-based films suggests their potential application as packaging for retarding beef deterioration [97].

Moreover, Sardinella protein isolate, obtained from blue crab and *Sardinella aurita* by-products, was used as a biopolymer in a study conducted by Azaza et al. [98]. The protein isolate was incorporated into the chitosan matrix to obtain an active packaging composite film with better characteristics. Incorporating the protein isolate into the chitosan matrix improved the UV–VIS light barrier due to the formation of links between the two polymers, and it decreased WVP due to the strengthening of the cross-linking in the composite films and the limitation of the mobility of the polymer matrix. Although the protein isolate incorporation into the chitosan matrix led to a slight decrease in mechanical properties, the results showed that the composite films had better biological characteristics than the control film [98].

In a recent study, chitosan was combined with two other biopolymers, sodium alginate and carboxymethyl cellulose, and the characteristics of the new composite films were assessed [99]. These three biopolymers were used due to the potential antibacterial activity of the chitosan films, the high strength of the CMC, and the flexibility and film-forming capacity of sodium alginate. The composite film presented improved mechanical properties and also good antibacterial activity, with a 96% antibacterial rate against *E. coli* and a 93% antibacterial rate against *S. aureus*; therefore, the composite film has potential for use as an active packaging [99].

3.2.2. Combined Chitosan with Polyphenols/Extracts Rich in Polyphenols

In an attempt to increase the biological and functional properties of chitosan-based films, several studies have been performed and assessed on the effects of the incorporation of various kinds of natural extracts rich in polyphenols or phenolic compounds into the chitosan matrix (Table 5). Plants react to stress by producing polyphenolic compounds as secondary metabolites. Polyphenolic compounds present in plants are phenolic acids, flavonoids, and anthocyanins. These compounds minimize oxidation and cell damage and act as powerful antibacterial and antioxidant agents. When they are utilized in film or coating formulations, the polyphenolic compounds display synergy and increase the composite films' total antibacterial and antioxidant activity [17]. Several studies have found that incorporating phenolic compounds or plant extracts rich in polyphenolic content into chitosan films improves their mechanical properties (Table 5).

The mechanical characteristics of a film are highly reliant on intra-molecular bonding, the type of chitosan matrix, the microstructure of the chitosan network, and the presence of crystalline phase inside the film [101,102]. The literature indicates that modifications of the mechanical characteristics of chitosan films by the incorporation of various phenols/extracts rich in polyphenols are not similar. Some compounds increase the mechanical strength, whereas others reduce it.

The incorporation of propolis extract (5%–20%) increases the TS and EAB in chitosan–propolis composite film [103,104]. Additionally, it has been described in several studies that the incorporation of phenolic compounds, such as gallic acid [105], epigallocatechin gallate nanocapsules (2.5%, 4.5%, 6.0% (w/v)) [106], ellagic acid (0.5%, 1.0%, 2.5%, 5.0% w/w) [107], protocatechuic acid (0.8, 1.6, 2.4, 3.2 g) [108], proanthocyanidins (5, 10, 15, 20 wt%) [109], syringic acid (0.25%, 0.5%) [110], phenolic acids (ferulic, caffeic, tannic, gallic, 1%) [81], or curcumin (1%) [111], into chitosan-based films, reinforces the mechanical strength of the chitosan composite. In chitosan-based films, the incorporation of olive leaves extract (10%, 20%, 30% w/w) [112], or purple rice and black rice extracts ((1, 3, 5 wt%) [113], achieved an increase in the film's TS and EAB, whereas the addition of pomegranate peel extract (10 g/L) [114], thyme extract (0.15% w/w) [115], turmeric extract [116], mango leaf extract (1%–5%) [117], or purple-fleshed sweet potato extract (5, 10, 15 wt%) [118] improved only the TS, and the addition of pomegranate peel extract (1%, 2%, 3%) [119], grapefruit seed extract (0.5, 1.0, 1.5% v/v) [120], *Berberis crataegina* fruit extract (1 g) [121], and *Nigella sativa* seed-cake extract (2.5%, 5%, 7.5%) [122] improved only the EAB. On the other hand, several studies have reported that the incorporation of polyphenols/extracts rich in polyphenols may decrease the mechanical characteristics [84,101,102,114,123–126].

Usually, an increase in the strength of chitosan–polyphenols films is associated with a strong bond between the phenolic compounds and the chitosan matrix that explains the enhanced stiffness. On the contrary, a decline in strength is related to the reduction of intermolecular interactions between chitosan chains in the presence of polyphenolic compounds [116]. Riaz et al. [102] reported that the decrease in mechanical characteristics is due to the reduction of intermolecular interactions between chitosan chains and the loss of crystalline phase inside the film.

The barrier properties of chitosan-based films are helpful in maintaining the preservation and nutritional value, and in prolonging the shelf life, of food products [127].

A recent study indicated that chitosan–propolis films' water vapor and oxygen permeability decreased with increased amounts of propolis incorporation into the chitosan matrix [103]. In one study, blueberry, parsley, and red grape extracts [128] were incorporated into the chitosan matrix. Oxygen permeability was reduced by an average of 21% for films containing 5% blueberry extract, by 16% for films containing 5% parsley extract, and by 14% percent for films containing 5% red grapes extract. The decrease in oxygen permeability of the chitosan–extract films is caused by the potential of cross-linking between the polyphenolic content of the extracts and the chitosan matrix [128].

Furthermore, the incorporation of epigallocatechin gallate nanocapsules [106], protocatechuic acid [108], turmeric extract [116], and *Sonneratia caseolaris* leaf extract [84] into chitosan-based films reported UV–VIS light barrier properties for the chitosan composite films. The UV–VIS light barrier property of a film is a significant aspect, because the packaged food's resistance to oxidative degradation might be enhanced by the UV–VIS light barrier characteristics, which could prevent nutritional loss, color changes in the food, and off flavors [108]. The UV–VIS barrier property could be related to the incorporation of polyphenolic compounds into the inter-molecular pores of the chitosan matrix, which might block UV–VIS light transmission, and to the aromatic groups present in polyphenolic compounds, which might absorb the UV–VIS radiations [106,116].

Since one of the main purposes of a film is to prevent moisture transfer between the food and the environment in order to avoid or postpone deterioration, WVP should be reduced as low as necessary to keep products fresh [123]. Decreased WVP was reported in several studies for chitosan-based films incorporating various natural extracts abundant in polyphenols, such as pomegranate peel extract [119], tea extract [124], *Lycium barbarum* fruit extract [125], honeysuckle flower extract [126], *Nigella sativa* seedcake extract [122], mango leaf extract [117], *Herba Lophatheri* extract [129], Chinese chive (*Allium tuberosum*) root extract [102], and olive leaves extract [112]. Similar results were reported when phenolic compounds were incorporated into the chitosan matrix. For instance, the incorporation of syringic acid in the chitosan matrix improves WVP [110]. The results show that chitosan–syringic acid films significantly decreased WVP in comparison to control films, and WVP decreased as the syringic acid amounts increased [110].

Chitosan-based films may be used as active packaging in order to inhibit food oxidation. Antioxidant activity of the chitosan-based films was improved by the addition of propolis extract [103,104]. With an increased amount of propolis being integrated into the chitosan matrix, the DPPH radical scavenging capacity improved. This result might be explained by the presence of phenolic compounds in the propolis extract [104]. Chitosan-based films incorporating different phenolic compounds have higher antioxidant activity than simple chitosan films [81,108]. The enhancement of the antioxidant activity of chitosan-based films was noticed with the addition of blueberry, parsley, and red grapes containing polyphenols extracts [128]. Recently, Rambabu et al. [117] reported that chitosan-based film incorporating mango leaf extract presented higher antioxidant activity compared with control films. The antioxidant activity of mango leaf extract is due to the presence of polyphenolic content and some compounds with antioxidant potential, such as mangiferin [117].

It is well known that polyphenols are natural compounds with a variety of biological properties. There are various phenolic compounds that possess antimicrobial activity; therefore, antimicrobial properties of chitosan films are expected to increase with the addition of polyphenols to their composition [130]. Several researchers evaluated the biological properties of chitosan-based films enriched with polyphenols/extracts rich in polyphenols (Table 5). The combination of chitosan-based films with polyphenols/extracts rich in polyphenols augmented the antimicrobial activity against Gram-negative and Gram-positive bacteria [84,101,102,107,114,116,121,129,131]. Recently, Sun et al. [123] reported that polyphenol compounds extracted from thinned young apple increased the antimicrobial activity of chitosan-based films against three molds (*Colletotrichum fructicola*, *Botryosphaerial dothidea*, and *Alternaria tenuissima*). Moreover, the composite films did not have activity against yeasts (*S. cerevisiae*, Baker's yeast, and *Tropical candida*) [123].

Table 5. The characteristics of chitosan–polyphenol mixtures.

Formulation	Physical/Chemical/Mechanical Characteristics	Biological Properties	Applications	References
Chitosan, propolis extract	↑ TS, EAB, ↓ WVP, oxygen permeability color changes of the films	Antioxidant activity (DPPH), antimicrobial activity against *S. aureus*, *Salmonella Enteritidis*, *E. coli*, and *P. aeruginosa*	Active packaging	[103]
Chitosan, propolis extract	↑ thickness, thermal stability, TS ↓ transparency, EAB, WS color change	Antioxidant activity (DPPH (49.8%–94.5%), ABTS (20.3%–83.6%)), antimicrobial activity against *Staphylococcus hominis*, *Pantoea* sp., *Arthrobacter* sp., *Erwinia* sp., *B. cereus*, *E. coli*, *S. aureus*, *Metschnikowia rancensis*, *Cladosporium* sp., *Penicillium brevicompactum*, *Botrytis cinerea*, and *Alternaria* sp.	Active packaging	[104]
Chitosan, gallic acid	↑ TS (for chitosan:gallic acid ratio 1:0.1, 1:0.5) ↓ EAB and WVP (for chitosan:gallic acid ratio 1:0.1)	Antioxidant activity (DPPH, ABTS), antimicrobial activity against *E. coli* and *L. monocytogenes*	Active food packaging	[105]
Chitosan, epigallocatechin gallate nanocapsules (with zein)	↑ TS, EAB, VIS-light protection	Antioxidant activity (DPPH)	Active packaging	[106]
Chitosan, ellagic acid	↑ EAB, WVP ↓ TS, Young's modulus, UV-light protection good thermal stability	Antioxidant activity (DPPH), antimicrobial activity against *P. aeruginosa* and *S. aureus*, prevent photo-oxidation of light-sensitive foods	Active packaging	[107]

Table 5. Cont.

Formulation	Physical/Chemical/Mechanical Characteristics	Biological Properties	Applications	References
Chitosan, protocatechuic acid	↑ thickness, opacity, WS, UV-light barrier ↓ moisture content, WVP, EAB, color change TS increased up to 1% acid incorporation, afterwards decreased	Antioxidant activity (DPPH)	Active packaging	[108]
Chitosan, thinned young apple polyphenols	↑ thickness, density, WS ↓ WVP, TS, EAB, water content	Antioxidant activity (DPPH 68%–92%), antimicrobial activity against E. coli, S aureus, L. monocytogenes, Colletotrichum fructicola, Botryosphaerial dothidea, and Alternaria tenuissima	Active packaging	[123]
Chitosan, apple peel polyphenols	↑ thickness, density, WS, ↓ thermal stability, WVP, TS, EAB, moisture content, transparency color change	Antioxidant activity (DPPH 30%–67%, ABTS 70%–90%), antimicrobial activity against E. coli, B. cereus, S. aureus, and S. typhimurium	Active packaging	[101]
Chitosan, proanthocyanidins	↓ thermal stability	Antioxidant activity (DPPH, ABTS), antimicrobial activity against M. luteus, B. subtilis, E. coli, S. aureus, Proteus vulgaris, and P. aeruginosa	Active packaging	[132]
Chitosan, proanthocyanidins	↑ thickness, opacity, thermal stability, WS, WVP, TS, UV–VIS light barrier ↓ moisture content, EAB, oxygen permeability color change	Antioxidant activity (DPPH), antimicrobial activity against E. coli, Salmonella, S. aureus, and L. monocytogenes	Active packaging	[109]
Chitosan, syringic acid	↑ thickness, density, WS, opacity, TS when the amount of syringic acid was under 0.5% and EAB when the amount of syringic acid was 0.25%, ↓ moisture content, thermal stability and WVP color change	Antimicrobial activity against S. aureus and E. coli	Preservation of quail egg Active packaging	[110]
Chitosan, phenolic acids (ferulic acid, caffeic acid, tannic acid, gallic acid)	↑ TS, EAB, Young's modulus, thermal stability, WVP color change	Antioxidant activity (DPPH 17%–89%)	Active packaging	[81]
Chitosan, curcumin	↑ TS ↓ EAB, WVP,	Antimicrobial activity against S. aureus and Rhizoctonia solani	Active packaging	[111]
Chitosan, carvacrol	↓ WVP, TS, EAB, thickness and transparency, change color to yellow	Antioxidant activity (FRAP), antimicrobial activity against E. coli and S. aureus	Active packaging	[114]
Chitosan, pomegranate peel extract	↑ thickness, TS ↓ EAB and transparency change color	Antioxidant activity (FRAP), antimicrobial activity against S. aureus	Active packaging	[114]
Chitosan, pomegranate peel extract	↑ EAB ↓ TS, WVP	Antioxidant activity (DPPH 21%–57%)	Active packaging	[119]
Chitosan, thyme extract	↑ TS, EM, opacity decreased: EAB, color change	Antioxidant activity (DPPH)	Active packaging	[115]
Chitosan, turmeric extract	↑ TS, Young's modulus, WVP, UV–VIS barrier property	Antimicrobial activity against S. aureus and Salmonella	Active packaging	[116]
Chitosan, tea extract	↑ thickness, WS ↓ water content, WVP, TS, EAB	Antioxidant activity (DPPH)	Active packaging	[124]
Chitosan, grapefruit seed extract	↑ thickness, EAB, ↓TS	Antifungal activity	Bread preservation	[120]
Chitosan, maqui berry extract (Aristotelia chilensis)	not determined	Antioxidant activity (DPPH, FRAP), antimicrobial activity against Serratia marcescens, Alcaligenes faecalis, Aeromonas hydrophila, Pseudomonas fluorescens, Citrobacter freundii, Achromobacter denitrifican, S. putrefaciens	Active packaging	[131]
Chitosan, Lycium barbarum fruit extract	↑ density ↓ TS, EAB, WVP, WS, moisture content	Antioxidant activity (DPPH)	Active packaging	[125]
Chitosan, honeysuckle flower extract (Lonicera japonica Thunb)	↑ WS, density ↓ WVP, TS, EAB, moisture content	Antioxidant activity (DPPH), antimicrobial activity against E. coli	Active packaging	[126]

Table 5. Cont.

Formulation	Physical/Chemical/Mechanical Characteristics	Biological Properties	Applications	References
Chitosan, *Berberis crataegina* fruit extract	↑ thickness, EAB ↓ transparency, TS, WS, Young's modulus	Antioxidant activity (DPPH 86%), antimicrobial activity against *E. coli*, *S. thympurium*, *Proteus microbilis*, *Proteus vulgaris*, *P. aeruginosa*, *Enterobacter aerogenes*, *S. aureus*, *Streptococcus mutans*, *Bacillus thuringiensis*	Active packaging	[121]
Chitosan, *Nigella sativa* seedcake extract	↑ thickness, EAB, ↓ moisture content, WVP, TS color change	Antioxidant activity (DPPH, FRAP)	Active packaging	[122]
Chitosan, mango leaf extract	↑ thickness, TS, EM, ↓ moisture content, WS, WVP, EAB	Antioxidant activity (DPPH, FRAP, ABTS)	Cashew nuts preservation	[117]
Chitosan, *Herba Lophatheri* extract from dried leaves of *Lophatherum gracile* Brongn	↑ opacity, density, ↓ WS, WVP, moisture content color change, higher oil resistance	Antioxidant activity (DPPH), antimicrobial activity against *E. coli* and *S. aureus*	Active packaging	[129]
Chitosan, Chinese chive (*Allium tuberosum*) root extract	↑ thickness, thermal stability ↓ TS, EAB, WS, WVP, moisture content	Antioxidant activity (DPPH 20%–47%, ABTS 28%–57%), antimicrobial activity against *B. cereus*, *S. aureus*, *E. coli*, and *S. typhimurium*	Soybean oil packaging	[102]
Chitosan, *Sonneratia caseolaris* (L.) Engl. leaf extract	↑ light barrier property, WS, WVP ↓ TS, EAB, moisture content change color	Antimicrobial activity against *S. aureus* and *P. aeruginosa*	Vietnamese banana preservation	[84]
Chitosan, olive leaves extract	↑ WS, TS, and EAB, ↓ WVP	Antioxidant activity (ABTS), antimicrobial activity against *L. monocytogenes* and *Campylobacter jejuni*	Active packaging	[112]
Chitosan, blueberry extract by-products	↑ thickness, WVP ↓ oxygen permeability, water content	Antioxidant activity (DPPH, ABTS, FRAP)	Active packaging	[128]
Chitosan, parsley extract by-products	↑ thickness, WVP ↓ oxygen permeability, water content	Antioxidant activity (DPPH, ABTS, FRAP)	Active packaging	[128]
Chitosan, red grapes extract by-products	↑ thickness, WVP ↓ oxygen permeability, water content	Antioxidant activity (DPPH, ABTS, FRAP), antimicrobial activity against *E. coli*	Active packaging	[128]
Chitosan, purple-fleshed sweet potato extract	↑ thickness, WS, WVP when the extract exceeded 5 wt% and TS when the extract was 5 wt% ↓ EAB, WVP when the extract was 5 wt%, TS when the extract exceeded 5 wt%, moisture content and light transmittance	Antioxidant activity (DPPH), color variations of films to pH, pink-red (pH 3.0–6.0), purple-brown (pH 7.0–8.0), and greenish–green (pH 9.0–10.0)	Monitoring food spoilage	[118]
Chitosan, purple rice extract	↑ thickness, EAB, TS, light barrier property, and WVP when the extract exceeded 1 wt% ↓ moisture content, change color	Antioxidant activity (DPPH), pH-sensitive in different buffer solutions	Monitor pork spoilage	[113]
Chitosan, black rice extract	↑ thickness, EAB, light barrier property ↓ moisture content, TS when the extract exceeded 1 wt% change color	Antioxidant activity (DPPH)	Active packaging	[113]

↑-increased values of the tested characteristics, ↓-decreased values of the tested characteristics.

3.2.3. Combined Chitosan and Essential Oil

Essential oils are secondary plant metabolites with strong fragrance and great antioxidant and antimicrobial properties. The main content is represented by bioactive compounds, such as polyphenolic compounds, alkaloids, aldehydes, carotenoids, and monoterpenes [133]. In order to decrease their volatility, and enhance antioxidant and antimicrobial activities, essential oils may be incorporated in polymer matrices such as chitosan [17]. A unique advantage of essential oils utilization seems to be the synergistic effects of their constituents, as contrasted to the sum of the activities of the separate bioactive compounds [134]. Several studies have reported that chitosan films incorporating essential oils have improved characteristics (Table 6).

Shen and Kamden [135] reported that the incorporation of citronella and cedarwood essential oils into the chitosan matrix affected the mechanical characteristics. The TS of composite films was reduced when the amounts of the essential oils increased. They observed that incorporating low amounts of essential oils led to increased EAB. However, they noticed that by incorporating increased amounts of essential oils, a decrease in EAB was obtained. This result may be described as the substitution of stronger polymer–polymer bonds with weaker polymer–oil bonds in the film network [135].

Similar results were reported by other researchers. Priyardashi et al. [136] observed an increase in EAB for chitosan–apricot kernel essential oil films with the incorporation of low amounts of the essential oil; however, when the ratio of chitosan: essential oil exceeded 1: 0.5, the EAB decreased. Moreover, they observed an increase in the TS with an increase in the essential oil amount being incorporated into the film [136].

A similar pattern for the EAB of chitosan–*piper betle* Linn oil films was reported in the study conducted by Nguyen et al. [137].

The addition of basil essential oil and thyme essential oil to chitosan-based films improved the mechanical characteristics of the composite film. The TS and EM were increased, whereas the EAB decreased [138].

An important property of the composite films is water vapor permeability. Decreased WVP was reported in several studies for chitosan-based films incorporating various essential oils, such as citronella (10%, 20%, 30% w/w) [135], cedarwood (10%, 20%, 30% w/w) [135], basil (0.5 g, 1 g/100 g) [138], thyme (0.5 g, 1 g/100 g) [138], and apricot kernel (0.125, 0.25, 0.5, 1%) [136]. Hydrogen and covalent bonds between the chitosan matrix and the bioactive compounds may result in reduced WVP of chitosan–essential oil films. These interactions might minimize the capacity of hydrophilic groups to establish hydrophilic linkages and, therefore, minimize the interactions with water, resulting in a composite film with better moisture resistance. Additionally, even at low concentrations, the existence of a hydrophobic dispersion causes discontinuity in the hydrophilic phase that leads to decreasing WVP [135,136].

Several studies reported that by incorporating essential oils into a chitosan matrix, the biological activities may be improved. For instance, Liu et al. [139] prepared chitosan–peppermint essential oil and chitosan–fennel essential oil films and evaluated the antioxidant activity of the composite films using DPPH scavenging method. The chitosan-based film had the lowest antioxidant activity among all the films (55%). The chitosan-based film's antioxidant activity could be associated with the NH_2 units in the chitosan matrix, units that interacted with DPPH and generated stable molecules. The incorporation of essential oils into chitosan-based films enhanced their ability to scavenge DPPH. The chitosan–peppermint essential oil film had higher antioxidant activity (67%) than the chitosan-based film, due to the peppermint composition with antioxidant properties. Moreover, the trans-anethole molecule in fennel essential oil could be responsible for the greater antioxidant activity of the chitosan–fennel essential oil film (68%) compared to the chitosan–peppermint essential oil film [139].

In another study, Hafsa et al. [140] prepared chitosan–*Eucalyptus globulus* essential oil films with different essential oil content and evaluated its antioxidant and antimicrobial activities. The authors reported that the DPPH scavenging ability of the composite increased with increasing essential oil content. The highest antioxidant activity was 44% (chitosan incorporating 4% (v/v) essential oil), which was substantially higher than that of chitosan-based film (only 10%). Furthermore, the antimicrobial activity of the composite films was tested against three bacteria, *E. coli*, *S. aureus*, and *P. aeruginosa*, and two fungi, *C. albicans* and *Candida parapsilosis*. The results of the study showed that all composite films showed antimicrobial activity against all strains tested and the antimicrobial activity increased with increasing essential oil content [140].

Similar results were obtained in a study conducted by Priyadarshi et al. [136]. They incorporated different amounts of Apricot kernel essential oil into chitosan-based films, and they tested the biological activities for all the composite films and also for the chitosan-based film. It was observed that the antioxidant and antimicrobial activities increased with increasing essential oil content. Moreover, the chitosan–apricot kernel essential oil films were tested for antifungal activity. The authors evaluated the potential of the composite films for the inhibition the of growth of *Rhizopus stolonifer* on bread slices. The films were observed to successfully limit the growth of fungi on bread, hence extending its shelf life [136].

Table 6. The characteristics of chitosan–essential oil mixtures.

Formulation	Physical/Chemical/Mechanical Characteristics	Biological Properties	Applications	References
Chitosan, citronella essential oil	↑ EAB (low essential oil content), thermal stability ↓ WVP, TS, moisture content	Not determined	Packaging	[135]
Chitosan, cedarwood essential oil	↑ EAB (low essential oil content), thermal stability ↓ WVP, TS, moisture content	Not determined	Packaging	[135]
Chitosan, basil essential oil	↑ thickness, TS, EM ↓ WVP, EAB	Tested for antifungal activity, but the film did not inhibit the growth of *A. niger*, *Botrytis cinerea*, and *R. stolonifer*	Packaging	[138]
Chitosan, thyme essential oil	↑ thickness, TS, EM ↓ WVP, EAB	Tested for antifungal activity, but the film did not inhibit the growth of *A. niger*, *B. cinerea*, and *R. stolonifer*	Packaging	[138]
Chitosan, fennel essential oil	↑ density, thermal stability, and opacity ↓ WS, water swelling, thickness, and moisture content color change	Antioxidant activity (DPPH 68%)	Active packaging	[139]
Chitosan, peppermint essential oil	↑ density, thermal stability, and opacity ↓ WS, water swelling, and thickness color change	Antioxidant activity (DPPH 66%)	Active packaging	[139]
Chitosan, *Eucalyptus globulus* essential oil	↑ opacity ↓ moisture content, WS	Antioxidant activity (DPPH 23%–43%), antimicrobial activity against *E. coli*, *S. aureus*, *P. aeruginosa*, *C. albicans*, *C. parapsilosis*	Active packaging	[140]
Chitosan, apricot kernel essential oil	↑ opacity, TS ↓ moisture content, WS, WVP EAB first increased, and then when the ratio of chitosan:essential oil exceeded 1: 0.5 decreased	Antioxidant activity (DPPH 26%–35%), antimicrobial activity against *S. aureus* and *B. subtillis*, antifungal activity against *R. stolonifer*	Inhibited the growth of fungi on bread, active food packaging	[136]
Chitosan, *piper betle* Linn oil	↑ UV-light barrier, EAB (at 0.4 and 1% oil incorporation), ↓ thermal stability, TS, EM, and EAB (at 1.2% oil incorporated)	Antioxidant activity (DPPH), antimicrobial activity against *S. aureus*, *E. coli*, *P. aeruginosa*, and *S. typhimurium*	King orange preservation	[137]

↑-increased values of the tested characteristics, ↓-decreased values of the tested characteristics.

4. Conclusions

Gelatin and chitosan are two of the most important biopolymers for the production of films and coatings. Although gelatin exhibits excellent gas barrier and swelling properties, it has poor mechanical resistance and is permeable to water vapor molecules. The chitosan-based films have some disadvantages, including low water resistance, low UV–VIS light barrier properties, and reduced mechanical characteristics.

These limitations could be improved by combining gelatin and chitosan with other biopolymers or with bioactive compounds, such as polyphenols, natural extracts, and essential oils. Combining gelatin and chitosan with other biopolymers leads to improved mechanical properties, water vapor, and UV–VIS light barriers, and greater thermal stability of the obtained films. Additionally, incorporating different polyphenolic compounds, natural extracts rich in phenolic content, and essential oils into gelatin-based and chitosan-based films leads to increased physicochemical and mechanical properties and, even more relevant for the food industry, improved biological properties, and antioxidant and antimicrobial activities.

All these improved characteristics of composite films help in maintaining the quality, reducing lipid oxidation, avoiding microbial growth, and extending the shelf life of the packed products.

On the other hand, the implementation of alternative biodegradable materials in the existing infrastructure, and the technological transfer of the findings from a laboratory scale to industrial levels, represents a serious economic effort and a great challenge for both stakeholders and the scientific community.

Author Contributions: Conceptualization, B.E.Ș.; validation, B.E.Ș., C.S. and D.C.V.; writing—original draft preparation, B.E.Ș.; writing—review and editing, C.S. and D.C.V.; supervision, C.S. and D.C.V. All authors have read and agreed to the published version of the manuscript.

Funding: This research was funded by UEFISCDI-MCDI, project number PD 7/2022, PN-III-P1-1.1-PD-2021-0444.

Institutional Review Board Statement: Not applicable.

Informed Consent Statement: Not applicable.

Data Availability Statement: The data presented in this study are available on request from the corresponding author.

Acknowledgments: We kindly thank Bernadette E. Teleky for image support.

Conflicts of Interest: The authors declare no conflict of interest. The funders had no role in the design of the study; in the collection, analyses, or interpretation of data; in the writing of the manuscript; or in the decision to publish the results.

References

1. Iwata, T. Biodegradable and bio-based polymers: Future prospects of eco-friendly plastics. *Angew. Chem. Int. Ed.* **2015**, *54*, 3210–3215. [CrossRef] [PubMed]
2. Garavand, F.; Rouhi, M.; Razavi, S.H.; Cacciotti, I.; Mohammadi, R. Improving the integrity of natural biopolymer films used in food packaging by crosslinking approach: A review. *Int. J. Biol. Macromol.* **2017**, *104*, 687–707. [CrossRef] [PubMed]
3. Groh, K.J.; Backhaus, T.; Carney-Almroth, B.; Geueke, B.; Inostroza, P.A.; Lennquist, A.; Leslie, H.A.; Maffini, M.; Slunge, D.; Trasande, L.; et al. Overview of known plastic packaging-associated chemicals and their hazards. *Sci. Total Environ.* **2019**, *651*, 3253–3268. [CrossRef] [PubMed]
4. Cazón, P.; Velazquez, G.; Ramírez, J.A.; Vázquez, M. Polysaccharide-based films and coatings for food packaging: A review. *Food Hydrocoll.* **2017**, *68*, 136–148. [CrossRef]
5. Dehghani, S.; Hosseini, S.V.; Regenstein, J.M. Edible films and coatings in seafood preservation: A review. *Food Chem.* **2018**, *240*, 505–513. [CrossRef]
6. Kumar, V.A.; Hasan, M.; Mangaraj, S.; Pravitha, M.; Verma, D.K.; Srivastav, P.P. Trends in Edible Packaging Films and its Prospective Future in Food: A Review. *Appl. Food Res.* **2022**, *2*, 100118. [CrossRef]
7. Hahladakis, J.N.; Velis, C.A.; Weber, R.; Iacovidou, E.; Purnell, P. An overview of chemical additives present in plastics: Migration, release, fate and environmental impact during their use, disposal and recycling. *J. Hazard. Mater.* **2018**, *344*, 179–199. [CrossRef] [PubMed]
8. Mahmud, N.; Islam, J.; Tahergorabi, R. Marine biopolymers: Applications in food packaging. *Processes* **2021**, *9*, 2245. [CrossRef]
9. Kaur, J.; Rasane, P.; Singh, J.; Kaur, S. Edible Packaging: An Overview. In *Edible Food Packaging*; Springer Nature: Singapore, 2022; pp. 3–25. ISBN 9789811623837.
10. Ubeda, S.; Aznar, M.; Rosenmai, A.K.; Vinggaard, A.M.; Nerín, C. Migration studies and toxicity evaluation of cyclic polyesters oligomers from food packaging adhesives. *Food Chem.* **2020**, *311*, 125918. [CrossRef] [PubMed]
11. Samsudin, H.; Auras, R.; Burgess, G.; Dolan, K.; Soto-Valdez, H. Migration of antioxidants from polylactic acid films, a parameter estimation approach: Part I—A model including convective mass transfer coefficient. *Food Res. Int.* **2018**, *105*, 920–929. [CrossRef] [PubMed]
12. Nechita, P.; Roman, M. Review on Polysaccharides Used in Coatings for Food. *Coatings* **2020**, *10*, 566. [CrossRef]
13. Gómez-Estaca, J.; Gavara, R.; Catalá, R.; Hernández-Muñoz, P. The Potential of Proteins for Producing Food Packaging Materials: A Review. *Packag. Technol. Sci.* **2016**, *29*, 203–224. [CrossRef]
14. Lu, Y.; Luo, Q.; Chu, Y.; Tao, N.; Deng, S.; Wang, L.; Li, L. Application of Gelatin in Food Packaging: A Review. *Polymers* **2022**, *14*, 436. [CrossRef] [PubMed]
15. Díaz-montes, E.; Castro-muñoz, R. Trends in chitosan as a primary biopolymer for functional films and coatings manufacture for food and natural products. *Polymers* **2021**, *13*, 767. [CrossRef] [PubMed]
16. Calinoiu, L.F.; Ștefănescu, B.E.; Pop, I.D.; Muntean, L.; Vodnar, D.C. Chitosan coating applications in probiotic microencapsulation. *Coatings* **2019**, *9*, 194. [CrossRef]
17. Nair, M.S.; Tomar, M.; Punia, S.; Kukula-Koch, W.; Kumar, M. Enhancing the functionality of chitosan- and alginate-based active edible coatings/films for the preservation of fruits and vegetables: A review. *Int. J. Biol. Macromol.* **2020**, *164*, 304–320. [CrossRef] [PubMed]
18. Menezes, M.d.L.L.R.; da Rocha Pires, N.; da Cunha, P.L.R.; de Freitas Rosa, M.; de Souza, B.W.S.; de Andrade Feitosa, J.P.; de Souza, M.D.S.M. Effect of tannic acid as crosslinking agent on fish skin gelatin-silver nanocomposite film. *Food Packag. Shelf Life* **2019**, *19*, 7–15. [CrossRef]
19. Nurdiani, R.; Ma'rifah, R.D.A.; Busyro, I.K.; Jaziri, A.A.; Prihanto, A.A.; Firdaus, M.; Talib, R.A.; Huda, N. Physical and functional properties of fish gelatin-based film incorporated with mangrove extracts. *PeerJ* **2022**, *10*, e13062. [CrossRef] [PubMed]

20. Hanani, Z.A.N. Gelatin. In *Encyclopedia of Food and Health*; Elsevier: Amsterdam, The Netherlands, 2015; pp. 191–195. ISBN 9780123849533.
21. Kuan, Y.H.; Nafchi, A.M.; Huda, N.; Ariffin, F.; Karim, A.A. Comparison of physicochemical and functional properties of duck feet and bovine gelatins. *J. Sci. Food Agric.* **2017**, *97*, 1663–1671. [CrossRef]
22. Karayannakidis, P.D.; Zotos, A. Fish Processing By-Products as a Potential Source of Gelatin: A Review. *J. Aquat. Food Prod. Technol.* **2016**, *25*, 65–92. [CrossRef]
23. Ahmad, T.; Ismail, A.; Ahmad, S.A.; Khalil, K.A.; Kumar, Y.; Adeyemi, K.D.; Sazili, A.Q. Recent advances on the role of process variables affecting gelatin yield and characteristics with special reference to enzymatic extraction: A review. *Food Hydrocoll.* **2017**, *63*, 85–96. [CrossRef]
24. Said, N.S.; Sarbon, N.M. Physical and Mechanical Characteristics of Gelatin-Based Films as a Potential Food Packaging Material: A Review. *Membranes* **2022**, *12*, 442. [CrossRef] [PubMed]
25. Fakhreddin Hosseini, S.; Rezaei, M.; Zandi, M.; Ghavi, F.F. Preparation and functional properties of fish gelatin-chitosan blend edible films. *Food Chem.* **2013**, *136*, 1490–1495. [CrossRef]
26. Fakhouri, F.M.; Martelli, S.M.; Caon, T.; Velasco, J.I.; Mei, L.H.I. Edible films and coatings based on starch/gelatin: Film properties and effect of coatings on quality of refrigerated Red Crimson grapes. *Postharvest Biol. Technol.* **2015**, *109*, 57–64. [CrossRef]
27. Cao, N.; Fu, Y.; He, J. Preparation and physical properties of soy protein isolate and gelatin composite films. *Food Hydrocoll.* **2007**, *21*, 1153–1162. [CrossRef]
28. Nur Hazirah, M.A.S.P.; Isa, M.I.N.; Sarbon, N.M. Effect of xanthan gum on the physical and mechanical properties of gelatin-carboxymethyl cellulose film blends. *Food Packag. Shelf Life* **2016**, *9*, 55–63. [CrossRef]
29. Loo, C.P.Y.; Sarbon, N.M. Chicken skin gelatin films with tapioca starch. *Food Biosci.* **2020**, *35*, 100589. [CrossRef]
30. Jridi, M.; Abdelhedi, O.; Salem, A.; Kechaou, H.; Nasri, M.; Menchari, Y. Physicochemical, antioxidant and antibacterial properties of fish gelatin-based edible films enriched with orange peel pectin: Wrapping application. *Food Hydrocoll.* **2020**, *103*, 105688. [CrossRef]
31. Zhong, C.; Hou, P.F.; Li, Y.X.; Yang, W.Y.; Shu, M.; Wu, G.P. Characterization, antioxidant and antibacterial activities of gelatin film incorporated with protocatechuic acid and its application on beef preservation. *LWT* **2021**, *151*, 112154. [CrossRef]
32. Wang, Q.; Cao, J.; Yu, H.; Zhang, J.; Yuan, Y.; Shen, X.; Li, C. The effects of EGCG on the mechanical, bioactivities, cross-linking and release properties of gelatin film. *Food Chem.* **2019**, *271*, 204–210. [CrossRef]
33. Zhao, Y.; Li, Z.; Yang, W.; Xue, C.; Wang, Y.; Dong, J.; Xue, Y. Modification of gelatine with galla chinensis extract, a natural crosslinker. *Int. J. Food Prop.* **2016**, *19*, 731–744. [CrossRef]
34. Zhou, Q.; Li, P.; Fang, S.; Liu, W.; Mei, J.; Xie, J. Preservative effects of gelatin active coating enriched with eugenol emulsion on Chinese seabass (*Lateolabrax maculatus*) during superchilling (−0.9 °C) storage. *Coatings* **2019**, *9*, 489. [CrossRef]
35. Adilah, A.N.; Jamilah, B.; Noranizan, M.A.; Hanani, Z.A.N. Utilization of mango peel extracts on the biodegradable films for active packaging. *Food Packag. Shelf Life* **2018**, *16*, 1–7. [CrossRef]
36. Li, X.; Tu, Z.C.; Sha, X.M.; Ye, Y.H.; Li, Z.Y. Flavor, antimicrobial activity and physical properties of gelatin film incorporated with of ginger essential oil. *J. Food Sci. Technol.* **2022**, *59*, 815–824. [CrossRef]
37. Yang, S.Y.; Lee, K.Y.; Beak, S.E.; Kim, H.; Song, K. Bin Antimicrobial activity of gelatin films based on duck feet containing cinnamon leaf oil and their applications in packaging of cherry tomatoes. *Food Sci. Biotechnol.* **2017**, *26*, 1429–1435. [CrossRef]
38. Martucci, J.F.; Gende, L.B.; Neira, L.M.; Ruseckaite, R.A. Oregano and lavender essential oils as antioxidant and antimicrobial additives of biogenic gelatin films. *Ind. Crops Prod.* **2015**, *71*, 205–213. [CrossRef]
39. Lee, K.Y.; Lee, J.H.; Yang, H.J.; Song, K. Bin Production and characterisation of skate skin gelatin films incorporated with thyme essential oil and their application in chicken tenderloin packaging. *Int. J. Food Sci. Technol.* **2016**, *51*, 1465–1472. [CrossRef]
40. Tongnuanchan, P.; Benjakul, S.; Prodpran, T. Properties and antioxidant activity of fish skin gelatin film incorporated with citrus essential oils. *Food Chem.* **2012**, *134*, 1571–1579. [CrossRef]
41. Tongnuanchan, P.; Benjakul, S.; Prodpran, T. Physico-chemical properties, morphology and antioxidant activity of film from fish skin gelatin incorporated with root essential oils. *J. Food Eng.* **2013**, *117*, 350–360. [CrossRef]
42. Ahmad, M.; Benjakul, S.; Prodpran, T.; Agustini, T.W. Physico-mechanical and antimicrobial properties of gelatin film from the skin of unicorn leatherjacket incorporated with essential oils. *Food Hydrocoll.* **2012**, *28*, 189–199. [CrossRef]
43. Kavoosi, G.; Rahmatollahi, A.; Mohammad Mahdi Dadfar, S.; Mohammadi Purfard, A. Effects of essential oil on the water binding capacity, physico-mechanical properties, antioxidant and antibacterial activity of gelatin films. *LWT—Food Sci. Technol.* **2014**, *57*, 556–561. [CrossRef]
44. Teixeira, B.; Marques, A.; Pires, C.; Ramos, C.; Batista, I.; Saraiva, J.A.; Nunes, M.L. Characterization of fish protein films incorporated with essential oils of clove, garlic and origanum: Physical, antioxidant and antibacterial properties. *LWT—Food Sci. Technol.* **2014**, *59*, 533–539. [CrossRef]
45. Mihaly Cozmuta, A.; Turila, A.; Apjok, R.; Ciocian, A.; Mihaly Cozmuta, L.; Peter, A.; Nicula, C.; Galić, N.; Benković, T. Preparation and characterization of improved gelatin films incorporating hemp and sage oils. *Food Hydrocoll.* **2015**, *49*, 144–155. [CrossRef]
46. Gomez-Guillen, M.C.; Gimenez, B.; Lopez-Caballero, M.E.; Montero, M.P. Functional and bioactive properties of collagen and gelatin from alternative sources: A review. *Food Hydrocoll.* **2011**, *25*, 1813–1827. [CrossRef]

7. Karim, A.A.; Bhat, R. Fish gelatin: Properties, challenges, and prospects as an alternative to mammalian gelatins. *Food Hydrocoll.* **2009**, *23*, 563–576. [CrossRef]
8. Jongjareonrak, A.; Benjakul, S.; Visessanguan, W.; Prodpran, T.; Tanaka, M. Characterization of edible films from skin gelatin of brownstripe red snapper and bigeye snapper. *Food Hydrocoll.* **2006**, *20*, 492–501. [CrossRef]
9. Mhd Sarbon, N.; Badii, F.; Howell, N.K. Preparation and characterisation of chicken skin gelatin as an alternative to mammalian gelatin. *Food Hydrocoll.* **2013**, *30*, 143–151. [CrossRef]
10. Abedinia, A.; Ariffin, F.; Huda, N.; Mohammadi Nafchi, A. Preparation and characterization of a novel biocomposite based on duck feet gelatin as alternative to bovine gelatin. *Int. J. Biol. Macromol.* **2018**, *109*, 855–862. [CrossRef]
11. Rahman, M.N.; Jamalulail, S.A.S.K.A. Extraction, Physicochemical Characterizations and Sensory. *Borneo Sci.* **2012**, *30*, 1–13.
12. Nik Muhammad, N.A.; Huda, N.; Karim, A.A.; Mohammadi Nafchi, A. Effects of acid type extraction on characterization and sensory profile of duck feet gelatin: Towards finding bovine gelatin alternative. *J. Food Meas. Charact.* **2018**, *12*, 480–486. [CrossRef]
13. Ninan, G.; Joseph, J.; Aliyamveettil, Z.A. A comparative study on the physical, chemical and functional properties of carp skin and mammalian gelatins. *J. Food Sci. Technol.* **2014**, *51*, 2085–2091. [CrossRef] [PubMed]
14. Muyonga, J.H.; Cole, C.G.B.; Duodu, K.G. Extraction and physico-chemical characterisation of Nile perch (Lates niloticus) skin and bone gelatin. *Food Hydrocoll.* **2004**, *18*, 581–592. [CrossRef]
15. da Trindade Alfaro, A.; Balbinot, E.; Weber, C.I.; Tonial, I.B.; Machado-Lunkes, A. Fish Gelatin: Characteristics, Functional Properties, Applications and Future Potentials. *Food Eng. Rev.* **2015**, *7*, 33–44. [CrossRef]
16. Howell, N.K. Elucidation of protein protein interactions in gels and foams. In *Gums and Stabilizers for the Food Industry 7*; Oxford University Press: Wales, UK, 1994; pp. 77–89.
17. Sarbon, N.M.; Badii, F.; Howell, N.K. The effect of chicken skin gelatin and whey protein interactions on rheological and thermal properties. *Food Hydrocoll.* **2015**, *45*, 83–92. [CrossRef]
18. Ngarize, S.; Adams, A.; Howell, N. A comparative study of heat and high pressure induced gels of whey and egg albumen proteins and their binary mixtures. *Food Hydrocoll.* **2005**, *19*, 984–996. [CrossRef]
19. Kehoe, J.J.; Foegeding, E.A. The characteristics of heat-induced aggregates formed by mixtures of β-lactoglobulin and β-casein. *Food Hydrocoll.* **2014**, *39*, 264–271. [CrossRef]
20. Nazmi, N.N.; Isa, M.I.N.; Sarbon, N.M. Preparation and characterization of chicken skin gelatin/CMC composite film as compared to bovine gelatin film. *Food Biosci.* **2017**, *19*, 149–155. [CrossRef]
21. Alias, S.A.; Mhd Sarbon, N. Rheological, physical, and mechanical properties of chicken skin gelatin films incorporated with potato starch. *NPJ Sci. Food* **2019**, *3*, 26. [CrossRef]
22. Jahit, I.S.; Nazmi, N.N.M.; Isa, M.I.N.; Sarbon, N.M. Preparation and physical properties of gelatin/CMC/chitosan composite films as affected by drying temperature. *Int. Food Res. J.* **2016**, *23*, 1068–1074.
23. Bakry, N.F.; Isa, M.I.N.; Sarbon, N.M. Effect of sorbitol at different concentrations on the functional properties of gelatin/carboxymethyl cellulose (CMC)/chitosan composite films. *Int. Food Res. J.* **2017**, *24*, 1753–1762.
24. Fatyasari Nata, I.; Irawan, C.; Ramadhan, L.; Rizky Ramadhani, M. Influence of soy protein isolate on gelatin-based edible film properties. *MATEC Web Conf.* **2018**, *156*, 01014. [CrossRef]
25. Fu, S.; Wu, C.; Wu, T.; Yu, H.; Yang, S.; Hu, Y. Preparation and characterisation of Chlorogenic acid-gelatin: A type of biologically active film for coating preservation. *Food Chem.* **2017**, *221*, 657–663. [CrossRef] [PubMed]
26. Tammineni, N.; Ünlü, G.; Rasco, B.; Powers, J.; Sablani, S.; Nindo, C. Trout-Skin Gelatin-Based Edible Films Containing Phenolic Antioxidants: Effect on Physical Properties and Oxidative Stability of Cod-Liver Oil Model Food. *J. Food Sci.* **2012**, *77*, E342–E347. [CrossRef] [PubMed]
27. Zhang, X.; Ma, L.; Yu, Y.; Zhou, H.; Guo, T.; Dai, H.; Zhang, Y. Physico-mechanical and antioxidant properties of gelatin film from rabbit skin incorporated with rosemary acid. *Food Packag. Shelf Life* **2019**, *19*, 121–130. [CrossRef]
28. Halim, A.L.A.; Kamari, A.; Phillip, E. Chitosan, gelatin and methylcellulose films incorporated with tannic acid for food packaging. *Int. J. Biol. Macromol.* **2018**, *120*, 1119–1126. [CrossRef]
29. Wu, J.; Chen, S.; Ge, S.; Miao, J.; Li, J.; Zhang, Q. Preparation, properties and antioxidant activity of an active film from silver carp (*Hypophthalmichthys molitrix*) skin gelatin incorporated with green tea extract. *Food Hydrocoll.* **2013**, *32*, 42–51. [CrossRef]
70. Li, J.H.; Miao, J.; Wu, J.L.; Chen, S.F.; Zhang, Q.Q. Preparation and characterization of active gelatin-based films incorporated with natural antioxidants. *Food Hydrocoll.* **2014**, *37*, 166–173. [CrossRef]
71. Hanani, Z.A.N.; Yee, F.C.; Nor-Khaizura, M.A.R. Effect of pomegranate (*Punica granatum* L.) peel powder on the antioxidant and antimicrobial properties of fish gelatin films as active packaging. *Food Hydrocoll.* **2019**, *89*, 253–259. [CrossRef]
72. Bessaleh, S.; Jebahi, S.; Abbassi, R.; Ben Belgecem, S.; Faraz, A. Bioactive Gelatin-based Date By-Product for Packaging Applications: Physico-Chemical and Biological Characterization. *J. Mater. Appl.* **2022**, *11*, 10–16. [CrossRef]
73. Luo, Q.; Hossen, M.A.; Zeng, Y.; Dai, J.; Li, S.; Qin, W.; Liu, Y. Gelatin-based composite films and their application in food packaging: A review. *J. Food Eng.* **2022**, *313*, 110762. [CrossRef]
74. Ge, L.; Zhu, M.; Li, X.; Xu, Y.; Ma, X.; Shi, R.; Li, D.; Mu, C. Development of active rosmarinic acid-gelatin biodegradable films with antioxidant and long-term antibacterial activities. *Food Hydrocoll.* **2018**, *83*, 308–316. [CrossRef]
75. Liu, J.; Yong, H.; Liu, Y.; Qin, Y.; Kan, J.; Liu, J. Preparation and characterization of active and intelligent films based on fish gelatin and haskap berries (*Lonicera caerulea* L.) extract. *Food Packag. Shelf Life* **2019**, *22*, 100417. [CrossRef]

76. Zhao, Y.; Sun, Z. Effects of gelatin-polyphenol and gelatin–genipin cross-linking on the structure of gelatin hydrogels. *Int. J. Food Prop.* **2018**, *20*, S2822–S2832. [CrossRef]
77. Ruiz-Navajas, Y.; Viuda-Martos, M.; Sendra, E.; Perez-Alvarez, J.A.; Fernández-López, J. In vitro antibacterial and antioxidant properties of chitosan edible films incorporated with Thymus moroderi or Thymus piperella essential oils. *Food Control* **2013**, *30*, 386–392. [CrossRef]
78. Atarés, L.; Chiralt, A. Essential oils as additives in biodegradable films and coatings for active food packaging. *Trends Food Sci. Technol.* **2016**, *48*, 51–62. [CrossRef]
79. Burt, S. Essential oils: Their antibacterial properties and potential applications in foods—A review. *Int. J. Food Microbiol.* **2004**, *94*, 223–253. [CrossRef] [PubMed]
80. Sun, J.; Li, Y.; Cao, X.; Yao, F.; Shi, L.; Liu, Y. A Film of Chitosan Blended with Ginseng Residue Polysaccharides as an Antioxidant Packaging for Prolonging the Shelf Life of Fresh-Cut Melon. *Coatings* **2022**, *12*, 468. [CrossRef]
81. Kaczmarek-Szczepańska, B.; Zasada, L.; Grabska-Zielińska, S. The Physicochemical, Antioxidant, and Color Properties of Thin Films Based on Chitosan Modified by Different Phenolic Acids. *Coatings* **2022**, *12*, 126. [CrossRef]
82. Salgado-Cruz, M.d.l.P.; Salgado-Cruz, J.; García-Hernández, A.B.; Calderón-Domínguez, G.; Gómez-Viquez, H.; Oliver-Espinoza, R.; Fernández-Martínez, M.C.; Yáñez-Fernández, J. Chitosan as a coating for biocontrol in postharvest products: A bibliometric review. *Membranes* **2021**, *11*, 421. [CrossRef] [PubMed]
83. Aranaz, I.; Acosta, N.; Civera, C.; Elorza, B.; Mingo, J.; Castro, C.; Gandía, M.d.l.L.; Caballero, A.H. Cosmetics and cosmeceutical applications of chitin, chitosan and their derivatives. *Polymers* **2018**, *10*, 213. [CrossRef] [PubMed]
84. Nguyen, T.T.; Thi Dao, U.T.; Thi Bui, Q.P.; Bach, G.L.; Ha Thuc, C.N.; Ha Thuc, H. Enhanced antimicrobial activities and physiochemical properties of edible film based on chitosan incorporated with *Sonneratia caseolaris* (L.) Engl. leaf extract. *Prog. Org. Coatings* **2020**, *140*, 105487. [CrossRef]
85. Tokatlı, K.; Demirdöven, A. Effects of chitosan edible film coatings on the physicochemical and microbiological qualities of sweet cherry (*Prunus avium* L.). *Sci. Hortic. (Amst.)* **2020**, *259*, 108656. [CrossRef]
86. Hu, W.; Sarengaowa; Feng, K. Effect of Edible Coating on the Quality and Antioxidant Enzymatic Activity of Postharvest Sweet Cherry (*Prunus avium* L.) during Storage. *Coatings* **2022**, *12*, 581. [CrossRef]
87. Bigi, F.; Haghighi, H.; Siesler, H.W.; Licciardello, F.; Pulvirenti, A. Characterization of chitosan-hydroxypropyl methylcellulose blend films enriched with nettle or sage leaf extract for active food packaging applications. *Food Hydrocoll.* **2021**, *120*, 106979. [CrossRef]
88. Kumari, S.; Kishor, R. Chitin and chitosan: Origin, properties, and applications. In *Handbook of Chitin and Chitosan*; Volume 1: Preparation and Properties; Elsevier: Amsterdam, The Netherlands, 2020; pp. 1–33. ISBN 9780128179703.
89. Lisitsyn, A.; Semenova, A.; Nasonova, V.; Polishchuk, E.; Revutskaya, N.; Kozyrev, I.; Kotenkova, E. Approaches in animal proteins and natural polysaccharides application for food packaging: Edible film production and quality estimation. *Polymers* **2021**, *13*, 1592. [CrossRef]
90. Younes, I.; Rinaudo, M. Chitin and chitosan preparation from marine sources. Structure, properties and applications. *Mar. Drugs* **2015**, *13*, 1133–1174. [CrossRef] [PubMed]
91. Abdou, E.S.; Nagy, K.S.A.; Elsabee, M.Z. Extraction and characterization of chitin and chitosan from local sources. *Bioresour. Technol.* **2008**, *99*, 1359–1367. [CrossRef] [PubMed]
92. Sagheer, F.A.A.; Al-Sughayer, M.A.; Muslim, S.; Elsabee, M.Z. Extraction and characterization of chitin and chitosan from marine sources in Arabian Gulf. *Carbohydr. Polym.* **2009**, *77*, 410–419. [CrossRef]
93. Ren, L.; Yan, X.; Zhou, J.; Tong, J.; Su, X. Influence of chitosan concentration on mechanical and barrier properties of corn starch/chitosan films. *Int. J. Biol. Macromol.* **2017**, *105*, 1636–1643. [CrossRef] [PubMed]
94. Escamilla-García, M.; Reyes-Basurto, A.; García-Almendárez, B.E.; Hernández-Hernández, E.; Calderón-Domínguez, G.; Rossi-Márquez, G.; Regalado-González, C. Modified starch-chitosan edible films: Physicochemical and mechanical characterization. *Coatings* **2017**, *7*, 224. [CrossRef]
95. Kaya, M.; Akyuz, I.; Sargin, I.; Mujtaba, M.; Salaberria, A.M.; Labidi, J.; Cakmak, Y.S.; Koc, B.; Baran, T.; Ceter, T. Incorporation of sporopollenin enhances acid–base durability, hydrophobicity, and mechanical, antifungal and antioxidant properties of chitosan films. *J. Ind. Eng. Chem.* **2017**, *47*, 236–245. [CrossRef]
96. Younis, H.G.R.; Zhao, G. Physicochemical properties of the edible films from the blends of high methoxyl apple pectin and chitosan. *Int. J. Biol. Macromol.* **2019**, *131*, 1057–1066. [CrossRef]
97. Costa, S.M.; Ferreira, D.P.; Teixeira, P.; Ballesteros, L.F.; Teixeira, J.A.; Fangueiro, R. Active natural-based films for food packaging applications: The combined effect of chitosan and nanocellulose. *Int. J. Biol. Macromol.* **2021**, *177*, 241–251. [CrossRef] [PubMed]
98. Azaza, Y.B.; Hamdi, M.; Charmette, C.; Jridi, M.; Li, S.; Nasri, M.; Nasri, R. Development and characterization of active packaging films based on chitosan and sardinella protein isolate: Effects on the quality and the shelf life of shrimps. *Food Packag. Shelf Life* **2022**, *31*, 100796. [CrossRef]
99. Lan, W.; He, L.; Liu, Y. Preparation and properties of sodium carboxymethyl cellulose/sodium alginate/chitosan composite film. *Coatings* **2018**, *8*, 291. [CrossRef]
100. Mao, H.; Wei, C.; Gong, Y.; Wang, S.; Ding, W. Mechanical and water-resistant properties of eco-friendly chitosan membrane reinforced with cellulose nanocrystals. *Polymers* **2019**, *11*, 166. [CrossRef] [PubMed]

101. Riaz, A.; Lei, S.; Akhtar, H.M.S.; Wan, P.; Chen, D.; Jabbar, S.; Abid, M.; Hashim, M.M.; Zeng, X. Preparation and characterization of chitosan-based antimicrobial active food packaging film incorporated with apple peel polyphenols. *Int. J. Biol. Macromol.* **2018**, *114*, 547–555. [CrossRef] [PubMed]
102. Riaz, A.; Lagnika, C.; Luo, H.; Dai, Z.; Nie, M.; Hashim, M.M.; Liu, C.; Song, J.; Li, D. Chitosan-based biodegradable active food packaging film containing Chinese chive (*Allium tuberosum*) root extract for food application. *Int. J. Biol. Macromol.* **2020**, *150*, 595–604. [CrossRef] [PubMed]
103. Siripatrawan, U.; Vitchayakitti, W. Improving functional properties of chitosan films as active food packaging by incorporating with propolis. *Food Hydrocoll.* **2016**, *61*, 695–702. [CrossRef]
104. De Carli, C.; Aylanc, V.; Mouffok, K.M.; Santamaria-Echart, A.; Barreiro, F.; Tomás, A.; Pereira, C.; Rodrigues, P.; Vilas-Boas, M.; Falcão, S.I. Production of chitosan-based biodegradable active films using bio-waste enriched with polyphenol propolis extract envisaging food packaging applications. *Int. J. Biol. Macromol.* **2022**, *213*, 486–497. [CrossRef]
105. Wu, C.; Tian, J.; Li, S.; Wu, T.; Hu, Y.; Chen, S.; Sugawara, T.; Ye, X. Structural properties of films and rheology of film-forming solutions of chitosan gallate for food packaging. *Carbohydr. Polym.* **2016**, *146*, 10–19. [CrossRef] [PubMed]
106. Liang, J.; Yan, H.; Zhang, J.; Dai, W.; Gao, X.; Zhou, Y.; Wan, X.; Puligundla, P. Preparation and characterization of antioxidant edible chitosan films incorporated with epigallocatechin gallate nanocapsules. *Carbohydr. Polym.* **2017**, *171*, 300–306. [CrossRef] [PubMed]
107. Vilela, C.; Pinto, R.J.B.; Coelho, J.; Domingues, M.R.M.; Daina, S.; Sadocco, P.; Santos, S.A.O.; Freire, C.S.R. Bioactive chitosan/ellagic acid films with UV-light protection for active food packaging. *Food Hydrocoll.* **2017**, *73*, 120–128. [CrossRef]
108. Liu, J.; Liu, S.; Wu, Q.; Gu, Y.; Kan, J.; Jin, C. Effect of protocatechuic acid incorporation on the physical, mechanical, structural and antioxidant properties of chitosan film. *Food Hydrocoll.* **2017**, *73*, 90–100. [CrossRef]
109. Bi, F.; Zhang, X.; Bai, R.; Liu, Y.; Liu, J.; Liu, J. Preparation and characterization of antioxidant and antimicrobial packaging films based on chitosan and proanthocyanidins. *Int. J. Biol. Macromol.* **2019**, *134*, 11–19. [CrossRef] [PubMed]
110. Yang, K.; Dang, H.; Liu, L.; Hu, X.; Li, X.; Ma, Z.; Wang, X.; Ren, T. Effect of syringic acid incorporation on the physical, mechanical, structural and antibacterial properties of chitosan film for quail eggs preservation. *Int. J. Biol. Macromol.* **2019**, *141*, 876–884. [CrossRef]
111. Liu, Y.; Cai, Y.; Jiang, X.; Wu, J.; Le, X. Molecular interactions, characterization and antimicrobial activity of curcumin-chitosan blend films. *Food Hydrocoll.* **2016**, *52*, 564–572. [CrossRef]
112. Musella, E.; El Ouazzani, I.C.; Mendes, A.R.; Rovera, C.; Farris, S.; Mena, C.; Teixeira, P.; Poças, F. Preparation and characterization of bioactive chitosan-based films incorporated with olive leaves extract for food packaging applications. *Coatings* **2021**, *11*, 1339. [CrossRef]
113. Yong, H.; Liu, J.; Qin, Y.; Bai, R.; Zhang, X.; Liu, J. Antioxidant and pH-sensitive films developed by incorporating purple and black rice extracts into chitosan matrix. *Int. J. Biol. Macromol.* **2019**, *137*, 307–316. [CrossRef]
114. Yuan, G.; Lv, H.; Yang, B.; Chen, X.; Sun, H. Physical properties, antioxidant and antimicrobial activity of chitosan films containing carvacrol and pomegranate peel extract. *Molecules* **2015**, *20*, 11034–11045. [CrossRef]
115. Talón, E.; Trifkovic, K.T.; Nedovic, V.A.; Bugarski, B.M.; Vargas, M.; Chiralt, A.; González-Martínez, C. Antioxidant edible films based on chitosan and starch containing polyphenols from thyme extracts. *Carbohydr. Polym.* **2017**, *157*, 1153–1161. [CrossRef]
116. Kalaycıoğlu, Z.; Torlak, E.; Akın-Evingür, G.; Özen, İ.; Erim, F.B. Antimicrobial and physical properties of chitosan films incorporated with turmeric extract. *Int. J. Biol. Macromol.* **2017**, *101*, 882–888. [CrossRef]
117. Rambabu, K.; Bharath, G.; Banat, F.; Show, P.L.; Cocoletzi, H.H. Mango leaf extract incorporated chitosan antioxidant film for active food packaging. *Int. J. Biol. Macromol.* **2019**, *126*, 1234–1243. [CrossRef]
118. Yong, H.; Wang, X.; Bai, R.; Miao, Z.; Zhang, X.; Liu, J. Development of antioxidant and intelligent pH-sensing packaging films by incorporating purple-fleshed sweet potato extract into chitosan matrix. *Food Hydrocoll.* **2019**, *90*, 216–224. [CrossRef]
119. Fan, J.; Zhang, Z.H.; Qin, Y.Y.; Zhao, T.R.; Cheng, C.S. Characterization of antioxidant chitosan film incorporated with pomegranate peel extract. *Adv. Mater. Res.* **2013**, *706*, 24–27. [CrossRef]
120. Tan, Y.M.; Lim, S.H.; Tay, B.Y.; Lee, M.W.; Thian, E.S. Functional chitosan-based grapefruit seed extract composite films for applications in food packaging technology. *Mater. Res. Bull.* **2015**, *69*, 142–146. [CrossRef]
121. Kaya, M.; Ravikumar, P.; Ilk, S.; Mujtaba, M.; Akyuz, L.; Labidi, J.; Salaberria, A.M.; Cakmak, Y.S.; Erkul, S.K. Production and characterization of chitosan based edible films from Berberis crataegina's fruit extract and seed oil. *Innov. Food Sci. Emerg. Technol.* **2018**, *45*, 287–297. [CrossRef]
122. Kadam, D.; Shah, N.; Palamthodi, S.; Lele, S.S. An investigation on the effect of polyphenolic extracts of Nigella sativa seedcake on physicochemical properties of chitosan-based films. *Carbohydr. Polym.* **2018**, *192*, 347–355. [CrossRef] [PubMed]
123. Sun, L.; Sun, J.; Chen, L.; Niu, P.; Yang, X.; Guo, Y. Preparation and characterization of chitosan film incorporated with thinned young apple polyphenols as an active packaging material. *Carbohydr. Polym.* **2017**, *163*, 81–91. [CrossRef] [PubMed]
124. Peng, Y.; Wu, Y.; Li, Y. Development of tea extracts and chitosan composite films for active packaging materials. *Int. J. Biol. Macromol.* **2013**, *59*, 282–289. [CrossRef]
125. Wang, Q.; Tian, F.; Feng, Z.; Fan, X.; Pan, Z.; Zhou, J. Antioxidant activity and physicochemical properties of chitosan films incorporated with *Lycium barbarum* fruit extract for active food packaging. *Int. J. Food Sci. Technol.* **2015**, *50*, 458–464. [CrossRef]
126. Wang, L.; Wang, Q.; Tong, J.; Zhou, J. Physicochemical Properties of Chitosan Films Incorporated with Honeysuckle Flower Extract for Active Food Packaging. *J. Food Process Eng.* **2017**, *40*, e12305. [CrossRef]

127. Mujtaba, M.; Morsi, R.E.; Kerch, G.; Elsabee, M.Z.; Kaya, M.; Labidi, J.; Khawar, K.M. Current advancements in chitosan-based film production for food technology; A review. *Int. J. Biol. Macromol.* **2019**, *121*, 889–904. [CrossRef] [PubMed]
128. Dordevic, S.; Dordevic, D.; Sedlacek, P.; Kalina, M.; Tesikova, K.; Antonic, B.; Tremlova, B.; Treml, J.; Nejezchlebova, M.; Vapenka, L.; et al. Incorporation of natural blueberry, red grapes and parsley extract by-products into the production of chitosan edible films. *Polymers* **2021**, *13*, 3388. [CrossRef] [PubMed]
129. Wang, L.; Guo, H.; Wang, J.; Jiang, G.; Du, F.; Liu, X. Effects of Herba Lophatheri extract on the physicochemical properties and biological activities of the chitosan film. *Int. J. Biol. Macromol.* **2019**, *133*, 51–57. [CrossRef]
130. Zhu, F. Polysaccharide based films and coatings for food packaging: Effect of added polyphenols. *Food Chem.* **2021**, *359*, 129871. [CrossRef] [PubMed]
131. Genskowsky, E.; Puente, L.A.; Pérez-Álvarez, J.A.; Fernandez-Lopez, J.; Muñoz, L.A.; Viuda-Martos, M. Assessment of antibacterial and antioxidant properties of chitosan edible films incorporated with maqui berry (*Aristotelia chilensis*). *LWT* **2015**, *64*, 1057–1062. [CrossRef]
132. Jing, Y.; Huang, J.; Yu, X. Preparation, characterization, and functional evaluation of proanthocyanidin-chitosan conjugate. *Carbohydr. Polym.* **2018**, *194*, 139–145. [CrossRef]
133. Shiekh, K.A.; Ngiwngam, K.; Tongdeesoontorn, W. Polysaccharide-Based Active Coatings Incorporated with Bioactive Compounds for Reducing Postharvest Losses of Fresh Fruits. *Coatings* **2022**, *12*, 8. [CrossRef]
134. Elsabee, M.Z.; Morsi, R.E.; Fathy, M. *Chitosan-Oregano Essential Oil Blends Use as Antimicrobial Packaging Material*; Elsevier Inc.: Amsterdam, The Netherlands, 2016; ISBN 9780128007235.
135. Shen, Z.; Kamdem, D.P. Development and characterization of biodegradable chitosan films containing two essential oils. *Int. J. Biol. Macromol.* **2015**, *74*, 289–296. [CrossRef] [PubMed]
136. Priyadarshi, R.; Sauraj; Kumar, B.; Deeba, F.; Kulshreshtha, A.; Negi, Y.S. Chitosan films incorporated with Apricot (Prunus armeniaca) kernel essential oil as active food packaging material. *Food Hydrocoll.* **2018**, *85*, 158–166. [CrossRef]
137. Nguyen, T.T.; Nguyen, T.T.T.; Van Tran, T.; Van Tan, L.; Danh, L.T.; Than, V.T. Development of antibacterial, antioxidant, and uv-barrier chitosan film incorporated with piper betle linn oil as active biodegradable packaging material. *Coatings* **2021**, *11*, 351. [CrossRef]
138. Perdones, Á.; Chiralt, A.; Vargas, M. Properties of film-forming dispersions and films based on chitosan containing basil or thyme essential oil. *Food Hydrocoll.* **2016**, *57*, 271–279. [CrossRef]
139. Liu, T.; Wang, J.; Chi, F.; Tan, Z.; Liu, L. Development and characterization of novel active chitosan films containing fennel and peppermint essential oils. *Coatings* **2020**, *10*, 936. [CrossRef]
140. Hafsa, J.; ali Smach, M.; Khedher, M.R.B.; Charfeddine, B.; Limem, K.; Majdoub, H.; Rouatbi, S. Physical, antioxidant and antimicrobial properties of chitosan films containing Eucalyptus globulus essential oil. *LWT* **2016**, *68*, 356–364. [CrossRef]

Review

Applicability of Agro-Industrial By-Products in Intelligent Food Packaging

Silvia Amalia Nemes [1], Katalin Szabo [1] and Dan Cristian Vodnar [1,2,*]

1. Institute of Life Sciences, University of Agricultural Sciences and Veterinary Medicine, Calea Mănăştur 3–5, 400372 Cluj–Napoca, Romania; amalia.nemes@usamvcluj.ro (S.A.N.); katalin.szabo@usamvcluj.ro (K.S.)
2. Faculty of Food Science and Technology, University of Agricultural Sciences and Veterinary Medicine, Calea Mănăştur 3–5, 400372 Cluj–Napoca, Romania
* Correspondence: dan.vodnar@usamvcluj.ro; Tel.: +40-747341881

Received: 20 May 2020; Accepted: 1 June 2020; Published: 8 June 2020

Abstract: Nowadays, technological advancement is in continuous development in all areas, including food packaging, which tries to find a balance between consumer preferences, environmental safety, and issues related to food quality and control. The present paper concretely details the concepts of smart, active, and intelligent packaging and identifies commercially available examples used in the food packaging market place. Along with this purpose, several bioactive compounds are identified and described, which are compounds that can be recovered from the by-products of the food industry and can be integrated into smart food packaging supporting the "zero waste" activities. The biopolymers obtained from crustacean processing or compounds with good antioxidant or antimicrobial properties such as carotenoids extracted from agro-industrial processing are underexploited and inexpensive resources for this purpose. Along with the main agro-industrial by-products, more concrete examples of resources are presented, such as grape marc, banana peels, or mango seeds. The commercial and technological potential of smart packaging in the food industry is undeniable and most importantly, this paper highlights the possibility of integrating the by-products derived compounds to intelligent packaging elements (sensors, indicators, radio frequency identification).

Keywords: smart packaging; by-products; antioxidant properties; indicators; sensors; zero-waste; food quality; shelf-life

1. Introduction

The food packaging concept arose with the desire of humans to conserve food for a longer time, and it was adapted gradually to the industrialization and commercialization processes [1].

Since the earliest times, people consumed fresh food on the same days when raw materials were hunted or reaped from the garden without any food preservation issues. With the evolution through time and the trend of the population to live in communities, the need for food preservation appeared and food packaging solutions were found. Glass is found as the first material for food packaging in written history, as a precedent of paper, which is now widely used in the food industry [1]. Moving to one of the most debated packagings of the contemporary era, plastic material is found in specific studies from the year 1870 [1]. Back then, brothers John W. Hyatt and Isaiah S. Hyatt had patented the first commercially available plastic material, which was a mixture containing pyroxylin and camphor used in the manufacture of objects such as dental plates or shirt collars [1]. The evolution of this packaging material is spectacular, reaching 380 million metric tons in 2015 globally, covering 40% of the materials used for packaging. From this ratio, 60% is used only in the food packaging industry, and the rest is used in areas such as healthcare, cosmetics, or household [2].

Regardless of specific domains (e.g., medical, pharmaceutical, automobile, construction), constant progress can be observed in all fields, as well in the food packaging industry, and a significant evolution is perceived, according to global digitalization. If in the past few years researchers were focused on developing new packaging materials, in the present moment, the spotlight relies on developing new packaging concepts, such as electronic devices (e.g., indicators, sensors) incorporated in the package, providing more information about the food inside.

Nowadays, the challenge in the food industry is to fulfill consumer demand, which is focused on healthy, fresh, and the least processed food products. With the need to have relevant options that enhance their lifestyle, concerning the shorter time for shopping in the supermarket, and less time spent cooking the food, consumers are also aware of the effects of packaging on the environment. Globalization has allowed many products from all around the world to be more accessible, and consumers are used to obtaining these products at any time and relatively at a low cost [3]. As a result of these reasons and others such as safety, marketing, or cost-effectiveness, producers pay much attention to the packaging they choose for the products.

At a global level, the food packaging area is one of the most advanced industries, which makes people think about the economic impact of this manufacturing sector [4]. Advancing this idea, the price of the packaging is found in the final price of the food, and consumers know this aspect. This is one of the reasons why it is difficult to introduce a new concept of packaging in the market place, only if its cost is lower or if the consumer is informed and knows its benefits [5].

In 2009, the food packaging industry was estimated at the value of US$380 billion, and it is one of the largest packaging industries, representing more than 50% of all packaging industries at a global level. Plastic is the most imported packaging material, and it is evaluated at $9.5 billion, followed by paper with $4 billion, glass with $1.6 billion, and wood $0.3 billion. The food packaging market has a huge economic impact also on the developing countries taken in the study by The Food and Agricultural Organization of the United Nations (FAO), and its value was estimated at US$15.4 billion [4].

Generally, the package of a product has a substantial role in its journey, beginning with transport, the distribution process, retailing, and most important, protection and preservation. With all these efforts, a report published by the World Packaging Organization shows that more than 25% of food products are wasted because of improper packagings such as inappropriate dosage, an absence of reclosing function, insufficient storage protection, or smaller package sizes [5]. To solve these specific problems strongly related to the global concern of food waste, the technologies of food packaging keep advancing. Furthermore, producers from the packaging market are permanently challenged to develop new packaging models in line with consumers' request, which is in a continuous changing process.

Consumer requests are in a continuously changing process; therefore, producers from the packaging market are permanently challenged to develop new packaging models. The latest trends on the supermarket's shelves are the ready-to-eat products with improved shelf life based on the newest methods, such as active, intelligent, and smart packaging concepts.

Besides the safety improvements for food quality or marketing solutions that smart packaging offers for consumers and producers, it has an impact also for the economic domain, especially due to the global market which is expected to be doubled until 2021 [6], and to reach US$26.7 billion until 2024 [7]. For example, active packaging, which is characterized by an active function to extend the shelf life of the food product through bioactive compounds, was estimated in 2013 to occupy 26.9% from the global market of the packaging industry, which is evaluated at US$6.4 billion, while intelligent packagings were estimated at US$2.3 billion [8]. The most relevant countries that develop active and intelligent packaging are the US, Japan, and Australia. Figure 1 presents the growth rate predicted by 2026 [7].

The main purpose of food packaging is to prolong the shelf life of the food by protecting it from the outside environment [9]. Improving this aspect and maintaining the quality of the products supposed to use new strategic methods of fabrication or a new strategy of packaging. Various strategies can be used for improving the shelf life of a product, such as time–temperature devices, nanomaterials, the addition of chemicals, carbon dioxide emitters/absorbers, oxygen indicators, and barcode label biosensors [3,10–12]. This kind of packing can be part of two categories, active or intelligent packing, and if these two methods are combined, it can be called smart packaging [13].

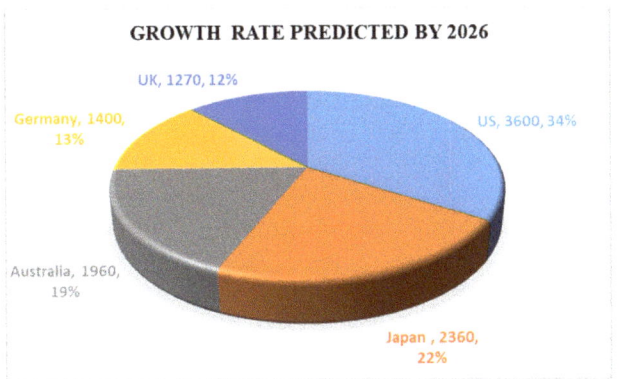

Figure 1. Global growth rate predicted by 2026 for the largest market (Data Sources from Ref. [7]).

The present review paper is focused on bringing answers and opening future research directions for the most common problems of the food packaging industry, such as food safety and control managing, food waste, and foodborne diseases. In addition, further attention is needed to the outstanding, unresolved, research issues such as reusability, biodegradability, and environmental protection against the waste resulted from food packaging industries. In this context, the present review points out possible solutions for developing alternative food packaging materials and recovering bioactive compounds from agro-industrial by-product sources that can support the "zero waste" agenda by integrating it in innovative food packaging.

The paper also aimed to make a classification of the newest smart packaging, including how this technology can be a solution for food waste and how compounds recovered from agro-industrial wastes can be used in active and intelligent packaging.

2. Concepts Definitions

Beginning with the middle of the 20th century, important evolution and major changes have been made in the development of new food packaging concepts, such as aseptic processing and packaging, modified atmosphere packaging, microwaveable packaging, and smart, active, and intelligent packaging systems [1]. Besides, the three advanced packaging technologies are all derived from the artificial intelligence technology associated with recent advances of computer science, which are found to serve in several practical applications, for example, porous materials [14], medical science [15,16], and the food industry [17,18].

Innovative packaging can be a way to transfer and apply intelligent science from the food industry to consumers. The newest concepts, active, intelligent, and smart packing exceed the usual attributes of a food pack. Intelligent packaging is considered a system that communicates with the consumer [19]. Active packaging is considered one of the most innovative concepts due to its precise functions that enable reading, feeling, seeing, or smelling the food inside [20]. Smart packaging is considered an update of the active pack to grow the industry of food products, especially the food packaging industry, with precious functions such as chemical and electrical-driven functions such as electronic displays storage temperature, self-heating or self-cooling containers, updated nutritional data, and the application of high-voltage pulsed electric fields [20–22]. Innovative packaging is a hot topic with an increased research interest; therefore, the concepts are described differently by scientists, and clear definitions have not been established yet. Some of these definitions are presented in Table 1.

Table 1. Definitions found in the literature for active, intelligent, and smart packaging.

Innovative Packaging	Definitions	Reference
Active packaging	- "active food contact materials and articles (hereinafter referred to as active materials and articles) means materials and articles that are intended to extend the shelf life or to maintain or improve the condition of packaged food. They are designed to deliberately incorporate components that would release or absorb substances into or from the packaged food or the environment surrounding the food".	[23]
	- A system that makes the interaction between the food, packaging, and the environment.	[24]
	- Packages that increase the shelf life of the food product by using natural compounds incorporated in the packaging.	
Intelligent packaging	- "'intelligent food contact materials and articles' (hereinafter referred to as intelligent materials and articles) means materials and articles which monitor the condition of packaged food or the environment surrounding the food"	[23]
	- Systems that provide the user with information on the conditions of food and should not release their constituents into the food.	[25]
	- A system adept to "sense, detect or record external or internal changes in the product"	[19]
	- A system added to packaging with the purpose of monitoring (food quality, critical control points) and giving information about the supply chain.	[3]
	- "The term "intelligent" involves an "ON/OFF" switching function on the package in response to changing external/internal stimuli, to communicate the product's status to its consumers or end users"	[26]
Smart packaging	- "Innovative packaging system that combines the benefits of measuring, estimating, or predicting different aspects of food quality or safety with the release of an active substance that extends the product shelf life".	[27]
	- The packaging capable of monitoring the changes that can appear in the product, packaging, or environment.	[24]
	- Upgraded packaging system with functional attributes that bring benefits to the food products and consumers.	[21]

The packaging has a role in preventing contamination of the food product and maintaining its freshness also. The food validity period can be determined depending on food product specifications such as the content of saturated and unsaturated fatty acids, enzyme activity, water activity, pH, or protein content [10,28–30], and the packaging type used as a barrier layer from contamination [19].

The shelf life of a food product is determined as the time between when the product was packaged and the last day that it can be consumed without any health risks and at the same quality. The expiration date can be limited by numerous intrinsic and extrinsic factors that are presented in Table 2, which can influence sensorial, textural, and microbial characteristics [13].

Table 2. Intrinsic and extrinsic factors that can influence the shelf life.

Intrinsic Factors	Extrinsic Factors	Reference
Water activity	Time–temperature profile	[9]
pH value	Temperature control	[31]
Redox potential	Relative humidity	[13]
Available oxygen	Exposure to light	[32]
Nutrients	Microbial environment	[32]
Natural microflora	Environment in packaging	[9,13]
Biochemical products	Heat treatment	[13]
Preservative	Purchaser handling	[13]

Distance between the place of production and place of sale is another issue raised with globalization that accentuated the drawbacks related to shelf life. Most of the foods are perishable and for this reason, producers have chosen different methods to improve the product's shelf life such as cooling, heat treatment, or modified atmosphere [33]. With all these, the microorganisms are not eliminated, and they stimulate reducing the quality and food safety. To solve and control this issue, the concept of active, intelligent, and smart packaging was developed.

3. Smart Packaging

According to many studies, smart packaging is defined as the packaging that includes both active and intelligent systems acting synergistically, as illustrated in Figure 2. It is capable of monitoring the changes during storage (increases/decreases in temperature or humidity) and acts to slow down the quality degradation. Using the compounds from active packagings, such as antioxidants, emitters of carbon dioxide, antibacterial agents, humidity, ethylene, and oxygen scavengers together with intelligent devices has obtained the concept named "smart packaging" [7,24,34,35].

Smart packaging includes devices that are capable of heating or cooling food inside and show in real time the nutritional information on the electronic display [21]. Some of the smart packaging belonging to the canning and beverage industries has been shown in a study [36]. A device incorporated in the package that can change the temperature of the food inside was developed for bottles, cans, or carton packages. It can lower the product's temperature with 18 °C in a short time (two or three minutes) before consumption. The principle of this packaging is based on absorbing the heat from the liquid inside, using the vapors obtained by releasing from a vinyl bag a quantity of pressurized water that evaporates immediately [37].

Another example of smart packaging is dedicated to heat coffee, tea, soup, and hot chocolate cups. In this case, the exothermic reaction between water and calcium oxide is the basis. Inside the cup, calcium and water are placed separately. Consumers are asked to invert the cup and mix the components, thus activating an exotherm reaction. The material that the cup is made of allows keeping the temperature for approximately 20 min [36].

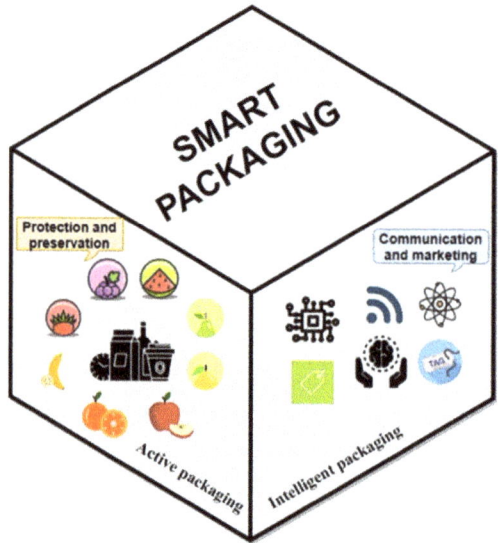

Figure 2. Smart packaging concept.

A new principle improved is an electronic tracking tag, which helps to have better control on the food distribution chain. Appling the Radio Frequency Identification (RFID) technology was advanced as a pack that is useful to prevent fraud, decrease economical losses, and enhance the fundamental operation such as storage, distribution, and marketing. RFID packaging can be considered smart packaging only if it is used alongside a sensor or biosensor that can provide the specific location of the product [35].

Oxygen-absorbing technology involves the combination of oxidation of certain compounds (iron powder, ascorbic acid, photosensitive polymers, enzymes), which are able to reduce the oxygen level inside the package, from the regular value between 0.3% and 3%, which is found in the conventional packaging systems such as modified atmosphere, vacuum, or substitution of internal atmosphere to an oxygen value of 0.01% [30]. Ascorbic acid and sodium bicarbonate were used for the development of a smart oxygen-absorbing label used in the meat industry. It is working to release carbon dioxide while absorbing the oxygen from the packaging, which is a reaction that preferably happens only when it is required. This action offers protection and prolongs the shelf life of the food inside. It can be made to operate at different temperatures (refrigeration or freezing) and a specific humidity [38].

The area of smart packaging is in full development and can provide solutions to reduce foodborne diseases and also help to reduce environmental problems. Research in this field is growing, evolves, and matures continuously, and smart packing is expected to improve food safety and quality soon.

4. Active Packaging

The concept of active food packaging is defined as innovative packaging that interacts with the food and absorbs derived chemicals to prolong the shelf life and ensure safety and quality at the same time. Active packaging creates a barrier between food and environmental space and stays in touch with the food inside, offering protection [29]. An important characteristic of active packaging is the capability of maintaining a low concentration in oxygen to slow the lipid oxidation. A low oxygen atmosphere can be achieved by the incorporation of antioxidant compounds (that act as protectors against oxidation processes) such as vitamins C and E [11], propolis, tocopherols, or plant extracts [29].

Overall, active packaging involves active compounds from different sources that can be in contact with the food or can be incorporated in the coating material [8]. Active compounds should be safe for human health; if combined active compounds are used, the mixture must be evaluated and authorized by the European Food Safety Authority [25].

4.1. Nanomaterials

The latest trends in the active packaging industry are to develop new active materials for preventing degradation and maintaining the quality for a longer time [28]. Good candidates to evolve in this direction are nanomaterials given their mechanical, optical, thermal, and antimicrobial attributes [11].

Nano packaging materials have an advantage in the food industry due to the protection they can assure against the foodborne pathogens as *Escherichia coli*, *Staphylococcus aureus*, and *Salmonella*, which are known for food poisoning [39]. Researchers have claimed that nanotechnology can solve the most common food poisoning symptoms such as fever, diarrhea, nausea, vomiting, and abdominal pain, which can even cause death in the case of children, pregnant women, and old people [40–42]. According to many research papers, nano packaging can also come with solutions for reducing environmental waste by using bionanocomposites (chitosan, starch, alginate, carboxymethyl cellulose, pectin) [43–45]. To make biocomposite materials an efficient solution for the food packaging industry, they should be improved for having high antimicrobial activity, better mechanical proprieties, and gas barrier functions. These improvements are possible using nanotechnology and active and intelligent packaging solutions [46]. A good mechanical amendment and antimicrobial activity for the most common pathogens found present in food were identified as zinc oxide (ZnO) nanoparticles and titanium dioxide (TiO_2) nanoparticles [39,43]. Both biocompatible materials have been tested for their efficacy against *Salmonella typhi*, *Klebsiella pneumoniae*, and *Shigella flexneri*. The results showed that TiO_2 and ZnO have inhibitory action against the bacteria mentioned above [39,47].

4.2. Polymers

The research on sustainability and the circular economy bring to the front the zero-waste processes that require the re-integration of by-products as a priority for the zero-waste agenda [48]. In this context, many studies have been made for developing new methods to integrate compounds extracted from by-products in usual technologies, such as food packaging.

The most important fact is that those compounds can be recovered from food by-products, such as seafood by-products (heads, gills, skin, trimmings) [49,50] for chitosan extraction, and fruit and vegetable industries by-products (pumpkin seeds and peels, grapefruit peel, sunflower head, sisal waste, pomegranate peel, eggplant peel, sour orange peel) for alginate and pectin [51–54]. They present high proprieties of biocompatibility, bioadhesion, and biodegradability, which is why their applicability is persistently growing [44].

As a coating material, a mixture of chitosan with 0.5% apple peel polyphenols or chitosan–proanthocyanidins combination has shown better mechanical proprieties and tensile strength [55,56]. Better flexibility and water resistance presented the films based on chitosan and grape seed extract or chitosan and apricot kernel essential oil in a ratio of 1:1 [57,58]. The application of starch in the packaging industry is limited by the disadvantage of reduced plasticity [59]. A mixture form by urea and ethanolamine can act as a plasticizer to prepare thermoplastic starch and succeeds to growth its mechanical properties and solve the issue of low plasticity [60]. Packaging mixtures presented behind do not cause environmental problems, are biodegradable and not toxic, and can be obtained from sustainable sources [61,62].

4.3. Antioxidants

Polyphenols are parting from the class of phytochemicals, and they can be found in plant-derived products such as coffee, tea, wine, fruits, vegetables, or chocolate [63,64]. They can be recovered from plant residues such as peels, bran, husks, and skins, which usually are wasted in their processing [65,66]. Many studies highlight the benefits that polyphenols can bring to human health, beginning with the preventive effects against cardiovascular disease [67–69], anti-inflammatory, anticoagulant, anticancer, and antioxidant proprieties [70–76]. For those functions and others (e.g., the prebiotic effect [64]), polyphenols are used in several areas such as drugs and the food industry or packaging area.

However, active packaging is still a studied area with great potential, which can improve the food packaging industry considerably.

5. Intelligent Packaging

The food industry is in continuous progress and one of the most revolutionary concepts developed recently is intelligent packaging [26]. Intelligent packaging can be represented by a small tag that is capable of monitoring the quality of food and can notice the consumer if there is a contamination problem with the food product [77]. In comparison with active packaging, intelligent packaging has the advantage to communicate directly with the consumers through an incorporated device [78].

With the technology evolution, indicators used in intelligent packaging were classified into two categories: indirect indicators and direct indicators, as presented in Figure 3. Researchers are now focused to develop more the second category (direct indicators) because of their ability to maintain the quality of the product and also to give more targeted information about volatile compounds of microbial origin, biogenic amines, toxins, or pathogenic bacterias [3].

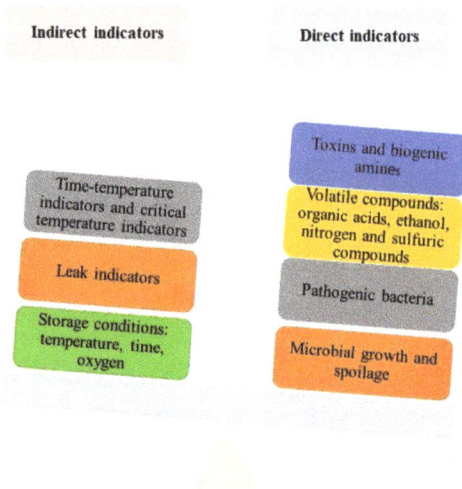

Figure 3. Food quality and safety indicators used in intelligent packaging.

The latest consumer demands of the 21st century are the products that can help them saving time, such as "ready-to-eat" and "heat and eat" meals. Another requirement registered is represented by the products' microwavable, easy opening, reusable characteristics [79]. An easy-to-use food package label (Figure 4a) for pork was developed using pH indicators (bromocresol purple, bromothymol blue, and a mixture of bromothymol blue and methyl red) for monitoring freshness and shelf life [80]. A self-adhesive label, named Fresh-Check (Figure 4b), was developed to ensure consumers about the freshness of the perishable food products. "The active center circle of the Fresh-Check darkens irreversibly, faster at higher temperatures and slower at a lower temperature"; in this way, consumers can know for sure if the product is fresh or not [81]. Another study has developed an on-package colorimetric sensor label (Figure 4c) for monitoring the ripeness of the apple after packing. This sensor was able to detect the aldehyde emission, in solution or vapor, of apples based on Methyl Red. As the apples mature, the sensor label changes its color in yellow, orange, and red in the end [82].

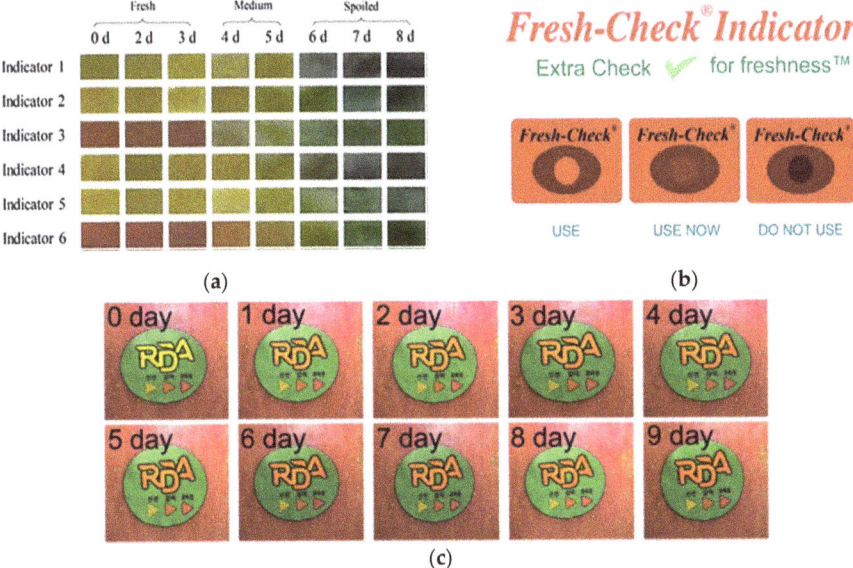

Figure 4. (**a**) Color changes of indicator labels for lean pork, (**b**) Fresh-Check label, (**c**) Color changes of a sensor label after exposure to apple flavor.

However, carefully watching the action of the intelligent labels, an important aspect to note is that these labels used as indicators are in direct contact with the food, which can be dry (coffee, fruits, and vegetables, bakery ware), liquid (beer, beverage), or semisolid (meat, fish). The migration of compounds from food contact labels according to food safety needs to be tested agreeing to the specific European directives (Regulation 450/2009) that regulate that active and intelligent packaging materials require authorization [3].

Intelligent packaging can give information about the state of food inside and can present the entire lifecycle of the product beginning with packing and distribution up to selling [78,83]. It can be used along with active packaging for watching the effect of the active compounds and their efficacy.

Intelligent packaging has a specific role: to improve safety and quality issues and to control the traceability of the food products. To fulfill this task, systems such as sensors, indicators, and Radio Frequency Identification (RFID) are used [26].

5.1. Sensors Used in Food Packaging

Sensors are one of the most studied electronic devices from the intelligent packaging field. Their main function is to detect and convert a signal form to another, using a transducer [19]. Sensors can be classified into active or passive sensors. If the transducer needs external power for measurement, the sensor is active; if it measures without help, the sensor is passive [24]. A traditional sensor is capable of measuring parameter changes such as temperature, pH, humidity, light exposure, or color changes. Research is advancing in the improvement of chemical sensors that can monitor package integrity and food quality and safety [19,24,84]. Among the chemical sensors found in the scientific literature are those detecting volatile organic compounds, compounds with high sensitivity, and gas molecules (H_2, CO, NO_2, O_2, H_2S, NH_3, CO_2, CH_4), which have high importance to the food sector [24].

Biosensors are part of chemical sensors that differ by the biological components used as detectors such as cells, antibodies, bacteria, yeast, fungi, plant and animal cells, biological tissue or enzymes, which are obtained by isolation and purification from biotechnological processes [24]. The most successful type of biosensor is the glucose sensor for diabetics, which is from the medical part, but they are found also in the pharmaceutical industry, food and process control, environmental monitoring, defense, and the security area [85].

In the intelligent packaging industry, chemical sensors, biosensors, and others can be used and incorporated in films. Flex Alert Company Ltd. with Vancouver-based partners has developed a commercially available biosensor that is capable of detecting pathogens (*E. coli*, *Salmonella*, and aflatoxins) in coffee beans, dried nuts, seeds, wine barrels, and fresh fruit. An active biosensor, using wireless communication, alerts manufacturers, consumers, and distributors to some toxin presence by sending information in real-time during storage and in packaged products [86].

Following the potential of sensors in the food industry, it can be seen that they may be a solution for combating food waste and also for reducing the risk of diseases caused by altered foods.

5.2. Indicators Used in Food Packaging

Indicators are part of the intelligent packaging group of devices. They have a different method of action than sensors do, by providing immediate qualitative visual information such as color change, color intensity change, or diffusion of colors, which are irreversible changes about the product [24,87]. Indicators are the most commonly used in commercial form, and a few of them are presented in Table 3.

Table 3. Commercially available indicators for intelligent packaging.

Type	Trade Name	Manufacturer	Information	Reference
Gas indicators	O$_2$ Sense	Freshpoint Lab	- aims to alert consumers, producers, or sellers if the integrity of the packaging is damaged by detecting oxygen inside, which can lead to the degradation of quality and safety of the product inside. - the indication is based on a color change.	[24,87,88]
	Novas	Insignia Technologies Ltd.	- specially made for products packed in plastic material, in a controlled atmosphere. - when the packaging is degraded, the pigment used as an indicator changes its color.	[24,87]
	Ageless Eye	Mitsubishi Gas Chemical Inc.	- an in-packaging indicator that monitors the presence/absence of oxygen. - if the oxygen level is 0.1% or less, the color of the indicator is pink; when the oxygen level attains 0.5% or more, the color indicated will be blue.	[88–91]
Freshness indicator	Freshtag	COX Technologies	- fish and seafood products are emitting a special odor (caused by volatile amines) when they lose their freshness. - the dye-based indicator interacts with the odor-causing chemicals and produces a change color reaction. - the intense pink color created indicates the lack of freshness.	[24,92]
	Sensorq	DSM NV and Food Quality Sensor International	- it is placed inside of the packaging. - is applied for meat and poultry products. - detect the gaseous by-products of bacterias that cause food poisoning.	[87,93]
Time-temperature	Timestrip	Timestrip UK Ltd.	- it monitors time and temperature for products where temperature can be a critical control point. - the adhesive label is applied to the food packaging and once that is activated, it starts to monitor the temperature (days, weeks or months). - the device records the time when the temperature recommended was not respected. - it is applicable for fresh seafood, fresh produce, airline catering, school meals, home delivery diets, food retailing, restaurants, hub, and spoke production.	[87,94]
	Monitormark	3M	- it is used as part of the secondary packaging and monitors the storage and transportation condition. - it is an adhesive label that easily attaches to the packaging and visually shows exposure and relative time over which exposure happened. - it has an irreversible action, even if the temperature during storage or delivery returns to a normal value. - it is applied for food products such as bakery, beverage, confectionery, and meat products that have critical temperature points beginning with –15 to 26 °C.	[19,87,95,96]

Table 3. Cont.

Type	Trade Name	Manufacturer	Information	Reference
	Onvu	Ciba Specialty Chemicals and FreshPoint	- the system monitors the freshness of a food product and is specially made for products sensitive to temperature. - after the label is activated, it becomes dark and then grows progressively lighter with time if the temperature rises. - the product has reached the end of the shelf life when the color reaches the reference color tone. - is commercially available for products as meat, fish, and dairy products with a shelf life of 5–6 days at 5 °C.	[97,98]
	L5-8 Smart TTI Seafood Label	Vitsab	- specially made for seafood products and is improved for the recognition of *Clostridium Botulinum* toxin formation; - the second most common pathogen (25%) in seafood is *Clostridium Botulinum*, and its growth and multiplication are directly influenced by temperature, for example, at 60 °F, multiplication is 10–12 times more rapid than at 40 °F or 72 °F, it is even 25–27 times more rapid than at 40 °F. - this label is based on an enzymatic reaction, which is given by the enzyme mix and substrate from the center of the label; the mixture is activated by applying moderate pressure on the "window" and is recognized by a homogenous green color in the "window". - the shade of green color can be changed in four ways (25%, 40%, 65%, 85%), and this is correlated with the ending of the product's validity. - if the green color changes and is replaced by an orange color (100%) or red (120%), it means that the product is no longer safe for consumption.	[24,99,100]
Time–temperature indicators	Cook-Chex	Pymah Corp	- it comes in the form of a cardboard tag. - the label contains a purple color chemical indicator (chromium–chloride complex) which, at various conditions of time and temperature in a pure steam atmosphere, changes its color to green. - this type of indicator is used commercially to verify the sterilization or autoclaving operation in the canning industry.	[87,101]
	Evigence sensors	EVIGENCE SENSORS	- it is a visual indicator that undergoes color changes after exposure at a higher temperature than the one established by the manufacturer. - the speed of changing the color of the indicator (from silver to white) increases directly in proportion to the increase in the temperature of the product's the environment. - indicator labels can be calibrated for products with a shelf life from some hours to several years. - labels are improved so that their application and activation is automatically done on the food packaging line. - using the SMART DOT™ app for mobile, consumers can scan the label before, during, and post-acquisition, for having more details (the remaining shelf life, time to repurchase).	[102,103]

Table 3. Cont.

Type	Trade Name	Manufacturer	Information	Reference
Time–temperature indicator/freshness indicator	Fresh-Check	Temptime Corp	- monitors the food in the store or at home. - it is an adhesive label, which has a central reactive part, represented by a circle; the circle darkens at an accelerated rate if the product is not stored at the appropriate temperature, and consumers can see the freshness of the food products in real time.	[19,104]
Disinfection Indicator	Thermostrip DL	LCR HALLCREST LLC	- a waterproof and self-adhesive label that irreversibly changes its color depending on the temperature of dishware in dishwashers. its main purpose is to indicate if the proper dishwasher temperature has been reached for sanitation. - it can be used also as proof for HACCP (Hazard Analysis and Critical Control Point). The temperature range covered by the indicator is beginning with 29 to 290 °C.	[105]

5.3. Radio Frequency Identification

Radio Frequency Identification (RFID) is an automatic identification technology based on remote data storage and retrieval using devices named labels, which are improved with wireless sensors for product identification. The RFID system is made of specific elements presented in Figure 5. In the food industry, RFID technology is part of the intelligent packaging sector. An RFID tag tracks food traceability in real time and monitors the entire cold chain for food applications (e.g., intercontinental fresh fish logistic chain) [106].

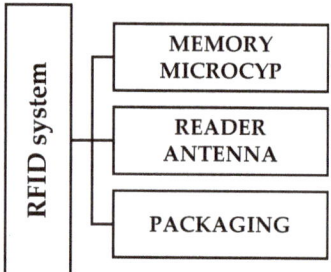

Figure 5. Radio Frequency Identification system elements.

RFID tags can be part of two categories, passive and active tags; the category depends on the use or not of external or integrated batteries for the transfer of information in the memory microchip. Passive tags are using a magnetic field created between the reader antenna and tag; thus, the intelligent tag takes the energy and transfers the encoded data to the memory chip of the label. Active RFID tags are using batteries for energy power to make the circuit between the microchip and the reader antenna [87]. Taking back to first principles, the working process of the RFID system is explained in Figure 6.

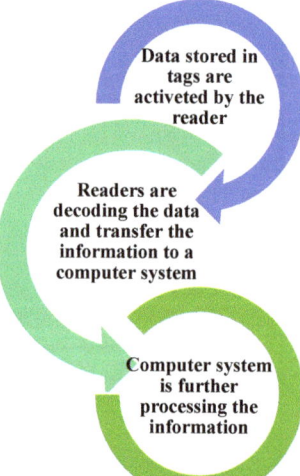

Figure 6. The working principles of an RFID system.

Therefore, RFID tags can increase the efficiency of supply chains and provide suppliers with an advanced method of food product monitoring until they become available to consumers.

6. Food Packaging Materials

6.1. Typical Packaging Materials

Since the Industrial Revolution from the 18th–19th century, industries had lost the value of reusing the packaging or the packaging material [24]. Over the years, this fact caused unwanted waste, which created much damage to the environment. The Food and Agricultural Organization of the United Nations (FAO) made a short report and a SWOT (Strengths, Weaknesses, Opportunities, and Threats) analysis of the food packaging industry in developing countries. They have presented a global view of this industry, with special attention in its size, structure, and the materials used. The most applied materials for packing are represented by paper, rigid plastic, metal, and glass, as seen in Figure 7 [4].

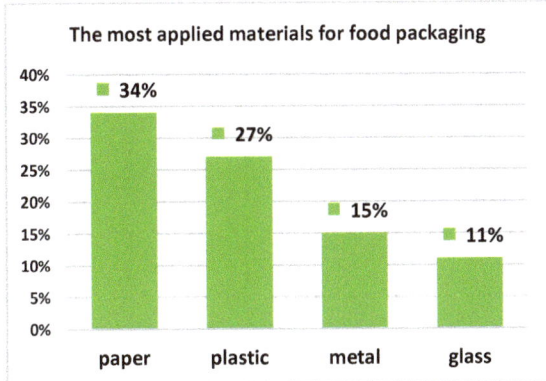

Figure 7. The most applied materials for food packaging (Data Sources from Ref. [4]).

In 2015, an amount of 322 M tons of plastic waste was registered, of which 49 M tons was from the food packaging industry [43]. Packaging materials used in the modern era such as glass, metals, paper, plastic, and alternative biomaterials are presented in Table 4 with their advantageous and disadvantageous characteristics.

Table 4. The advantages and disadvantages of typical and alternative packagings.

	Advantage	Disadvantage	References
Typical Packaging Materials			
Glass	Inorganic and chemically inert; Transparent, recyclable and it does not pollute; Easy to model; It can be sterilized at high temperatures; It offers protection against water and gas vapor; It does not deteriorate by oxidation or other forms; It does not react with food or the environment.	High manufacturing temperatures; Big energy consumption at manufacturing; High net mass; Breaks easily and it is nonbiodegradable.	[1,13]
Metals (steel, tin, chromium, and aluminum)	It is a barrier to gasses and humidity; It can be sterilized and it is recyclable; Covers a larger area of the food industry; Low toxicity; Good mechanical strength and resistance to working.	Has high costs of manufacturing; Energy intensive;	[1,13]
Paper and cardboard	Low costs, low weight, recyclable, and biodegradable; The resulting wastes can be reused for energy recovery; It has good printing properties.	Barriers to moisture and vapor are weak; Shock protection is low.	[1,13]
Plastic	Economically efficient; Good mechanical properties and a barrier to gasses and humidity; It is flexible, transparent, or can be colored.	Replaces glass, metal, and paper packaging; Pollute the soil; It is not degradable and harms marine life; The debris cannot be eliminated and can be ingested from seafood.	[1,13]
Alternative Packaging Materials			

Table 4. *Cont.*

	Advantage	Disadvantage	References
Chitosan-based biofilms	Non-toxic and biodegradable; Environmental friendly; Antimicrobial and bacteriostatic proprieties; Preserve the taste; Improve physical, mechanical, and chemical properties; Maintain tissue firmness; inhibit the increase of respiration rate.	Restricted scope.	[13,107–109]
Starch-based biofilms	Frequency in nature, renewable; Environmentally friendly; Low cost, non-toxic; Biodegradable; Suitable for foods with low humidity (confectionery and biscuit trays).	High water solubility; Insufficient water barrier properties; Weak mechanical properties; Reduced plasticity.	[1,59,107,110]
Films and coatings based on mango by-product	Good properties of permeability; Color stability; Antioxidant activity; Hydrophobicity; Decreases in manufacturing cost.	Low surface tension.	[111,112]

6.2. Alternative Materials

Alternative biomaterials can refer to different compounds (e.g., poly-lactic acid, starch, cellulose) incorporated in coating materials. The most important aspect of the food industry is the material used for packaging. To solve the issues of plastic pollution caused by the food packaging or pharmaceutical industries, the researchers on this domain are focused on finding alternatives packaging solutions [43].

The latest trends are to use bioplastic such as biopolymers and also plant extract incorporated for their antimicrobial properties [107]. Natural biodegradable polymers are obtained in a complex metabolic process. Biopolymers belong to the class of proteins, polysaccharides, lipids, phenolic compounds, and other classes, and they can be extracted from biomass or obtained from the vegetal, animal, or microbial sources presented in Figure 8 [45].

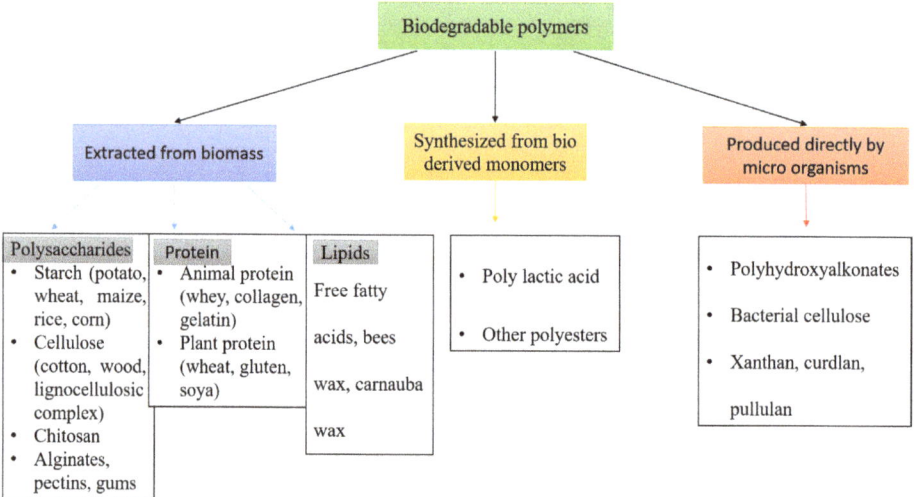

Figure 8. Classification of natural biodegradable polymers.

Starch is one of the most used natural biodegradable polymers at the moment, and it is being part of the carbohydrates class; its chemical structure is presented in Figure 9. The most important sources of starch are cereals (wheat, rice, corn) and tubers (potato) [107,113]. Starch contains two fundamental compounds: amylose, which is responsible for the structure and the rigid proprieties, and amylopectin, which helps digestion through its precise functions such as the enhancing of solubility of the polymer [114]. More than this, using starch-based biofilms is offering food package antioxidant activity, especially if gallic acid is added, 0.3 g/g starch [110]. A Systematic Review and Meta-Analysis has also shown recently the potential of using potato starch in the development of biofilms used for food packaging because it is an important raw material derived from many industries such as fruits and vegetable processing, has a low cost, and is the most wealthy biomaterial [115].

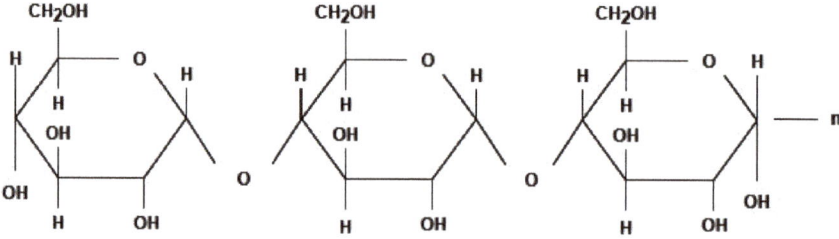

Figure 9. Chemical structure of polymeric starch.

Carotenoids are a part of the lipophilic pigment class. The most well-known sources are red, yellow, and orange fruits and vegetables, especially carrots, tomato, watermelon, and some species of fish such as salmon and crustaceans, principally cooked lobster and crab [61,116]. In addition to other pigments such as chlorophyll, β-carotene, anthocyanins, and lycopene, which are presented in Figure 10, β-carotene is added in the composition of biofilms used to make food packaging. The great attention of the researchers received the importance of β-carotene for its proprieties of excellent natural antioxidant and colorant products [117]. The encapsulation of β-carotene is used for obtaining active biodegradable packaging films. The presence of β-carotene has proved better thermal protection for the packaging films and greater protection against the oxidation process [61]. A recent study has developed a poly-lactic acid (PLA) film with three different compositions, using lycopene, β-carotene, and bixin. The standard curves of the three compounds were made in the following concentrations: β-carotene (λmax = 449 nm), from 2.2 to 105.4 µg mL^{-1} (R2 = 0.999); lycopene (λmax = 480 nm), from 0.3 to 13.9 µg mL^{-1} (R2 = 0.999); bixin (λmax = 457 nm), from 0.3 to 5.3 µg mL^{-1} (R2 = 0.996). All types were destined for sunflower oil oxidation protection [118]. The carotenoids were progressively released to the food simulant, the attainment of approximately 45% release of β-carotene and lycopene, and approximately 55% release of bixin. The PVA films improved with β-carotene and lycopene have proved a good barrier against light and oxygen, and PVA films with bixin presented the best antioxidant activity for sunflower oil [118].

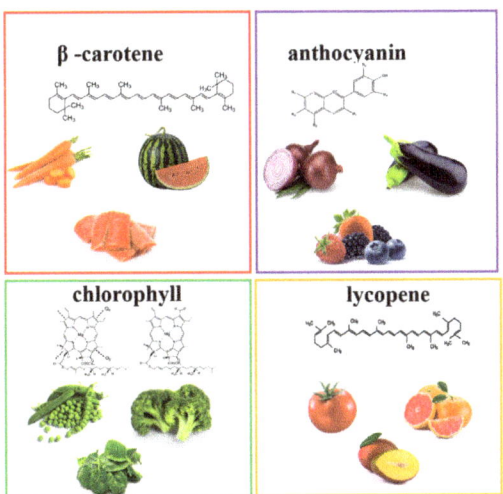

Figure 10. Bioactive pigments and their chemical structure.

Intelligent and active packaging has been obtained by poly(vinyl alcohol) (PVA) films mixed with chitosan, itaconic acid, and tomato by-products extract (TBE) [117]. TBE was used in another research study as a natural pigment due to its carotenoid content [119]. TBE is rich in carotenoids (lutein, lycopene, and β-carotene), an content of 3.273 mg/100 g DW, and phenolic compounds (e.g., caffeic acid–glucoside isomer, 5-caffeoy lquinic acid, quercetin-diglucoside), the total content of 80.596 mg/100 g DW [117]. This active film has proven to have increased physical properties (diameter, thickness, density, weight), inhibition effects against bacterias such as *S. aureus* and *P. aeruginosa*, and antimicrobial activity to *S. aureus*, *E. coli*, *P. aeruginosa*, *S. enterica* Enteritidis, and *S. enterica* Typhimurium [117].

The literature shows that chlorophyll is recommended to be used as a colorimetric temperature indicator for 50 to 70 degrees in chitosan films, along with anthocyanin, which is capable of detecting a temperature variation in intelligent packaging due to the property of changing its color depending on the temperature [34,120].

Another bioactive compound from the pigment class with high antioxidant activity that can be incorporated in active packaging is lycopene [116]. The most known natural sources are tomatoes, especially the peels [121], but it can be found also in guava or papaya [122]. An important aspect is that lycopene can be extracted from the tomato product waste [123]. A study shows that tomato processing by-products contain up to 0.1% lycopene and other bioactive compounds such as tocopherols and other carotenoids [124]. A research paper presented a method to replace synthetic oil-based stabilizers with naturals antioxidants compounds obtained from tomato and grape seeds recovered from the tomato and wine industries [125].

By-products can be recovered from food waste, which represents a high-value source of functional components such as proteins, fibers, polysaccharides, phytochemicals, lipids, and fatty acids [126]. One of the best-known by-products that is very frequently used in the world is bran resulting from the grain industry. Bran has a low price, great availability, and multiple functional proprieties (antioxidant and anti-inflammatory activity) due to the high content of phenolic compounds, fibers, and minerals [127].

A by-product that can be used from the winery waste is grape marc. [128]. Globally, the production of grapes, achieved in 2017 an amount of more than 77,000 tons and only the pulp is used for the wine production. The rest (seeds and peels) forms the grape marc abundant in epicatechin, catechin, gallic acid, procyanidins, and phenolic acids, which makes it have good antioxidant and antimicrobial activity [129].

Globally, banana production is registering an amount of 102 tons annually, of which 35% is only the peel, which is considered waste. One study developed an antioxidant chitosan biofilm improved with banana peels extract, which is fraught in bioactive compounds. The active biofilms were particularly made for maintaining the postharvest quality of apple during storage [109]. The by-product of mango is another example of a fruit that is used for developing active films for the food packaging industry. The seeds of mango had a large concentration of bioactive compounds, while the peel is rich in polysaccharides. The film created present good properties of permeability, color stability, antioxidant activity, and greater hydrophobicity [112].

7. Future Perspectives

Intelligent food packaging is an area with great potential for increasing the food packaging sector, providing fast, inexpensive, and efficient ways to monitor the environmental conditions of food in the supply chain. The next generation of food packaging developments and food packaging materials must be more environmentally friendly, and more importantly, they should be reusable, easy to use, and communicative with the consumers, to avoid the specific problems related to food waste, food quality managing, or foodborne diseases. For all this to become possible, supplementary attention should be driven to the innovative packaging and materials. Furthermore, recent strategies involving agro-industrial by-products showed that numerous bioactive compounds can be recovered and integrated into functional food packaging. Overall, future studies will be made on this topic with the purpose of continuing evolution in the food packaging industry.

8. Conclusions

This review paper presents the most important aspects associated with active and intelligent packaging and highlights many bioactive compounds recovered from agro-industrial by-products that may be useful in active or intelligent packaging. Smart packaging has a strong impact on food quality and control by allowing consumers to directly interpret the freshness and safety of the food inside. Intelligent packaging is an improved version of active packaging, simply by including devices from indicators and sensors classes for increased accuracy. The next future research can be made for by-product compounds integration in the elements of smart packaging (e.g., indicators, sensors) or for the design of new concepts of smart packaging that includes recovered bioactive compounds. Nevertheless, the trend of zero waste and the efforts made in the world of science to find environmentally friendly solutions to combat food waste and to capitalize raw materials at high capacity is constantly growing; accordingly to this tendency, shortly, more compounds will be integrated into food coatings.

Funding: This research was funded by a grant of the Romanian National Authority for Scientific Research and Innovation, CCDI-UEFISCDI, project No. 27/2018 CO FUND–MANUNET III-NON-ACT-2, within PNCDI III and the publication was supported by funds from the National Research Development Projects to finance excellence (PFE)-37/2018–2020 granted by the Romanian Ministry of Research and Innovation.

Conflicts of Interest: The authors declare no conflict of interest.

References

1. Robertson, G.L. History of Food Packaging. In *Reference Module in Food Science*; Elsevier: Brisbane, Australia, 2019.
2. Groh, K.J.; Backhaus, T.; Carney-Almroth, B.; Geueke, B.; Inostroza, P.A.; Lennquist, A.; Leslie, H.A.; Maffini, M.; Slunge, D.; Trasande, L.; et al. Overview of known plastic packaging-associated chemicals and their hazards. *Sci. Total Environ.* **2019**, *651*, 3253–3268. [CrossRef] [PubMed]
3. Dainelli, D.; Gontard, N.; Spyropoulos, D.; Zondervan-van den Beuken, E.; Tobback, P. Active and intelligent food packaging: Legal aspects and safety concerns. *Trends Food Sci. Technol.* **2008**, *19*, S103–S112. [CrossRef]
4. FAO. *Appropriate Food Packaging Solutions for Developing Countries*; Food and Agriculture Organization of the United Nations: Rome, Italy, 2014.
5. Wikström, F.; Williams, H.; Trischler, J.; Rowe, Z. The Importance of Packaging Functions for Food Waste of Different Products in Households. *Sustainability* **2019**, *11*, 2641. [CrossRef]
6. Fuertes, G.; Soto, I.; Carrasco, R.; Vargas, M.; Sabattin, J.; Lagos, C. Intelligent Packaging Systems: Sensors and Nanosensors to Monitor Food Quality and Safety. *J. Sens.* **2016**, *2016*, 1–8. [CrossRef]
7. Schaefer, D.; Cheung, W.M. Smart Packaging: Opportunities and Challenges. *Procedia CIRP* **2018**, *72*, 1022–1027. [CrossRef]
8. Renata Dobrucka, R.C. Active and Intelligent Packaging Food—Research and Development—A Review. *Pol. J. Food Nutr. Sci.* **2014**, *64*, 7–15. [CrossRef]
9. Robertson, G.L. *Food Packaging and Shelf Life*; CRC Press: Boca Raton, FL, USA, 2009.
10. Yam, K.L.; Takhistov, P.T.; Miltz, J. Intelligent Packaging: Concepts and Applications. *J. Food Sci.* **2005**, *70*, R1–R10. [CrossRef]
11. Madhusudan, P.; Chellukuri, N.; Shivakumar, N. Smart packaging of food for the 21st century—A review with futuristic trends, their feasibility and economics. *Mater. Today Proc.* **2018**, *5*, 21018–21022. [CrossRef]
12. Restuccia, D.; Spizzirri, U.G.; Parisi, O.I.; Cirillo, G.; Curcio, M.; Iemma, F.; Puoci, F.; Vinci, G.; Picci, N. New EU regulation aspects and global market of active and intelligent packaging for food industry applications. *Food Control* **2010**, *21*, 1425–1435. [CrossRef]
13. Ghoshal, G. Recent Trends in Active, Smart, and Intelligent Packaging for Food Products. In *Food Packaging and Preservation*; Academic Press: Cambridge, MA, USA, 2018; pp. 343–374.
14. Xiao, B.; Wang, S.; Wang, Y.A.N.; Jiang, G.; Zhang, Y.; Chen, H.; Liang, M.; Long, G.; Chen, X. Effective Thermal Conductivity of Porous Media with Roughened Surfaces by Fractal-Monte Carlo Simulations. *Fractals* **2020**, *28*, 2050029. [CrossRef]
15. Mesko, B. The Real Era of the Art of Medicine Begins with Artificial Intelligence. *J. Med. Internet Res.* **2019**, *21*, 5. [CrossRef] [PubMed]
16. Wang, C.H.; Zhu, X.F.; Hong, J.C.; Zheng, D.D. Artificial Intelligence in Radiotherapy Treatment Planning: Present and Future. *Technol. Cancer Res. Treat.* **2019**, *18*, 11. [CrossRef] [PubMed]
17. McClements, D.J. Future foods: A manifesto for research priorities in structural design of foods. *Food Funct.* **2020**, *11*, 1933–1945. [CrossRef] [PubMed]
18. Viejo, C.G.; Torrico, D.D.; Dunshea, F.R.; Fuentes, S. Emerging Technologies Based on Artificial Intelligence to Assess the Quality and Consumer Preference of Beverages. *Beverages* **2019**, *5*, 62. [CrossRef]
19. Kalpana, S.; Priyadarshini, S.R.; Maria Leena, M.; Moses, J.A.; Anandharamakrishnan, C. Intelligent packaging: Trends and applications in food systems. *Trends Food Sci. Technol.* **2019**, *93*, 145–157. [CrossRef]

20. Majid, I.; Thakur, M.; Nanda, V. Innovative and Safe Packaging Technologies for Food and Beverages: Updated Review. In *Innovations in Technologies for Fermented Food and Beverage Industries*; Springer: Cham, Switzerland, 2018; pp. 257–287.
21. Opara, U.L.; Mditshwa, A. A review on the role of packaging in securing food system: Adding value to food products and reducing losses and waste. *Afr. J. Agric. Res.* **2013**. [CrossRef]
22. Ortega-Rivas, E. *Non-Thermal Food Engineering Operations*; Springer: Chihuahua, Mexico, 2012.
23. Commission, T.E. On materials and articles intended to come into contact with food and repealing Directives 80/590/EEC and 89/109/EEC. In Proceedings of the Regulation (EC) No 1935/2004 of the European Parliament and of the Council, Strasbourg, France, 27 October 2004.
24. Vanderroost, M.; Ragaert, P.; Devlieghere, F.; De Meulenaer, B. Intelligent food packaging: The next generation. *Trends Food Sci. Technol.* **2014**, *39*, 47–62. [CrossRef]
25. Commission, E. Guidance to the commission regulation (EC) No 450/2009 of 29 May 2009 on active and intelligent materials and articles intended to come into contact with food. *Off. J. Eur. Union* **2009**.
26. Ghaani, M.; Cozzolino, C.A.; Castelli, G.; Farris, S. An overview of the intelligent packaging technologies in the food sector. *Trends in Food Sci. Technol.* **2016**, *51*, 1–11. [CrossRef]
27. Vilas, C.; Mauricio-Iglesias, M.; García, M.R. Model-based design of smart active packaging systems with antimicrobial activity. *Food Packag. Shelf Life* **2020**, *24*. [CrossRef]
28. Realini, C.E.; Marcos, B. Active and intelligent packaging systems for a modern society. *Meat Sci* **2014**, *98*, 404–419. [CrossRef] [PubMed]
29. Fang, Z.; Zhao, Y.; Warner, R.D.; Johnson, S.K. Active and intelligent packaging in meat industry. *Trends Food Sci. Technol.* **2017**, *61*, 60–71. [CrossRef]
30. Ramos, M.; Valdés, A.; Mellinas, A.; Garrigós, M. New Trends in Beverage Packaging Systems: A Review. *Beverages* **2015**, *1*, 248–272. [CrossRef]
31. Galić, K.; Ščetar, M.; Kurek, M. The benefits of processing and packaging. *Trends Food Sci. Technol.* **2011**, *22*, 127–137. [CrossRef]
32. Gnanaraj, J.; Welt, B.A.; Otwell, W.S.; Kristinsson, H.G. Influence of Oxygen Transmission Rate of Packaging Film on Outgrowth of Anaerobic Bacterial Spores. *J. Aquat. Food Prod. Technol.* **2005**, *14*, 51–69. [CrossRef]
33. Holley, R.A.; Patel, D. Improvement in shelf-life and safety of perishable foods by plant essential oils and smoke antimicrobials. *Food Microbiol.* **2005**, *22*, 273–292. [CrossRef]
34. Latos-Brozio, M.; Masek, A. The application of natural food colorants as indicator substances in intelligent biodegradable packaging materials. *Food Chem. Toxicol.* **2020**, *135*, 110975. [CrossRef]
35. Yucel, U. Intelligent Packaging. In *Reference Module in Food Science*; Elsevier: Amsterdam, The Netherlands, 2016. [CrossRef]
36. Jung H, H.; Colin H, L.H.; Rodrigues, E.T. *Intelligent Packaging*; Academic Press: Cambridge, MA, USA, 2005.
37. Brody, A.L. Packages that heat and cool themselves. *Food Technol Chic.* **2002**, *56*, 80–82.
38. YAM, K.L. *The Wiley Encyclopedia of Packaging Technology*; John Wiley & Sons: Hoboken, NJ, USA, 2009.
39. Vodnar, D.C.; Mitrea, L.; Călinoiu, L.F.; Szabo, K.; Ştefănescu, B.E. Removal of bacteria, viruses, and other microbial entities by means of nanoparticles. In *Advanced Nanostructures for Environmental Health*; Baia, L., Pap, Z., Hernadi, K., Baia, M., Eds.; Elsevier: Amsterdam, The Netherlands, 2020; pp. 465–491.
40. Rai, M.; Ingle, A.P.; Gupta, I.; Pandit, R.; Paralikar, P.; Gade, A.; Chaud, M.V.; dos Santos, C.A. Smart nanopackaging for the enhancement of food shelf life. *Environ. Chem. Lett.* **2018**, *17*, 277–290. [CrossRef]
41. Huayhongthong, S.; Khuntayaporn, P.; Thirapanmethee, K.; Wanapaisan, P.; Chomnawang, M.T. Raman spectroscopic analysis of food-borne microorganisms. *LWT* **2019**, *114*, 108419. [CrossRef]
42. Pal, M. Nanotechnology: A New Approach in Food Packaging. *J. Food Microbiol. Saf. Hyg.* **2017**, *2*. [CrossRef]
43. Al-Tayyar, N.A.; Youssef, A.M.; Al-Hindi, R. Antimicrobial food packaging based on sustainable Bio-based materials for reducing foodborne Pathogens: A review. *Food Chem.* **2020**, *310*, 125915. [CrossRef] [PubMed]
44. Martau, G.A.; Mihai, M.; Vodnar, D.C. The Use of Chitosan, Alginate, and Pectin in the Biomedical and Food Sector-Biocompatibility, Bioadhesiveness, and Biodegradability. *Polymers* **2019**, *11*, 1837. [CrossRef] [PubMed]
45. Lalit, R.; Mayank, P.; Ankur, K. Natural Fibers and Biopolymers Characterization: A Future Potential Composite Material. *Stroj. Cas. J. Mech. Eng.* **2018**, *68*, 33–50. [CrossRef]

46. Youssef, A.M.; Assem, F.M.; Abdel-Aziz, M.E.; Elaaser, M.; Ibrahim, O.A.; Mahmoud, M.; Abd El-Salam, M.H. Development of bionanocomposite materials and its use in coating of Ras cheese. *Food Chem.* **2019**, *270*, 467–475. [CrossRef]
47. Venkatasubbu, G.D.; Baskar, R.; Anusuya, T.; Seshan, C.A.; Chelliah, R. Toxicity mechanism of titanium dioxide and zinc oxide nanoparticles against food pathogens. *Colloids Surf. B Biointerfaces* **2016**, *148*, 600–606. [CrossRef]
48. Grillo, G.; Boffa, L.; Binello, A.; Mantegna, S.; Cravotto, G.; Chemat, F.; Dizhbite, T.; Lauberte, L.; Telysheva, G. Cocoa bean shell waste valorisation; extraction from lab to pilot-scale cavitational reactors. *Food Res. Int.* **2019**, *115*, 200–208. [CrossRef]
49. Vidanarachchi, J.K.; Ranadheera, C.S.; Wijerathne, T.D.; Udayangani, R.M.C.; Himali, S.M.C.; Pickova, J. *Applications of Seafood By-Products in the Food Industry and Human Nutrition*; Kim, S.K., Ed.; Springer: New York, NY, USA, 2014.
50. Salazar-Leyva, J.A.; Lizardi-Mendoza, J.; Ramirez-Suarez, J.C.; Valenzuela-Soto, E.M.; Ezquerra-Brauer, J.M.; Castillo-Yanez, F.J.; Pacheco-Aguilar, R. Acidic proteases from Monterey sardine (Sardinops sagax caerulea) immobilized on shrimp waste chitin and chitosan supports: Searching for a by-product catalytic system. *Appl. Biochem. Biotechnol.* **2013**, *171*, 795–805. [CrossRef]
51. Fleury, N.; Lahaye, M. Studies on by-products from the industrial extraction of alginate. *J. Appl. Phycol.* **1993**, *5*, 63–69. [CrossRef]
52. Masmoudi, M.; Besbes, S.; Abbes, F.; Robert, C.; Paquot, M.; Blecker, C.; Attia, H. Pectin Extraction from Lemon By-Product with Acidified Date Juice: Effect of Extraction Conditions on Chemical Composition of Pectins. *Food Bioprocess Technol.* **2010**, *5*, 687–695. [CrossRef]
53. Kazemi, M.; Khodaiyan, F.; Labbafi, M.; Hosseini, S.S. Ultrasonic and heating extraction of pistachio by-product pectin: Physicochemical, structural characterization and functional measurement. *J. Food Meas. Charact.* **2019**, *14*, 679–693. [CrossRef]
54. Lalnunthari, C.; Devi, L.M.; Badwaik, L.S. Extraction of protein and pectin from pumpkin industry by-products and their utilization for developing edible film. *J. Food Sci. Technol.* **2019**, *57*, 1807–1816. [CrossRef] [PubMed]
55. Riaz, A.; Lei, S.; Akhtar, H.M.S.; Wan, P.; Chen, D.; Jabbar, S.; Abid, M.; Hashim, M.M.; Zeng, X. Preparation and characterization of chitosan-based antimicrobial active food packaging film incorporated with apple peel polyphenols. *Int. J. Biol. Macromol.* **2018**, *114*, 547–555. [CrossRef] [PubMed]
56. Bi, F.; Zhang, X.; Bai, R.; Liu, Y.; Liu, J.; Liu, J. Preparation and characterization of antioxidant and antimicrobial packaging films based on chitosan and proanthocyanidins. *Int. J. Biol. Macromol.* **2019**, *134*, 11–19. [CrossRef] [PubMed]
57. Sogut, E.; Seydim, A.C. The effects of Chitosan and grape seed extract-based edible films on the quality of vacuum packaged chicken breast fillets. *Food Packag. Shelf Life* **2018**, *18*, 13–20. [CrossRef]
58. Priyadarshi, R.; Sauraj; Kumar, B.; Deeba, F.; Kulshreshtha, A.; Negi, Y.S. Chitosan films incorporated with Apricot (Prunus armeniaca) kernel essential oil as active food packaging material. *Food Hydrocoll.* **2018**, *85*, 158–166. [CrossRef]
59. Davoodi, M.; Kavoosi, G.; Shakeri, R. Preparation and characterization of potato starch-thymol dispersion and film as potential antioxidant and antibacterial materials. *Int. J. Biol. Macromol.* **2017**, *104*, 173–179. [CrossRef]
60. Ma, X.F.; Yu, J.G.; Wan, J.J. Urea and ethanolamine as a mixed plasticizer for thermoplastic starch. *Carbohydr. Polym.* **2006**, *64*, 267–273. [CrossRef]
61. Assis, R.Q.; Pagno, C.H.; Costa, T.M.H.; Flôres, S.H.; Rios, A.d.O. Synthesis of biodegradable films based on cassava starch containing free and nanoencapsulated β-carotene. *Packag. Technol. Sci.* **2018**, *31*, 157–166. [CrossRef]
62. Yong, H.; Wang, X.; Bai, R.; Miao, Z.; Zhang, X.; Liu, J. Development of antioxidant and intelligent pH-sensing packaging films by incorporating purple-fleshed sweet potato extract into chitosan matrix. *Food Hydrocoll.* **2019**, *90*, 216–224. [CrossRef]
63. Ma, G.; Chen, Y. Polyphenol supplementation benefits human health via gut microbiota: A systematic review via meta-analysis. *J. Funct. Foods* **2020**, *66*. [CrossRef]

64. Moorthy, M.; Chaiyakunapruk, N.; Jacob, S.A.; Palanisamy, U.D. Prebiotic potential of polyphenols, its effect on gut microbiota and anthropometric/clinical markers: A systematic review of randomised controlled trials. *Trends Food Sci. Technol.* **2020**. [CrossRef]
65. Dzah, C.S.; Duan, Y.; Zhang, H.; Serwah Boateng, N.A.; Ma, H. Latest developments in polyphenol recovery and purification from plant by-products: A review. *Trends Food Sci. Technol.* **2020**, *99*, 375–388. [CrossRef]
66. Calinoiu, L.F.; Vodnar, D.C. Thermal Processing for the Release of Phenolic Compounds from Wheat and Oat Bran. *Biomolecules* **2019**, *10*, 21. [CrossRef] [PubMed]
67. Castro-Barquero, S.; Shahbaz, M.; Estruch, R.; Casas, R. Cardiovascular Protection by Dietary Polyphenols. In *Reference Module in Food Science*; Elsevier: Amsterdam, The Netherlands, 2019.
68. Mendonca, R.D.; Carvalho, N.C.; Martin-Moreno, J.M.; Pimenta, A.M.; Lopes, A.C.S.; Gea, A.; Martinez-Gonzalez, M.A.; Bes-Rastrollo, M. Total polyphenol intake, polyphenol subtypes and incidence of cardiovascular disease: The SUN cohort study. *Nutr. Metab. Cardiovasc. Dis.* **2019**, *29*, 69–78. [CrossRef] [PubMed]
69. Sanches-Silva, A.; Testai, L.; Nabavi, S.F.; Battino, M.; Pandima Devi, K.; Tejada, S.; Sureda, A.; Xu, S.; Yousefi, B.; Majidinia, M.; et al. Therapeutic potential of polyphenols in cardiovascular diseases: Regulation of mTOR signaling pathway. *Pharmacol. Res.* **2020**, *152*, 104626. [CrossRef]
70. Burgos-Edwards, A.; Martín-Pérez, L.; Jiménez-Aspee, F.; Theoduloz, C.; Schmeda-Hirschmann, G.; Larrosa, M. Anti-inflammatory effect of polyphenols from Chilean currants (Ribes magellanicum and R. punctatum) after in vitro gastrointestinal digestion on Caco-2 cells. *J. Funct. Foods* **2019**, *59*, 329–336. [CrossRef]
71. Dzah, C.S.; Duan, Y.; Zhang, H.; Wen, C.; Zhang, J.; Chen, G.; Ma, H. The effects of ultrasound assisted extraction on yield, antioxidant, anticancer and antimicrobial activity of polyphenol extracts: A review. *Food Biosci.* **2020**, *35*, 100547. [CrossRef]
72. Hoskin, R.T.; Xiong, J.; Esposito, D.A.; Lila, M.A. Blueberry polyphenol-protein food ingredients: The impact of spray drying on the in vitro antioxidant activity, anti-inflammatory markers, glucose metabolism and fibroblast migration. *Food Chem.* **2019**, *280*, 187–194. [CrossRef]
73. Khouya, T.; Ramchoun, M.; Amrani, S.; Harnafi, H.; Rouis, M.; Couchie, D.; Simmet, T.; Alem, C. Anti-inflammatory and anticoagulant effects of polyphenol-rich extracts from Thymus atlanticus: An in vitro and in vivo study. *J. Ethnopharmacol.* **2020**, *252*, 112475. [CrossRef]
74. Le Sage, F.; Meilhac, O.; Gonthier, M.P. Anti-inflammatory and antioxidant effects of polyphenols extracted from Antirhea borbonica medicinal plant on adipocytes exposed to Porphyromonas gingivalis and Escherichia coli lipopolysaccharides. *Pharmacol. Res.* **2017**, *119*, 303–312. [CrossRef] [PubMed]
75. Sajadimajd, S.; Bahramsoltani, R.; Iranpanah, A.; Kumar Patra, J.; Das, G.; Gouda, S.; Rahimi, R.; Rezaeiamiri, E.; Cao, H.; Giampieri, F.; et al. Advances on Natural Polyphenols as Anticancer Agents for Skin Cancer. *Pharmacol. Res.* **2020**, *151*, 104584. [CrossRef] [PubMed]
76. Zhang, H.; Tsao, R. Dietary polyphenols, oxidative stress and antioxidant and anti-inflammatory effects. *Curr. Opin. Food Sci.* **2016**, *8*, 33–42. [CrossRef]
77. Han, J.-W.; Ruiz-Garcia, L.; Qian, J.-P.; Yang, X.-T. Food Packaging: A Comprehensive Review and Future Trends. *Compr. Rev. Food Sci. Food Saf.* **2018**, *17*, 860–877. [CrossRef]
78. Dobrucka, R.; Przekop, R. New perspectives in active and intelligent food packaging. *J. Food Process. Preserv.* **2019**, *43*. [CrossRef]
79. Yam, K.L.; Lee, D.S. Emerging food packaging technologies: An overview. In *Emerging Food Packaging Technologies*; Yam, K.L., Lee, D.S., Eds.; Woodhead Publishing: Cambridge, MA, USA, 2012.
80. Chen, H.; Zhang, M.; Bhandari, B.; Yang, C. Development of a novel colorimetric food package label for monitoring lean pork freshness. *LWT* **2019**, *99*, 43–49. [CrossRef]
81. Kuswandi, B.; Wicaksono, Y.; Abdullah, A.; Heng, L.Y.; Ahmad, M. Smart packaging: Sensors for monitoring of food quality and safety. *Sens. Food Qual. Saf.* **2011**, *5*, 137–146. [CrossRef]
82. Kim, Y.H.; Yang, Y.J.; Kim, J.S.; Choi, D.S.; Park, S.H.; Jin, S.Y.; Park, J.S. Non-destructive monitoring of apple ripeness using an aldehyde sensitive colorimetric sensor. *Food Chem.* **2018**, *267*, 149–156. [CrossRef]
83. Kerry, J.P.; O'Grady, M.N.; Hogan, S.A. Past, current and potential utilisation of active and intelligent packaging systems for meat and muscle-based products: A review. *Meat Sci.* **2006**, *74*, 113–130. [CrossRef]

84. Mahalik, N.; Kim, K. The Role of Information Technology Developments in Food Supply Chain Integration and Monitoring. In *Innovation and Future Trends in Food Manufacturing and Supply Chain Technologies*; Leadley, C.E., Ed.; Woodhead Publishing: Cambridge, MA, USA, 2016; pp. 21–37.
85. Turner, A.P. Biosensors: Sense and sensibility. *Chem. Soc. Rev.* **2013**, *42*, 3184–3196. [CrossRef] [PubMed]
86. Flex Alert Company Ltd. Available online: https://flex-alert.com/ (accessed on 26 March 2020).
87. Biji, K.B.; Ravishankar, C.N.; Mohan, C.O.; Srinivasa Gopal, T.K. Smart packaging systems for food applications: A review. *J. Food Sci. Technol.* **2015**, *52*, 6125–6135. [CrossRef] [PubMed]
88. Galanakis, C.M. *Food Quality and Shelf Life*; Elsevier Science: Amsterdam, The Netherland, 2019.
89. Yahia, E. *Postharvest Technology of Perishable Horticultural Commodities*; Woodhead Publishing: Cambridge, MA, USA, 2019.
90. Mitsubishi Gas Chemical. Available online: https://www.mgc.co.jp/eng/products/sc/rpsystem/metal/indicator.html (accessed on 4 April 2020).
91. Peng, S.-L.; Pal, S.; Huang, L. *Principles of Internet of Things (IoT) Ecosystem: Insight Paradigm*; Springer: Cham, Switzerland, 2020.
92. Dutra Resem Brizio, A.P. Use of Indicators in Intelligent Food Packaging. In *Reference Module in Food Science*; Elsevier: Amsterdam, The Netherland, 2016. [CrossRef]
93. DSM Invests in Food Freshness Device Company. Available online: https://www.dsm.com/markets/foodandbeverages/fr_FR/home.html (accessed on 6 April 2020).
94. Timestrip® Cold Chain Products for Food. Available online: https://timestrip.com/ (accessed on 6 April 2020).
95. 3M™ MonitorMark™ Time Temperature Indicators. Available online: https://www.3m.com/ (accessed on 7 April 2020).
96. Kour, H.; Wani, N.A.T.; Anisa Malik, R.K.; Chauhan, H.; Gupta, P.; Bhat, A.; Singh, J. Advances in food packaging—A review. *Stewart Postharvest Rev.* **2013**, *9*, 1–7. [CrossRef]
97. Brizio, A.P.D.R.; Prentice, C. Use of smart photochromic indicator for dynamic monitoring of the shelf life of chilled chicken based products. *Meat Sci.* **2014**, *96*, 1219–1226. [CrossRef] [PubMed]
98. OnVu time-temperature indicator (TTI) system. Available online: https://www.aiche.org/ (accessed on 7 April 2020).
99. Baptista, R.C.; Rodrigues, H.; Sant'Ana, A.S. Consumption, knowledge, and food safety practices of Brazilian seafood consumers. *Food Res. Int.* **2020**, *132*, 109084. [CrossRef] [PubMed]
100. TTI Label. Available online: http://vitsab.com/en/startpage/ (accessed on 8 April 2020).
101. U.S. Food & Drug Administration. Steam Activated Heat Sensitive Indicators. Available online: https://www.fda.gov/ (accessed on 8 April 2020).
102. Karst, T. Evigence Sensors Seeking Quality, Food Safety Uses in Fresh Produce. Available online: https://www.thepacker.com/ (accessed on 9 April 2020).
103. Time Temperature Indicators. Available online: https://evigence.com/ (accessed on 9 April 2020).
104. Fresh-Check® Indicator Temperature Intelligence™. Available online: http://fresh-check.com/ (accessed on 7 April 2020).
105. Thermostrip® DL. Available online: https://www.hallcrest.com/ (accessed on 9 April 2020).
106. Abad, E.; Palacio, F.; Nuin, M.; de Zarate, A.G.; Juarros, A.; Gomez, J.M.; Marco, S. RFID smart tag for traceability and cold chain monitoring of foods: Demonstration in an intercontinental fresh fish logistic chain. *J. Food Eng.* **2009**, *93*, 394–399. [CrossRef]
107. Munteanu, S.B.; Vasile, C. Vegetable Additives in Food Packaging Polymeric Materials. *Polymers* **2019**, *12*, 28. [CrossRef]
108. Gholami, R.; Ahmadi, E.; Ahmadi, S. Investigating the effect of chitosan, nanopackaging, and modified atmosphere packaging on physical, chemical, and mechanical properties of button mushroom during storage. *Food Sci. Nutr.* **2020**, *8*, 224–236. [CrossRef]
109. Zhang, W.; Li, X.; Jiang, W. Development of antioxidant chitosan film with banana peels extract and its application as coating in maintaining the storage quality of apple. *Int. J. Biol. Macromol.* **2019**. [CrossRef]
110. Zhao, Y.; Saldaña, M.D.A. Use of potato by-products and gallic acid for development of bioactive film packaging by subcritical water technology. *J. Supercrit. Fluids* **2019**, *143*, 97–106. [CrossRef]
111. Torres-León, C.; Rojas, R.; Contreras-Esquivel, J.C.; Serna-Cock, L.; Belmares-Cerda, R.E.; Aguilar, C.N. Mango seed: Functional and nutritional properties. *Trends Food Sci. Technol.* **2016**, *55*, 109–117. [CrossRef]

112. Torres-León, C.; Vicente, A.A.; Flores-López, M.L.; Rojas, R.; Serna-Cock, L.; Alvarez-Pérez, O.B.; Aguilar, C.N. Edible films and coatings based on mango (var. Ataulfo) by-products to improve gas transfer rate of peach. *LWT* **2018**, *97*, 624–631. [CrossRef]
113. Dufresne, A. *Nanocellulose: From Nature to High Performance Tailored Materials*, 2nd ed.; Walter de Gruyter GmbH & Co KG: Berlin, Germany, 2018.
114. Cerqueira, F.M.; Photenhauer, A.L.; Pollet, R.M.; Brown, H.A.; Koropatkin, N.M. Starch Digestion by Gut Bacteria: Crowdsourcing for Carbs. *Trends Microbiol.* **2020**, *28*, 95–108. [CrossRef] [PubMed]
115. Sadeghizadeh-Yazdi, J.; Habibi, M.; Kamali, A.A.; Banaei, M. Application of Edible and Biodegradable Starch-Based Films in Food Packaging: A Systematic Review and Meta-Analysis. *Curr. Res. Nutr. Food Sci. J.* **2019**, *7*, 624–637. [CrossRef]
116. Rodriguez-Amaya, D.B. Update on natural food pigments–A mini-review on carotenoids, anthocyanins, and betalains. *Food Res. Int.* **2019**, *124*, 200–205. [CrossRef] [PubMed]
117. Szabo, K.; Teleky, B.-E.; Mitrea, L.; Călinoiu, L.-F.; Martău, G.-A.; Simon, E.; Varvara, R.-A.; Vodnar, D.C. Active Packaging—Poly(Vinyl Alcohol) Films Enriched with Tomato By-Products Extract. *Coatings* **2020**, *10*, 141. [CrossRef]
118. Stoll, L.; Rech, R.; Flores, S.H.; Nachtigall, S.M.B.; de Oliveira Rios, A. Poly(acid lactic) films with carotenoids extracts: Release study and effect on sunflower oil preservation. *Food Chem.* **2019**, *281*, 213–221. [CrossRef]
119. Mitrea, L.; Calinoiu, L.F.; Martau, G.A.; Szabo, K.; Teleky, B.E.; Muresan, V.; Rusu, A.V.; Socol, C.T.; Vodnar, D.C. Poly(vinyl alcohol)-Based Biofilms Plasticized with Polyols and Colored with Pigments Extracted from Tomato By-Products. *Polymers* **2020**, *12*, 532. [CrossRef]
120. Maciel, V.B.V.; Yoshida, C.M.P.; Franco, T.T. Development of a prototype of a colourimetric temperature indicator for monitoring food quality. *J. Food Eng.* **2012**, *111*, 21–27. [CrossRef]
121. Giovannucci, E. A review of epidemiologic studies of tomatoes, lycopene, and prostate cancer. *Exp. Biol. Med.* **2002**, *227*, 852–859. [CrossRef]
122. Assis, R.Q.; Rios, P.D.A.; de Oliveira Rios, A.; Olivera, F.C. Biodegradable packaging of cellulose acetate incorporated with norbixin, lycopene or zeaxanthin. *Ind. Crops Prod.* **2020**, *147*. [CrossRef]
123. Szabo, K.; Catoi, A.F.; Vodnar, D.C. Bioactive Compounds Extracted from Tomato Processing by-Products as a Source of Valuable Nutrients. *Plant Foods Hum. Nutr.* **2018**, *73*, 268–277. [CrossRef] [PubMed]
124. Baranska, M.; Schutze, W.; Schulz, H. Determination of lycopene and beta-carotene content in tomato fruits and related products: Comparison of FT-Raman, ATR-IR, and NIR spectroscopy. *Anal. Chem.* **2006**, *78*, 8456–8461. [CrossRef] [PubMed]
125. Cerruti, P.; Malinconico, M.; Rychly, J.; Matisova-Rychla, L.; Carfagna, C. Effect of natural antioxidants on the stability of polypropylene films. *Polym. Degrad. Stab.* **2009**, *94*, 2095–2100. [CrossRef]
126. Calinoiu, L.F.; Farcas, A.; Socaci, S.; Vodnar, D.C. *Innovative Sources*; Academic Press Ltd-Elsevier Science Ltd.: London, UK, 2019; pp. 235–265.
127. Calinoiu, L.F.; Vodnar, D.C. Whole Grains and Phenolic Acids: A Review on Bioactivity, Functionality, Health Benefits and Bioavailability. *Nutrients* **2018**, *10*, 1615. [CrossRef] [PubMed]
128. Vandermeer, C.; Olejar, K.J.; Ricci, A.; Swift, S.; Versari, A.; Kilmartin, P.A. Effect of heat on grape marc extract. *Int. J. Nanotechnol.* **2018**, *15*, 792–797. [CrossRef]
129. Andrade, M.A.; Lima, V.; Sanches Silva, A.; Vilarinho, F.; Castilho, M.C.; Khwaldia, K.; Ramos, F. Pomegranate and grape by-products and their active compounds: Are they a valuable source for food applications? *Trends Food Sci. Technol.* **2019**, *86*, 68–84. [CrossRef]

© 2020 by the authors. Licensee MDPI, Basel, Switzerland. This article is an open access article distributed under the terms and conditions of the Creative Commons Attribution (CC BY) license (http://creativecommons.org/licenses/by/4.0/).

MDPI AG
Grosspeteranlage 5
4052 Basel
Switzerland
Tel.: +41 61 683 77 34

Coatings Editorial Office
E-mail: coatings@mdpi.com
www.mdpi.com/journal/coatings

Disclaimer/Publisher's Note: The statements, opinions and data contained in all publications are solely those of the individual author(s) and contributor(s) and not of MDPI and/or the editor(s). MDPI and/or the editor(s) disclaim responsibility for any injury to people or property resulting from any ideas, methods, instructions or products referred to in the content.

www.ingramcontent.com/pod-product-compliance
Lightning Source LLC
LaVergne TN
LVHW070650100526
838202LV00013B/931

www.ingramcontent.com/pod-product-compliance
Lightning Source LLC
LaVergne TN
LVHW070650100526
838202LV00013B/931